JAAatpl
JOINT AVIATION AUTHORITIES

Theoretical Training Manuals

Revised Edition

FLIGHT PERFORMANCE & PLANNING 1

This learning material has been approved as JAA compliant by the United Kingdom Civil Aviation Authority

OXFORD Aviation Training
Succeed through our experience

© Oxford Aviation Services Limited 2005
All Rights Reserved

This text book is to be used only for the purpose of private study by individuals and may not be reproduced in any form or medium, copied, stored in a retrieval system, lent, hired, rented, transmitted or adapted in whole or in part without the prior written consent of Oxford Aviation Services Limited.

Copyright in all documents and materials bound within these covers or attached hereto, excluding that material which is reproduced by the kind permission of third parties and acknowledged as such, belongs exclusively to Oxford Aviation Services Limited.

Certain copyright material is reproduced with the permission of the International Civil Aviation Organisation, the United Kingdom Civil Aviation Authority and the Joint Aviation Authorities (JAA).

This text book has been written and published as a reference work to assist students enrolled on an approved JAA Air Transport Pilot Licence (ATPL) course to prepare themselves for the JAA ATPL theoretical knowledge examinations. Nothing in the content of this book is to be interpreted as constituting instruction or advice relating to practical flying.

Whilst every effort has been made to ensure the accuracy of the information contained within this book, neither Oxford Aviation Services Limited nor the publisher gives any warranty as to its accuracy or otherwise. Students preparing for the JAA ATPL theoretical knowledge examinations should not regard this book as a substitute for the JAA ATPL theoretical knowledge training syllabus published in the current edition of 'JAR-FCL 1 Flight Crew Licensing (Aeroplanes)' (the Syllabus). The Syllabus constitutes the sole authoritative definition of the subject matter to be studied in a JAA ATPL theoretical knowledge training programme. No student should prepare for, or is currently entitled to enter himself/herself for the JAA ATPL theoretical knowledge examinations without first being enrolled in a training school which has been granted approval by a JAA-authorised national aviation authority to deliver JAA ATPL training.

Oxford Aviation Services Limited excludes all liability for any loss or damage incurred or suffered as a result of any reliance on all or part of this book except for any liability for death or personal injury resulting from Oxford Aviation Services Limited's negligence or any other liability which may not legally be excluded.

Cover photo by Chris Sheldon e-mail: fwog@mindless.com

First published by Jeppesen GmbH, Frankfurt, Germany: 2001
Second edition: Jeppesen GmbH, Frankfurt, Germany: 2002
This edition published by Transair (UK) Ltd, Shoreham, England: 2005
Printed in Malaysia by KHL Printing Co. Sdn. Bhd.

Contact Details:

Ground Training Department
Oxford Aviation Services Ltd
Oxford Airport
Kidlington
Oxford OX5 1RA
England

Tel: +44 (0)1865 844299
E-mail: ddd@oxfordaviation.net

Transair Pilot Shop
Transair (UK) Limited
Shoreham Airport
Shoreham-by-Sea
West Sussex BN43 5PA
England

Tel: +44 (0)1273 466000
E-mail: info@transair.co.uk

For further information on products and services from Oxford Aviation Training and Transair visit our websites at: www.oxfordaviation.net and www.transair.co.uk

FOREWORD

Joint Aviation Authorities (JAA) pilot licences were first introduced in 1999, and have now been adopted by nearly all member states. A steadily increasing number of non-European countries have also expressed the intention of aligning their training with JAA requirements, and some have already begun this process. The syllabi and the regulations governing the award and the renewal of licences are currently defined by the JAA's licensing agency, known as "Joint Aviation Requirements-Flight Crew Licensing", or JAR-FCL. Over the next few years, JAA responsibilities, including licensing, will gradually be transferred to the new European Aviation Safety Agency (EASA).

The JAR-FCL ATPL theoretical training requirements and associated ground examinations, although possibly similar in scope to those previously used by many national authorities, are inevitably different in a number of respects from the syllabi and examinations previously used under national schemes. Consequently, students who wish to train for the JAA ATPL licence need access to study material which has been specifically designed to meet the requirements of the new licensing system. This series of text books, prepared by Oxford Aviation Training (OAT) and published exclusively by Transair Pilot Shop, covers all JAR-FCL requirements and is specifically designed to help student pilots prepare for the ATPL theoretical knowledge examinations.

OAT is one of the world's leading professional pilot schools. Established for 40 years, Oxford has trained more than 14,000 professional pilots for over 80 airlines, world-wide. OAT was the first pilot school in the United Kingdom to be granted approval to train for the JAA ATPL and has been the leading contributor within Europe to the process of defining and improving the training syllabus. OAT led and coordinated the joint-European effort to produce the ATPL Learning Objectives which are now published by the JAA as the definitive guide to the theoretical knowledge requirements of ATPL training.

Since JAA ATPL training started in 1999, OAT has achieved an unsurpassed success rate in the JAA ATPL examinations. At the start of this year, OAT students had successfully passed more than 25,000 individual JAR-FCL examinations and currently more than 300 students a year graduate from Oxford's theoretical training programmes. The text books, together with an increasing range of Computer Based Training (CBT) products, are also now used by other Flight Training Organizations both in Europe and increasingly, throughout the world. Recognized by leading National Aviation Authorities as being fully compliant with JAR-FCL training requirements, the series has now effectively become the de-facto standard for JAR-FCL ATPL theoretical training. This achievement is the result of OAT's continued commitment to the development of the JAA licensing system. OAT's unrivalled experience and expertise make these books the best learning material available to any student who aspires to hold a JAA ATPL. The series is continually updated, with this year's edition incorporating specimen examination papers and "feedback" information, all specifically intended to help candidates prepare fully for their ATPL theoretical examinations

For those aspirant airline pilots who are not yet able to begin training but hope to do so in the future, these text books provide high-quality study material to help them prepare thoroughly for their formal training. The books also make excellent reading for general aviation pilots or for aviation enthusiasts who simply wish to further their knowledge of aeronautical subjects. We trust that your study of these books will not only be enjoyable but, for those of you currently undergoing ATPL training, will also lead to success in the JAA ATPL ground examinations.

Whatever your aviation ambitions, we wish you every success and above all, happy landings.

Mike Langley
Commercial Director

Oxford Aviation Training
March 2005

Textbook Series

Book	Title	JAR Ref. No.	Subject
1	010 Air Law	010	
2	020 Aircraft General Knowledge 1	021 01	Airframes & Systems
		021 01 01/04	Fuselage, Wings & Stabilising Surfaces
		021 01 07	Hydraulics
		021 01 05	Landing Gear
		021 01 06	Flight Controls
		021 01 08/09	Air Systems & Air Conditioning
		021 01 09/10	Anti-icing & De-icing
		021 04 00	Emergency Equipment
		021 01 11	Fuel Systems
3	020 Aircraft General Knowledge 2	021 02	Electrics – Electronics
		021 02 01	Direct Current
		021 02 02	Alternating Current
		021 02 05	Basic Radio Propagation.
4	020 Aircraft General Knowledge 3	021 00	Powerplant
		021 03 01	Piston Engines
		021 03 02	Gas Turbines
5	020 Aircraft General Knowledge 4	022	Instrumentation
		022 01	Flight Instruments
		022 03	Warning & Recording
		022 02	Automatic Flight Control
		022 04	Power Plant & System Monitoring Instruments
6	030 Flight Performance & Planning 1	031	Mass & Balance
		032	Performance
7	030 Flight Performance & Planning 2	033	Flight Planning & Monitoring
8	040 Human Performance & Limitations	040	
9	050 Meteorology	050	
10	060 Navigation 1	061	General Navigation
11	060 Navigation 2	062	Radio Navigation
12	070 Operational Procedures	070	
13	080 Principles of Flight	080	
14	090 Communications	091	VFR Communications
		092	IFR Communications
15	Reference Material		CAP 696, CAP 697, CAP 698 Aerodrome Information JAR FCL 1 Subpart J

FLIGHT PERFORMANCE AND PLANNING 1

TABLE OF CONTENTS

Subject Number   **Title**

1   **MASS AND BALANCE**

Chapter 1	Jar Ops-1 Subpart J
Chapter 2	Mass and Balance Definitions and Calculations
Chapter 3	Specimen Questions

2   **PERFORMANCE**

Chapter 1	Definitions
Chapter 2	General Principles: Take off
Chapter 3	General Principles: Climb and Descent
Chapter 4	General Principles: Cruise
Chapter 5	General Principles: Landing
Chapter 6	Single Engine Class B Aircraft: Take Off
Chapter 7	Single Engine Class B: Climb
Chapter 8	Single Engine Class B: Cruise
Chapter 9	Single Engine Class B: Landing
Chapter 10	Multi-Engined Class B: Take Off
Chapter 11	Multi Engined Class B: Climb and Cruise
Chapter 12	Multi-Engined Class B: Landing
Chapter 13	Class A Aircraft: Take Off
Chapter 14	Class A: Additional Take Off Procedures
Chapter 15	Class A: Take Off Climb
Chapter 16	Class A: En Route
Chapter 17	Class A: Landing
Chapter 18	Specimen Questions

JAAatpl
JOINT AVIATION AUTHORITIES

Theoretical Training Manuals

Revised Edition

FLIGHT PERFORMANCE & PLANNING 1

MASS & BALANCE

This learning material has been approved as JAA compliant by the United Kingdom Civil Aviation Authority

Oxford Aviation Training
Succeed through our experience

MASS AND BALANCE

PREFACE

The JAR Mass and Balance Examination

The JAR Mass and Balance Examination consists of approximately 30 questions. The time allowed is one hour. The questions can carry different marks depending on degree of difficulty and the mark will be indicated alongside the individual question.

The pass mark is 75%.

On sitting the examination you will be provided with :

    a.   The question paper
    b.   An answer proforma consisting of a series of rows of small rectangles marked 'a', 'b', 'c' and 'd' on which you score in your answer with a soft pencil.
    c.   Two pieces of plain paper on which to work out any calculations.
    d.   A simple electronic calculator (non scientific)
    e.   The Civil Aviation Authority Mass and Balance booklet (CAP696)
    f.   A copy of the Standard Mass Tables if required.(JAR - OPS Annex J)
    g.   A soft pencil.

You may take into the exam your navigation computer (CRP5) if desired.

All items have to be handed in at the end of the examination, including any scrap paper.

The Civil Aviation Mass and Balance Booklet CAP 696

This booklet is complete in itself and consists of :

    Section 1.   General notes
    Section 2.   Data for single engine piston/propeller (SEP.1) aeroplanes.
    Section 3.   Data for light twin engine piston/propeller aeroplane (MEP.1)
    Section 4.   Data for medium range twin jet (MRJT.1)

The examination questions will involve the data in the Mass and Balance booklet and general questions regarding mass and balance. Many of the answers required are printed in the booklet and will require nothing more than for you to locate them. Other questions will require you to use the data in the booklet together with additional information provided on the question paper to work out fairly simple mass and balance calculations. The booklet defines the mass and balance data and other general information for three separate aircraft types. Many of the question will involve the data regarding these three aircraft. A study of the course notes will enable you to carry out any mass and balance calculations required.

A thorough knowledge of the Mass and Balance booklet and the course notes is therefore essential.

MASS AND BALANCE

Standard Mass Tables

Details of the standard mass tables are given in the course notes Chapter one, which is a copy of the JAR - OPS-1 manual subpart J.

Example Questions

A study of the Mass and Balance booklet shows on page 5 that the Basic Empty Mass of a SEP 1. aeroplane is 2415 lb. A typical question may be as follows:

Question 1. Using the data provided for the SEP 1. Aeroplane. The Basic Empty Mass is?

- a)   2415 lb
- b)   1524 lb
- c)   4215 lb
- d)   1234 lb

As you can see, answering question one is simply a matter of looking through the data given for the SEP 1. and finding the relevant information.

MASS AND BALANCE

CHAPTER ONE - JAR OPS-1 SUBPART J

Contents

                                                                Page

JAR - OPS 1.605 GENERAL 1-1
JAR - OPS 1.607 TERMINOLOGY 1-1
JAR - OPS 1.610 LOADING, MASS AND BALANCE 1-2
JAR - OPS 1.615 MASS VALUES FOR CREW 1-2
JAR - OPS 1.620 MASS VALUES FOR PASSENGERS AND BAGGAGE 1-3
JAR - OPS 1.625 MASS AND BALANCE DOCUMENTATION 1-6

APPENDIX 1 TO JAR-OPS1.605
MASS AND BALANCE - GENERAL 1-7

APPENDIX TO JAR-OPS 1.620 (F)
DEFINITION TO THE AREA FOR FLIGHTS WITHIN THE EUROPEAN REGION.   1-11

APPENDIX 1 TO JAR - OPS 1.620 (G)
PROCEDURE TO ESTABLISHING REVISED STANDARD MASS VALUES FOR PASSENGERS AND THEIR BAGGAGE. 1-11

APPENDIX TO JAR - OPS 1.525
MASS AND BALANCE DOCUMENTATION 1-14

MASS AND BALANCE SELF ASSESSMENT QUESTIONS - THEORY . . . . 1 - 17

MASS AND BALANCE

MASS AND BALANCE **JAR OPS-1 SUBPART J**

SECTION 1

JAR - OPS 1.605 GENERAL

(See appendix 1 to JAR-OPS 1.605)

a) An operator shall ensure that during any phase of operation, the loading, mass and centre of gravity of the aeroplane complies with the limitations specified in the approved Aeroplane Flight Manual, or the Operations Manual if more restrictive.

b) An operator must establish the mass and the centre of gravity of any aeroplane by actual weighing prior to initial entry into service and thereafter at intervals of 4 years if individual aeroplane masses are used and 9 years if fleet masses are used. The accumulated effects of modifications and repairs on the mass and balance must be accounted for and properly documented. Furthermore, aeroplanes must be re-weighed if the effect of modifications on the mass and balance is not accurately known.

c) An operator must determine the mass of all operating items and crew members included in the aeroplane dry operating mass by weighing or by using standard masses. The influence of their position on the aeroplane centre of gravity must be determined.

d) An operator must establish the mass of the traffic load, including any ballast, by actual weighing or determine the mass of the traffic load in accordance with standard passenger and baggage masses specified in JAR-OPS 1.620.

e) An operator must determine the mass of the fuel load by using the actual density or, if not known, the density calculated in accordance with a method specified in the Operations Manual. (See IEM OPS 1.605 (e).)

JAR - OPS 1.607 TERMINOLOGY

(a) **Dry Operating Mass**.

The total mass of the aeroplane ready for a specific type of operation excluding all usable fuel and traffic load. The mass includes such items as:

(1) Crew and baggage;

(2) Catering and removable passenger service equipment; and

(3) Potable water and lavatory chemicals.

MASS AND BALANCE · JAR OPS-1 SUBPART J

b) **Maximum Zero Fuel Mass**

The maximum permissible mass of an aeroplane with no useable fuel. The mass of the fuel contained in the zero fuel mass when it is explicitly mentioned in the Aeroplane Flight Manual limitations.

c) **Maximum Structural Landing Mass.**

The maximum permissible total aeroplane mass on landing under normal circumstances.

d) **Maximum Structural Take-off Mass.**

The maximum permissible total aeroplane mass at the start of the take-off run.

e) **Passenger Classification.**

  i) Adults, male and female, are defined as persons of an age of 12 years and above.
  ii) Children are defined as persons of an age of two years and above but who are less than 12 years of age.
  iii) Infants are defined as persons who are less than 2 years of age.]

f) **Traffic Load**.

The total mass of passengers, baggage and cargo, including any non-revenue load.

JAR - OPS 1.610 LOADING, MASS AND BALANCE

An operator shall specify, in the Operations Manual, the principles and methods involved in the loading and in the mass and balance system that meets the requirements of JAR-OPS 1.605. This system must cover all types of intended operations.

JAR-OPS 1.615 MASS VALUES FOR CREW

a) An operator shall use the following mass values to determine the dry operating mass:

  i) Actual masses including any crew baggage; or
  ii) Standard masses, including hand baggage, of 85 kg for flight crew members and 75 kg for cabin crew members; or
  iii) Other standard masses acceptable to the Authority.

b) An operator must correct the dry operating mass to account for any additional baggage. The position of this additional baggage must be accounted for when establishing the centre of gravity of the aeroplane.

MASS AND BALANCE   **JAR OPS-1 SUBPART J**

JAR-OPS 1.620 Mass Values for Passengers and Baggage

a) An operator shall compute the mass of passengers and checked baggage using either the actual weighed mass of each person and the actual weighed mass of baggage or the standard mass values specified in Tables 1 to 3 below except where the number of passenger seats is less than 10. In such cases passenger mass may be established by use of a verbal statement by or on behalf of each passenger and adding to it a pre-determined constant to account for hand baggage and clothing (See AMC OPS 1.620 (a)). The procedure specifying when to select actual or standard masses and the procedure to be followed when using verbal statements must be included in the Operations Manual.

b) If determining the actual mass by weighing, an operator must ensure that passengers' personal belongings and hand baggage are included. Such weighing must be conducted immediately prior to boarding and at an adjacent location.

c) If determining the mass of passengers using standard mass values, the standard mass values in Tables 1 and 2 below must be used. The standard masses include hand baggage and the mass of any infant below 2 years of age carried by an adult on one passenger seat. Infants occupying separate passenger seats must be considered as children for the purpose of this sub-paragraph.

d) Mass values for passengers - 20 passengers seats or more.

    i) Where the total number of passenger seats available on an aeroplane is 20 or more, the standard masses of male and female in Table 1 are applicable. As an alternative, in cases where the total number of passenger seats available is 30 or more, the 'All Adult' mass values in Table 1 are applicable.

    ii) For the purpose of Table 1, holiday charter means a charter flight solely intended as an element of a holiday travel package. The holiday charter mass values apply provided that not more than 5% of passenger seats installed in the aeroplane are used for the non-revenue carriage of certain categories of passengers (See IEM OPS 1.620 (d) (2)).

MASS AND BALANCE **JAR OPS-1 SUBPART J**

Table 1

Passenger seats:	20 and more		30 and more All adult
	Male	Female	
All holiday flights except holiday charters	88 kg	70 kg	84 kg
Holiday charters	83 kg	69 kg	76 kg
Children	35 kg	35 kg	35 kg

e) Mass values for passengers - 19 passenger seats or less.

　i) Where the total number of passenger seats on an aeroplane is 19 or less, the standard masses in Table 2 are applicable.

　ii) On flights where no hand baggage is carried in the cabin is accounted for separately, 6 kg may be deducted from the above male and female masses. Articles such as an overcoat, an umbrella, a small handbag or purse, reading material or a small camera are not considered as hand baggage for the purpose of this sub-paragraph.

Table 2

Passenger seats	1-5	6-9	10-19
Male	104 kg	96 kg	92 kg
Female	86 kg	78 kg	74 kg
Children	35 kg	35 kg	35 kg

f) Mass values for baggage.

　i) Where the total number if passenger seats available on the aeroplane is 20 or more the standard mass values given in Table 3 are applicable for each piece of checked baggage. For aeroplanes with 19 passenger seats or less, the actual mass of checked baggage, determined by weighing, must be used.

　ii) For the purpose of Table 3:

　　a) Domestic flight means a flight with origin and destination within the borders of one State;

b) Flights within the European region means flights, other than Domestic flights, whose origin and destination are within the area specified in Appendix 1 to JAR-OPS 1.620 (f) ; and

c) Intercontinental flight, other than flights within the European region, means a flight with origin and destination in different continents.

Table 3-20 or more passenger seats

Type of flight	Baggage standard mass
Domestic	11 kg
Within the European region	13 kg
Intercontinental	15 kg
All other	13 kg

g) If an operator wishes to use standard mass values other than those contained in Tables 1 to 3 above, he must advise the Authority of his reasons and gain its approval in advance. He must also submit for approval a detailed weighing survey plan and apply the statistical analysis method given in Appendix 1 to JAR-OPS 1.620 (g). After verification and approval by the Authority of the results of the weighing survey, the revised standard mass values are only applicable to that operator. The revised standard mass values can only be used in circumstances consistent with those under which the survey was conducted. Where revised standard masses exceed those in Tables 1-3, then such higher values must be used. (See IEM OPS 1.620 (g)).

h) On any flight identified as carrying a significant number of passengers whose masses, including hand baggage, are expected to exceed the standard passenger mass, an operator must determine the actual mass of such passengers by weighing or by
Adding an adequate mass increment.
(See IEM OSP 1.620 (h) & (i)).

i) If standard mass values for checked baggage are used and a significant number of passengers check in baggage that is expected to exceed the standard baggage mass, an operator must determine the actual mass of such baggage by weighing or adding an adequate mass increment. (See IEM OPS 1.620 (h) & (i)).

j) An operator shall ensure that a commander is advised when a non-standard method has been used for determining the mass of the load and that this method is stated in the mass and balance documentation.

MASS AND BALANCE   **JAR OPS-1 SUBPART J**

JAR - OPS 1.625 MASS AND BALANCE DOCUMENTATION

(See Appendix 1 to JAR - OPS 1.625)

a) An operator shall establish mass and balance documentation prior to each flight specifying the load and its distribution. The mass and balance documentation must enable the commander to determine that the load and its distribution is such that the mass and balance limits of the aeroplane are not exceeded. The person preparing the mass and balance documentation must be named on the document. The person supervising the loading of the aeroplane must confirm by signature that the load and its distribution are in accordance with the mass and balance documentation. This document must be acceptable to the commander, his acceptance being indicated by countersignature or equivalent. (See also JAR-OPS 1.1055 (a) (12)).

b) An operator must specify procedures for Last Minute Changes to the load.

c) Subject to the approval of the Authority, an operator may use an alternative to the procedures required by paragraphs (a) and (b) above.

MASS AND BALANCE   **JAR OPS-1 SUBPART J**

APPENDIX 1 TO JAR-OPS1.605

MASS AND BALANCE - GENERAL

See JAR-OPS 1.605

(a) **Determination of the dry operating mass of an aeroplane.**

    (1) Weighing of an aeroplane.

        i) New aeroplanes are normally weighed at the factory and are eligible to be placed in operation without re-weighing if the mass and balance records have been adjusted for alterations or modifications to the aeroplane. Aeroplanes transferred from one JAA operator to another JAA operator with an approved programme need not be weighed prior to use by the receiving operator unless more than 4 years have elapsed since the last weighing.

        ii) The individual mass and centre of gravity (CG) position of each aeroplane shall be re-established periodically. The maximum interval between two weighings must be defined by the operator and must meet the requirements of JAR-OPS 1.605 9b). In addition, the mass and CG of each aeroplane shall be re-established either by:

            (a) Weighing; or

            (b) Calculation, if the operator is able to provide the necessary justification to prove the validity of the method of calculation chosen,

        Whenever the cumulative changes to the dry operating mass exceed ± 0.5% of the maximum landing mass or the cumulative change in CG position exceeds 0.5% of the mean aerodynamic chord.

    (2) Fleet mass and CG position.

        (i) For a fleet or group of aeroplanes of the same model and configuration, an average dry operating mass and CG position, provided that the dry operating masses and CG positions of the individual aeroplanes meet the tolerances specified in sub-paragraph (ii) below. Furthermore, the criteria specified in sub-paragraphs (iii), (iv) and (a) (3) below are applicable.

MASS AND BALANCE   **JAR OPS-1 SUBPART J**

(ii)   Tolerances.

        (A)   If the dry operating mass of any aeroplane weighed, or the calculated dry operating mass of any aeroplane of a fleet, varies by more than ± 0.5% of the maximum structural landing mass from the established dry operating fleet mass or the CG position varies by more than ± 0.5% of the mean aero-dynamic chord from the fleet CG, that aeroplane shall be omitted from that fleet. Separate fleets maybe established, each with differing fleet mean masses.

        (B)   In cases where the aeroplane mass is within the dry operating fleet mass tolerance but its CG position falls outside the permitted tolerance, the aeroplane may still be operated under the applicable dry operating mass but with an individual CG position.

        (C)   If an individual aeroplane has, when compared with other aeroplanes of the fleet, a physical, accurately accountable difference (e.g. galley or seat configuration), that causes exceedance of the fleet tolerances, this aeroplane may be maintained by the fleet provided that appropriate corrections are applied to the mass and/or CG position for that aeroplane.

        (D)   Aeroplanes for which no mean aerodynamic chord has been published must be operated with their individual mass and CG position values or must be subjected to a special study and approval.

iii)   Use of fleet values

        (A)   After the weighing of an aeroplane, or if any change occurs in the aeroplane equipment or configuration, the operator must verify that this aeroplane falls within the tolerances specified in sub-paragraph (2) (ii) above.

        (B)   Aeroplanes which have not been weighed since the last fleet mass evaluation can still be kept in a fleet operated with fleet values, provided that the individual values are revised by computation and stay within the tolerances defined in sub-paragraph (2) (ii) above. If these individual values no linger fall within the permitted tolerances, the operator must either determine new fleet values fulfilling the conditions of sub-paragraphs (2) (i) and (2) (ii) above, or operate the aeroplanes not falling within the limits with their individual values.

        (C)   To add an aeroplane to a fleet with fleet values, the operator must verify by weighing or computation that its actual values fall within the tolerances specified in sub-paragraph (2) (i) above.

MASS AND BALANCE — JAR OPS-1 SUBPART J

    (iv)   To comply with sub-paragraph (2) (i) above, the fleet values must be updated at least at the end if each fleet mass evaluation

(3)   Number of aeroplanes to be weighed to obtain fleet values.

    (i)   If 'n' is the number of aeroplanes in the fleet using fleet values, the operator must at least weigh, in the period between two mass fleet evaluations, a certain number of aeroplanes defined in the Table below:

Number of aeroplanes defined in the fleet	Minimum number of weighings
2 or 3	n
4 to 9	$\dfrac{n + 3}{2}$
10 or more	$\dfrac{n + 51}{10}$

    (ii)   In choosing the aeroplanes to be weighed, aeroplanes in the fleet which have not been weighed for the longest time shall be selected.

    (iii)   The interval between 2 fleet mass evaluations must not exceed 48 months.

(4)   Weighing procedure.

    (i)   The weighing must be accomplished by either the manufacturer or by an approved maintenance organisation.

    (ii)   Normal precautions must be taken consistent with good practices such as:

        a)   Checking for completeness of the aeroplane and equipment;
        b)   Determining that fluids are properly accounted for;
        c)   Ensuring that the aeroplane is clean; and
        d)   Ensuring that weighing is accomplished in an enclosed building.

    iii)   Any equipment used for weighing must be properly calibrated, zeroed, and used in accordance with the manufacturer's instructions. Each scale must be calibrated either by the manufacturer, by a civil department of weights and measures or by an appropriately authorised organisation within 2 years or within a time period defined by the manufacturer of the weighing equipment, whichever is less. The equipment must enable the mass of the aeroplane to be established accurately. (See AMC to Appendix 1 to JAR OPS 1.605 para (a) (4) (iii).)

MASS AND BALANCE   **JAR OPS-1 SUBPART J**

b) **Special standard masses for the traffic load.**

In addition to standard masses for passengers and checked baggage, an operator can submit for approval to the Authority standard masses for other load items.

c) **Aeroplane loading.**

1) An operator must ensure that the loading of its aeroplanes is performed under the supervision of qualified personnel.

2) An operator must ensure that the loading of the freight is consistent with the data used for the calculation of the aeroplane mass and balance.

3) An operator must comply with additional structural limits such as the floor length limitations, the maximum load per running metre, the maximum mass per cargo compartment, and/or the maximum seating limits.

d) **Centre of gravity limits.**

1) Operational CG envelope. Unless seat allocation is applied and the effects of the number of passengers per seat row, of cargo in individual cargo compartments and of fuel in individual tanks is accounted for accurately in the balance circulation, operational margins must be applied to the certified centre of gravity envelope. In determining CG margins, possible deviations from the assumed load distribution must be considered. If free seating is applied, the operator must introduce procedures to ensure corrective action by flight or cabin crew if extreme longitudinal seat selection occurs. The CG margins and associated operational procedures, including assumptions with regard to passenger seating, must be acceptable to the Authority. (See IEM to Appendix 1 to JAR-OPS 1.605 sub-paragraph (d).)

2) In-flight centre of gravity.

Further to sub-paragraph (d) (i) above, the operator must show that the procedures fully account for the extreme variation in CG travel during flight caused bu passenger/crew movement and fuel consumption/transfer.

MASS AND BALANCE **JAR OPS-1 SUBPART J**

APPENDIX TO JAR-OPS 1.620 (f)

DEFINITION TO THE AREA FOR FLIGHTS WITHIN THE EUROPEAN REGION.

For the purposes of JAR-OPS 1.620 (f), flights within the European region, other than domestic flights, are flights conducted within the area bounded by rhumb lines between the following points:

N7200 E04500
N4000 E04500
N3500 E03700
N3000 W00600
N2700 W00900
N2700 W03000
N6700 W03000
N7200 W01000
N7200 E04500

as depicted in Figure 1:

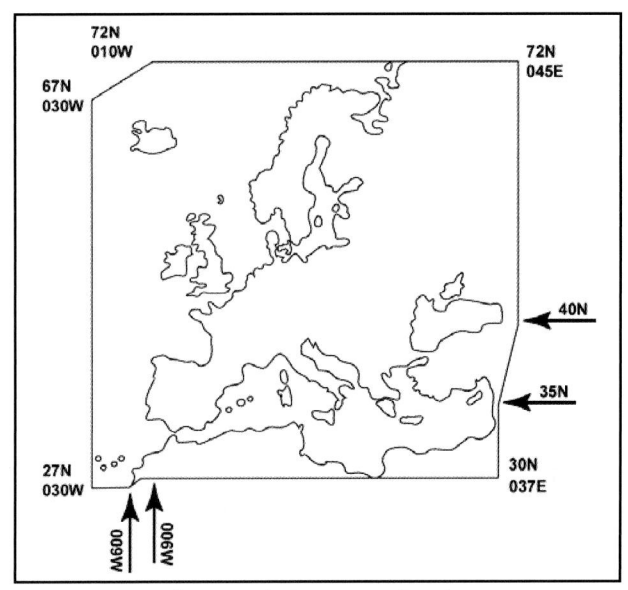

Figure 1.1 European Region

APPENDIX 1 TO JAR - OPS 1.620 (g)

PROCEDURE TO ESTABLISHING REVISED STANDARD MASS VALUES FOR PASSENGERS AND THEIR BAGGAGE.

(See IEM to Appendix 1 to JAR-OPS 1.620 (g)).

a) Passengers.

   1. Weight and sampling method.
      The average mass of passengers and their hand baggage must be determined by weighing, taking random samples. The selection of random samples must by nature and extent be representative of the passenger volume, considering the type of operation, the frequency of flights on various routes, in/outbound flights, applicable season and seat capacity of the aeroplane.

MASS AND BALANCE　　　　　　　　　　　　　　　　　　　　　**JAR OPS-1 SUBPART J**

2. Sample size. The survey plan must cover the weighing at least the greatest of:

   i) A number of passengers calculated from a pilot sample, using normal statistical procedures and based on a relative confidence range (accuracy) of 1% for all adult and 2% for separate male and female average masses (the statistical procedure, complemented with a worked example for determining the minimum sample size and the average mass, is included in IEM OPS 1.620 (g)); and

   ii) For aeroplanes:

   a) With a passenger seating capacity of 40 or more, a total of 2000 passengers; or

   b) With a passenger seating capacity of less than 40, a total number of 50 x (the passenger seating capacity).

3. Passenger masses. Passenger masses must include the mass of the passengers' belongings which are carried when entering the aeroplane. When taking random samples of passenger masses, infants shall be weighed together with the accompanying adult. (See also JAR-OPS 1.620 (c) (d) and (e).)]

4. Weighing location. The location for the weighing of passengers shall be selected as close as possible to the aeroplane, at a point where a change in the passenger mass by disposing of or by acquiring more personal belongings is unlikely to occur before the passengers board the aeroplane.

5. Weighing machine.
   The weighing machine to be used for passenger weighing shall have a capacity of at least 150 kg. The mass shall be displayed at minimum graduations of 500g. The weighing machine must be accurate to within 0.5% or 200 g whichever is the greater.

6. Recording of mass values.
   For each flight included in the survey, the mass of the passengers, the corresponding passenger category (i.e. male/female/children) and the flight number must be recorded.

(b) Checked baggage.

The statistical procedure for determining revised standard baggage mass values based on average baggage masses of the minimum required sample size is basically the same as for passengers and as specified in sub-paragraph (a) (1) (See also IEM OPS 1.620 (g)). For baggage, the relative confidence range (accuracy) amounts to 1%. A minimum of 2000 pieces of checked baggage must be weighed.

MASS AND BALANCE   **JAR OPS-1 SUBPART J**

(c)   Determination of revised standard mass values for passengers and checked baggage.

    1.   To ensure that, in preference to the use of actual masses determined by weighing, the use of revised standard mass values for passengers and checked baggage does not adversely affect operational safety, a statistical analysis (See IEM OPS 1.620(g)) must be carried out. Such an analysis will generate average mass values for passengers and baggage as well as other data.

    2.   On aeroplanes with 20 or more passenger seats, these averages apply as well as revised standard male and female mass values.

    3.   On smaller aeroplanes, the following increments must be added to the average passenger mass to obtain the revised standard mass values.

Number of passenger seats	Required mass increment
1 - 5 inc.	16 kg
6 - 9 inc.	8 kg
10 - 19 inc.	4 kg

Alternatively, all adult revised standard (average) mass values may be applied on aeroplanes with 30 or more passenger seats. Revised standard (average) checked baggage mass values are applicable to aeroplanes with 20 or more passenger seats.

    4.   Operators have the option to submit a detailed survey plan to the Authority for approval and subsequently a deviation from the revised standard mass value provided this deviating value is determined by use of the procedure explained in this appendix. Such deviations must be renewed at intervals not exceeding 5 years. (See AMC to Appendix 1 to JAR-OPS 1.620 9g), sub-paragraph (c) (4).

    5.   All adult revised standard mass values must be based on a male/female ratio of 80/20 in respect of all flights except holiday charters which are 50/50. If an operator wishes to obtain approval for use of a different ratio on specific routes or flights then data must be submitted to the Authority showing that the alternative male/female ratio is conservative and covers at least 84% of the actual male/female ratios on a sample of at least 100 representative flights.

    6.   The average mass values found are rounded to the nearest whole number in kg. Checked baggage mass values are rounded to the nearest 0.5 kg figure, as appropriate.

MASS AND BALANCE   **JAR OPS-1 SUBPART J**

APPENDIX TO JAR - OPS 1.525

MASS AND BALANCE DOCUMENTATION

See IEM to Appendix 1 to JAR - OPS 1.625

a)   Mass and balance documentation.

        1.   Contents.

               i)   The mass and balance documentation must contain the following information:

                       a)   The aeroplane registration and type;

                       b)   The flight identification number and date;

                       c)   The identity of the Commander;

                       d)   The identity of the person who prepared the document;

                       e)   The dry operating mass and the corresponding CG of the aeroplane;

                       f)   The mass of the fuel at take-off and the mass of trip fuel;

                       g)   The mass of consumables other than fuel;

                       h)   The components of the load including passengers, baggage, freight and ballast;

                       i)   The Take-off Mass, Landing Mass and Zero Fuel Mass;

                       j)   The load distribution;

                       k)   The applicable aeroplane CG positions; and

                       l)   The limiting mass and CG values.

               ii   Subject to the approval of the Authority, an operator may omit some of this data form the mass and balance documentation.

MASS AND BALANCE   **JAR OPS-1 SUBPART J**

    2.   Last minute change.

        If any last minute change occurs after the completion of the mass and balance documentation, this must be brought to the attention of the commander and at the last minute change must be entered on the mass and balance documentation. The maximum allowed change in the number of passengers or hold load acceptable as a last minute change must be specified in the Operations Manual. If this number is exceeded, new mass and balance documentation must be prepared.

b)   Computerised systems.

    Where mas and balance documentation is generated by a computerised mass and balance system, the operator must verify the integrity of the output data. He must establish a system to check that amendments of his input data are incorporated properly and in the system on a continuous basis by verifying the output data at intervals not exceeding 6 months.

c)   Onboard mass and balance systems.

    An operator must obtain the approval of the Authority if he wishes to use an onboard mass and balance computer system as a primary source for despatch.

d)   Datalink.

    When mass and balance documentation is sent to aeroplanes via datalink, a copy of the final mass and balance documentation as accepted by the commander must be available on the ground.

MASS AND BALANCE SELF ASSESSMENT QUESTIONS - THEORY

1. JAR Mass and Balance legislation can be found in:

   a. JAR OPS-1 subpart A
   b. JAR OPS-1 subpart D
   c. JAR OPS-1 subpart K
   d. JAR OPS-1 subpart J

2. The mass and centre of gravity of an aircraft must be established by actual weighing:

   a. by the pilot on entry of aircraft into service
   b. by the engineers before commencing service
   c. by the operator prior to initial entry of aircraft into service
   d. by the owner operator before the first flight of the day

3. The operator must establish the mass of the Traffic Load:

   a. prior to initial entry into service
   b. by actual weighing or determine the mass of the traffic load in accordance with standard masses as specified in JAR-OPS sub part J.
   c. prior to embarking on the aircraft
   d. by using an appropriate method of calculation as specified in the JAR Ops subpart J

4. The mass of the fuel load must be determined:

   a. by the operator using actual density or by density calculation specified in the Operations Manual.
   b. by the owner using actual density or by density calculation specified in JAR OPS - 1.
   c. by the pilot using actual density or by density calculation specified in the Operations Manual.
   d. by the fuel bowser operator using actual density or by density calculation specified in the Fuelling Manual.

5. The Dry Operating Mass is the total mass of the aeroplane ready for a specific type of operation and includes:

   a. Crew and passenger baggage, special equipment, water and chemicals
   b. .Crew and their hold baggage, special equipment, water and contingency fuel
   c. Crew baggage, catering and other special equipment, potage water and lavatory chemicals
   d. Crew and baggage, catering and passenger service equipment, potable water and lavatory chemicals.

MASS AND BALANCE **JAR OPS-1 SUBPART J**

6. The Maximum Zero Fuel Mass is the maximum permissible mass of the aeroplane:

   a. with no useable fuel
   b. with no useable fuel unless the Aeroplane Flight Manual Limitations explicitly include it.
   c. including the fuel taken up for take-off
   d. including all useable fuel unless the Aeroplane Flight Operations Manual explicitly excludes it.

7. The Maximum Structural Take-off Mass is:

   a. the maximum permissible total aeroplane mass on completion of the refuelling operation.
   b. the maximum permissible total aeroplane mass for take-off subject to the limiting conditions at the departure airfield.
   c. the maximum permissible total aeroplane mass for take-off but excluding fuel.
   d. the maximum permissible total aeroplane mass at the start of the take-off run.

8. The Regulated Take-off Mass:

   a. is the lower of maximum structural take-off mass and the performance limited take-off mass.
   b. is the higher of the maximum structural zero fuel mass and the performance limited take-off mass.
   c. the maximum structural take-off mass subject to any last minute mass changes.
   d. the maximum performance limited take-off mass subject to any last minute mass changes.

9. The Take-off mass

   a. the maximum permissible total aeroplane mass on completion of the refuelling operation.
   b. the mass of the aeroplane including everyone and everything contained within it at the start of the take-off run.
   c. the maximum permissible total aeroplane mass for take-off but excluding fuel.
   d. the maximum permissible total aeroplane mass at the start of the take-off run.

10. The Operating Mass:

    a. is the lower of the structural mass and the performance limited mass
    b. is the higher of the structural mass and the performance limited mass
    c. is the actual mass of the aircraft on take-off
    d. is the dry operating mass and the fuel load.

MASS AND BALANCE   **JAR OPS-1 SUBPART J**

11. The Basic Empty Mass is the mass of the aeroplane:

    a. plus non-standard items such as lubricating oil, fire extinguishers, emergency oxygen equipment etc.
    b. minus non-standard items such as lubricating oil, fire extinguishers, emergency oxygen equipment etc.
    c. plus standard items such as unusable fluids, fire extinguishers, emergency oxygen equipment, supplementary electronics etc.
    d. minus non-standard items such as unusable fluids, fire extinguishers, emergency oxygen and supplementary electronic equipment etc.

12. The Traffic Load:

    a. includes passenger masses and baggage masses but excludes any non-revenue load.
    b. includes passenger masses, baggage masses and cargo masses but excludes any non-revenue load.
    c. includes passenger masses, baggage masses, cargo masses and any non-revenue load.
    d. includes passenger masses, baggage masses and any non-revenue load but excludes cargo.

13. The Operating Mass:

    a. is the take-off mass minus the traffic load.
    b. is the landing mass minus the traffic load
    c. is the maximum zero fuel mass less the traffic load
    d. is the take-off mass minus the basic empty mass and crew mass.

14. The Traffic Load is:

    a. The Zero Fuel Mass minus the Dry operating Mass
    b. The Take-off Mass minus the sum of the Dry Operating Mass and the total fuel load.
    c. The landing Mass minus the sum of the Dry Operating Mass and the mass of the remaining fuel.
    d. all the above

15. The Basic Empty Mass is the:

    a. MZFM minus both traffic load and the fuel load
    b. Take-off mass minus the traffic load and the fuel load
    c. Operating mass minus the crew and fuel load
    d. Landing mass less traffic load

MASS AND BALANCE   JAR OPS-1 SUBPART J

16. Is it possible to fly a certified aircraft at a Regulated Take-off mass with both a full traffic load and a full fuel load?

    a. Some aircraft some of the time
    b. All aircraft all the time
    c. No, it is not possible!
    d. Only if the performance limited take-off mass is less than the structural limited take-off mass.

17. It is intended to fly a certified aircraft with both a full traffic load and a full fuel load.

    a. The CG might be in limits all of the flight.
    b. The CG limits will be in limits all of the flight.
    c. The CG might not be in limits any of the time during the flight.
    d. The CG will not be within the limits during the flight.

18. The term 'baggage' means:

    a. Excess frieght
    b. Any non-human, non-animal cargo
    c. any frieght or cargo not carried on the person
    d. personal belongings

19. Certified Transport category aircraft with less than 10 seats:

    a. may accept a verbal mass from or on behalf of each passenger.
    b. estimate the total mass of the passengers and add a pre-determined constant to account for hand baggage and clothing.
    c. may compute the actual mass of passengers and checked baggage.
    d. all the above.

20. When computing the mass of passengers and baggage:

    A. Personal belongings and hand baggage must be included
    B. Infants must be classed as children it they occupy a seat
    C. Standard masses include infants being carried by an adult.
    D. Table 1, Table 2 and Table 3 must be used as appropriate if using standard masses for passengers and freight.
    E. Weighing must be carried out immediately prior to boarding and at an adjacent location.

    a. A, B and E only
    b. B and D only
    c. A, B, C and E only
    d. All the above

MASS AND BALANCE   JAR OPS-1 SUBPART J

21. When computing the mass of passengers and baggage for an aircraft with 20 seats or more:

    A. Standard masses of male and female in Table 1 are applicable.
    B. If there are thirty seats or more, the 'All Adult' mass values in Table 1 may be used as an alternative.
    C. Holiday Charter masses apply to Table 1 and Table 3 if the charter is solely intended as an element of a holiday travel package.
    D. Holiday flights and holiday charters attract the same mass values.

    a. A, C and D only
    b. A and B only
    c. C and D only
    d. All the above

22. When computing the mass of passengers and baggage for an aircraft with 19 seats or less:

    A. The standard masses in Table 2 apply
    B. If hand baggage is accounted for separately, 6 kg may be deducted from the mass of each male and female.
    C. Table 2 masses vary with both the gender (male or female) of the seat occupant and the number of seats on the aircraft.
    D. Standard masses are not available for baggage.
    E. Standard masses are not available for freight.

    a. A only
    b. A, B and D only
    c. C and E only
    d. All the above

23. When computing the mass of checked baggage for an aircraft with twenty seats or more:

    A. Table 1 applies
    B. Table 2 applies
    C. Table 3 applies
    D. Mass is categorised by destination.
    E. Mass is categorised by gender

    a. A, C and D only
    b. B, C and E only
    c. C and D only
    d. All the above

MASS AND BALANCE   JAR OPS-1 SUBPART J

24. On any flight identified as carrying a significant number of passengers whose masses, including hand baggage, are expected to exceed the standard passenger mass the operator:

   a. must determine the actual mass of such passengers
   b. must add an adequate mass increment to each of such passengers
   c. must determine the actual masses of such passengers or add an adequate increment to each of such passengers.
   d. need only determine the actual masses or apply an increment if the Take-off mass is likely to be exceeded.

25. If standard mass tables are being used for checked baggage and a number of passengers check in baggage that is expected to exceed the standard baggage mass, the operator:

   a. determine the actual masses of such baggage
   b. must determine the actual mass of such baggage by weighing or by deducting an adequate mass increment.
   c. need may no alterations if the Take-off mass is not likely to be exceeded.
   d. must determine the actual mass of such baggage by weighing or adding an adequate mass increment.

26. Mass and balance documentation:

   A. must be established prior to each flight
   B. must enable the commander to determine that the load and its distribution is such that the mass and balance limits of the aircraft are not exceeded.
   C. must include the name of the person preparing the document.
   D. must be signed by the person supervising the loading to the effect that the load and its distribution is in accordance with the data on the document.
   E. must include the aircraft commander=s signature to signify acceptance of the document.

   a. All the above
   b. B, D and E only
   c. A, D and E only
   d. A and C only

27. Once the mass and balance documentation has been signed prior to flight:

   a. no load alterations are allowed.
   b. documented last minute changes to the load may be incorporated.
   c. the documentation is not signed prior to flight.
   d. acceptable last minute changes to the load must be documented.

MASS AND BALANCE   JAR OPS-1 SUBPART J

28. Aircraft must be weighed:

   A. on initial entry into service
   B. if the mass and balance records have not been adjusted for alterations or modifications.
   C. every four years after initial weigh
   D. whenever the cumulative changes to the dry operating mass exceed plus or minus 0.5% of the maximum landing mass.
   E. if the cumulative change in CG position exceeds 0.5% of the mean aerodynamic chord.

   a. A and C only
   b. A, B, C, D and E
   c. A, B and C only
   d. A, C and E only

29. Aeroplane loading:

   A. must be performed under the supervision of qualified personnel
   B. must be consistent with the data used for calculating the mass and balance.
   C. must comply with compartment dimension limitations
   D. must comply with the maximum load per running metre
   E. must comply with the maximum mass per cargo compartment

   a. A and B only
   b. A, B, D and E only
   c. A, B, C, D and E
   d. C, D and E only

30. An average dry operating mass and CG position may be used for a fleet or group of aeroplanes:

   A. if they are of the same model and configuration
   B. providing the individual masses and CG positions meet specific tolerances specified in JAR OPS section J.
   C. providing the dry operating mass of any aeroplane does not vary by more than 0.5% of the maximum structural landing mass of the fleet.
   D. providing that the CG position varies by more than 0.5% of the mean aerodynamic chord of the fleet.
   E. providing appropriate corrections to mass and CG position are applied to aircraft within the fleet which have a physical, accurately accountable difference.

   a. A, B, C, D and E
   b. A, B, C and E only
   c. B, C and D only
   d. A, D and E only

MASS AND BALANCE **JAR OPS-1 SUBPART J**

THEORY QUESTION ANSWERS

Question	Answer	Question	Answer
1	d	16	a
2	c	17	b
3	b	18	d
4	a	19	d
5	d	20	c
6	b	21	b
7	d	22	d
8	a	23	c
9	b	24	c
10	d	25	d
11	c	26	a
12	c	27	d
13	a	28	b
14	d	29	c
15	c	30	b

CHAPTER TWO - MASS AND BALANCE DEFINITIONS AND CALCULATIONS

Contents

		Page
2.1	INTRODUCTION	2-1
2.2	LIMITATIONS	2-1
2.3	EFFECTS OF OVERLOADING	2-1
2.4	EFFECTS OF OUT OF LIMIT CG POSITION	2-1
2.5	MOVEMENT OF CG IN FLIGHT	2-2
2.6	DEFINITIONS	2-2
2.7.	WEIGHING OF AIRCRAFT	2-7
2.8.	EQUIPMENT LIST	2-8
2.9.	CALCULATION OF FUEL MASS	2-8
2.10.	CALCULATION OF CENTRE OF GRAVITY	2-10
2.11.	CALCULATION OF CG FOR THE BASIC EMPTY MASS	2-12
2.12.	CALCULATION OF CG FOR LOADED MASS	2-13
2.13.	PROCEDURE FOR CALCULATING THE LOADED MASS	2-13
2.14.	COMPILING A MASS AND BALANCE DOCUMENT (LOAD SHEET)	2-14
2.15.	CG POSITION AS PERCENTAGE OF MEAN AERODYNAMIC CHORD	2 - 20
2.16.	RE-POSITIONING OF THE CENTRE OF GRAVITY	2 - 22
2.17.	RE-POSITIONING CENTRE OF GRAVITY BY REPOSITIONING MASS.	2 - 22
2.18.	RE-POSITIONING CENTRE OF GRAVITY BY ADDING OR SUBTRACTING MASS	2 - 26
2.19.	GRAPHICAL PRESENTATION	2 - 29
2.20.	CARGO HANDLING	2 - 29

2.21.	FLOOR LOADING	2 - 31
2.22.	LINEAR / RUNNING LOADS	2 - 31
2.23.	AREA LOAD LIMITATIONS	2 - 32
2.24.	SINGLE ENGINE PISTON/ PROPELLER AIRCRAFT (SEP 1)	2 - 33
2.25.	LIGHT TWIN PISTON/PROPELLER AIRCRAFT (MEP 1)	2 - 34
2.26.	MEDIUM RANGE JET TWIN (MRJT 1)	2 - 34
2.27.	MASS AND BALANCE CALCULATIONS (MRJT)	2 - 37
2.28.	LOAD AND TRIM SHEET MRJT	2 - 44
2.29.	SELF ASSESSMENT QUESTIONS FOR S.E.P.1; M.E.P.1 AND MRJT	2 - 52

MASS AND BALANCE   **LOADING AND CENTRE OF GRAVITY**

2.1 INTRODUCTION

JAR-OPS 1 Subpart J requires that during any phase of operation the loading, mass and centre of gravity of the aeroplane complies with the limitations specified in the approved Aeroplane Flight Manual, or the Operations Manual if more restrictive.

It is the responsibility of the commander of the aircraft to satisfy himself that this requirement is met.

2.2 LIMITATIONS

Limitations on mass are set to ensure adequate margins of strength and performance, and limitations on CG position are set to ensure adequate stability and control of the aircraft in flight.

2.3 EFFECTS OF OVERLOADING

If the limiting mass of an aircraft is exceeded:

a)   Performance is reduced:

        i)   Take off and landing distances will be increased.
        ii)   Rate of climb and ceiling height will be reduced.
        iii)   Range and endurance will be reduced.
        iv)   Maximum speed is reduced.

b)   Stalling speed is increased.

c)   Manoeuverability is reduced.

d)   Wear on tyres and brakes is increased.

e)   Structural safety margins will be reduced.

2.4 EFFECTS OF OUT OF LIMIT CG POSITION

The CG must lie between forward and aft limits. if the CG is out of limits control forces, stability, manoeuvrability and performance will all be affected. The following paragraphs give an indication of the effects that might occur if the CG is out of limits; students are advised to learn them well - they are frequently asked in the JAR exams.

1.   If the CG is outside the forward limit:

a)   A large balancing download is required from the tailplane, requiring large elevator deflections. This results in increased drag from the control surfaces, and because of the increased wing lift required to balance the tailplane download, increased induced drag. This causes reduced performance.

MASS AND BALANCE — LOADING AND CENTRE OF GRAVITY

b) Stalling speed will be increased because of the increased wing lift required.
c) Longitudinal stability is increased, leading to higher stick forces in pitch.
d) Range and endurance are decreased due to the increased trim drag as the elevators are used to trim the aircraft.
e) Nose up pitch is decreased because the elevators, having been used to trim the aircraft straight and level have less 'up' elevator range.

2. If the CG is outside the aft limit:

a) Longitudinal stability is reduced, and if the CG is too far aft, the aircraft will become unstable. Stick forces in pitch will be light, leading to the possibility of over stressing the aircraft by applying excessive 'g'.
b) Recovering from a spin may be more difficult because a flat spin may develop.
c) Range and endurance will usually decrease due to the extra drag.
e) Glide angle will be more difficult to sustain because of the tendency for the aircraft to pitch up.

2.5 MOVEMENT OF CG IN FLIGHT

It must be shown that the CG will be in limits for the entire flight. Any changes that occur in the aircraft's loading must be allowed for, as they may change the position of the CG. Changes will occur due to fuel and oil consumption, passenger movements within the aircraft and the dropping of loads, e.g. supplies or parachutists.

2.6 DEFINITIONS - (It is important that the student is fully familiar with each of these terms).

Centre of Gravity (CG).
The point through which the force of gravity is said to act on a mass (in C of G terms, the point on the aircraft through which the total mass is said to act in a vertically downward manner). The Centre of Gravity is also the point of balance and as such it affects the stability of the aircraft both on the ground and in the air.

Centre of Gravity limits.
The most forward and most aft positions of the CG at which the aircraft is permitted to fly. They are set by the manufacturer, defined in the Aeroplane Flight Manual and are mandatory. The CG limits are defined relative to the CG datum or may be expressed as a percentage of the Mean Aerodynamic Chord (MAC).

Centre of Gravity Datum.
A point designated by the manufacturer from which all measurements are taken in the calculation of the CG. The datum can be anywhere along the longitudinal axis of the aeroplane even if the axis is extended in front of or behind the aeroplane.

Arm.
The distance from the CG datum to the point at which the mass of a component acts (the CG of the component). By convention, all arms behind (aft) of the datum are positive and all arms forward (fwd) of the datum are negative - See fig.2.1.

MASS AND BALANCE   **LOADING AND CENTRE OF GRAVITY**

Moment.
The turning effect of a mass around the datum. It is the product of the mass multiplied by the arm. Since the arm may be positive or negative, a positive arm will give a positive moment and a negative arm will give a negative moment.

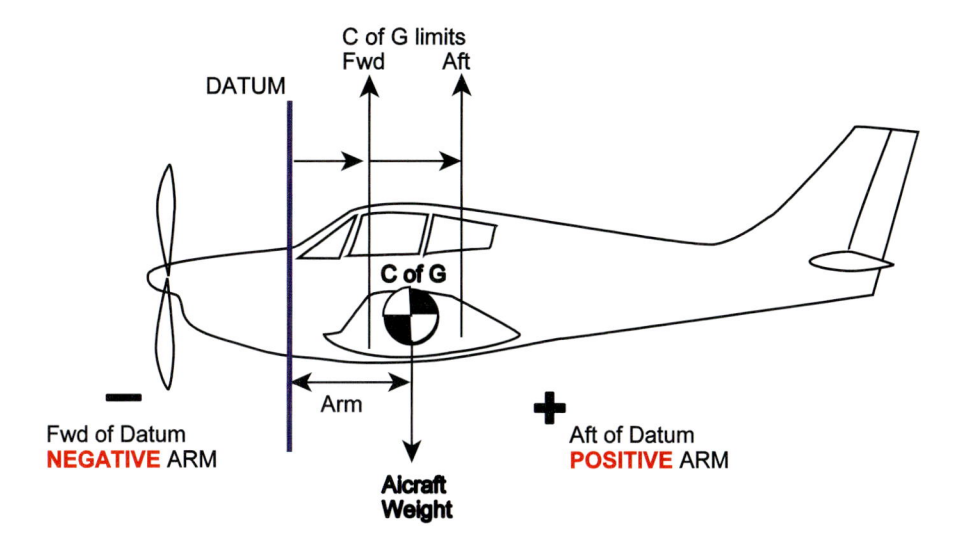

Figure 2.1 Positive and Negative Balance Arms

Loading Index.
A non dimensional figure which is a scaled down value of a moment used to simplify mass and balance calculations.

Dry Operating Index (DOI)
The index for the position of the centre of gravity at Dry Operating Mass.

Basic Empty Mass (BEM).
The mass of the aircraft with all its basic equipment plus a declared quantity of unusable fuel and oil.

Dry Operating Mass (DOM).
The total mass of the aeroplane ready for a specific type of operation, excluding all fuel and traffic load. This mass includes items such as :

       1.   Crew and crew baggage.
       2.   Catering and removable passenger service equipment.
       3.   Potable water and lavatory chemicals.

Operating Mass (OM)
Is the DOM plus fuel but without traffic load.

Traffic Load.
The total mass of passengers, baggage and freight.

MASS AND BALANCE — LOADING AND CENTRE OF GRAVITY

Useful Load
The total of Traffic Load plus Useable fuel.

Zero Fuel Mass (ZFM)
The total mass of the Dry Operating Mass plus the Traffic Load.

Maximum Zero Fuel Mass (MZFM).
The maximum permissible mass of an aeroplane with no useable fuel. The mass of fuel contained in particular tanks must be included in the ZFM when specified in the AFM limitations.

Take-Off Mass (TOM)
The mass of an aeroplane including everything and everyone in it at the start of the take-off run.

Maximum Structural Take Off Mass (MTOM).
The maximum permissible total aeroplane mass at the beginning of the take off run.

Maximum Structural Landing Mass (MLM).
The maximum permissible total aeroplane mass upon landing under normal circumstances.

Maximum Ramp Mass.(Max Structural Taxi Mass).
The maximum approved mass for commencement of ground manoeuvres. A mass greater than the Maximum Take Off Mass, to allow for fuel used in start up and taxi.

Examples of 'definition' questions:

1. The operating mass of an aircraft is:

    a. The dry operating mass plus the take-off fuel mass
    b. The empty mass plus the take-off fuel mass
    c. The empty mass plus crew, crew baggage and catering
    d. The empty mass plus the trip fuel mass

2. What effect has a centre of gravity close to the forward limit?

    a. A better rate of climb capability
    b. A reduction in the specific fuel consumption
    c. A reduce rate of climb
    d. A decreased induced drag

3. The DOM of an aeroplane is:

    a. TOM minus Operating Mass
    b. LM plus Trip Fuel
    c. Useful Load minus Operating Mass
    d. TOM minus Useful Load

4. The Traffic Load of an aeroplane is:

    a. TOM minus Operating Mass
    b. LM plus Trip Fuel
    c. Useful Load minus Operating Mass
    d. TOM minus Useful Load

MASS AND BALANCE   **LOADING AND CENTRE OF GRAVITY**

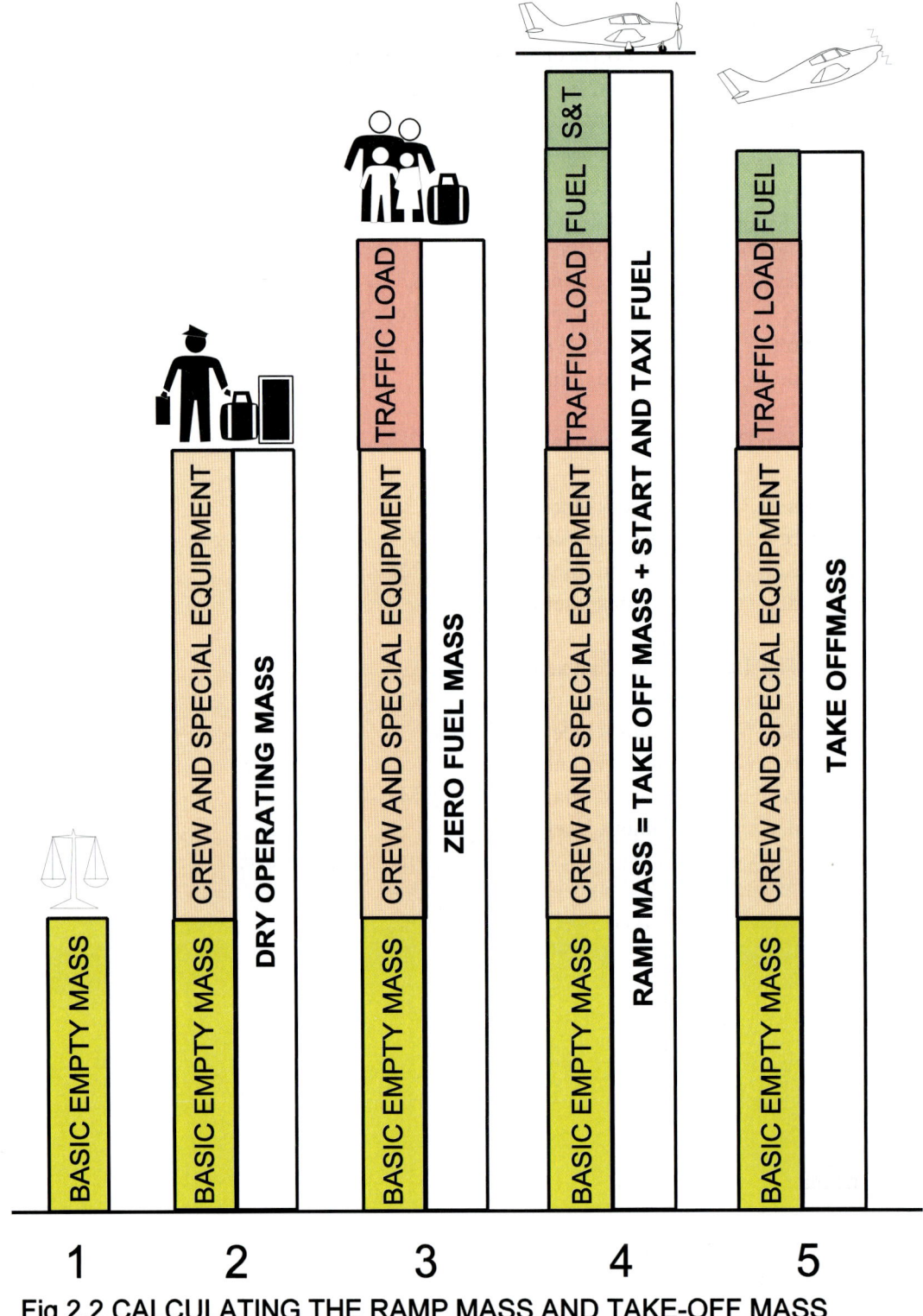

Fig 2.2 CALCULATING THE RAMP MASS AND TAKE-OFF MASS

2.7. WEIGHING OF AIRCRAFT

Aircraft are weighed in a draught free hangar and on each and every occasion of weighing a weighing schedule is compiled that states the basic equipment installed at the time together with the mass and CG position of the aircraft after completion of the weighing procedure. The weighing schedule is retained until the subsequent weighing operation and the information on it is used as the datum for the re-weigh.

The Basic Empty Mass (BEM) and Centre of Gravity (CG) of a new aeroplane are determined by the manufacturer using one of a number of weighing procedures. The BEM and CG position are subsequently checked by the operator at 4 yearly intervals or as specified in JAR-OPS 1 Subpart J.

Light aircraft generally use the BEM and CG as the start point for loading calculations but for larger aircraft, the masses of the crew, their baggage and special equipment are added to the BEM to arrive at the DOM and CG. This simplifies the more complicated "Load and Trim Sheet" calculations associated with such larger aircraft.

For a fleet or group of aeroplanes of the same model and configuration, an average DOM and CG position may be used as a Fleet Mass and CG position providing the requirements of Subpart J are met.

Weighing Equipment. The main types of weighing equipment in use are weigh-bridge scales, hydrostatic units, and electronic equipment.

  a) Weigh-bridge scales. This equipment consists of a separate weighing platform for each wheel or bogie of the aircraft. The mass at each platform is recorded directly on the balance arm.

  b) Hydrostatic units. The operation of these units depends on the hydraulic principle that pressure is proportional to the load applied. The units are interposed between the lifting jacks and the jacking points on the aircraft.

  c) Electronic equipment. This equipment uses strain gauges, which vary their electrical resistance with load.

When the empty mass of the aircraft has been determined by weighing, its fully loaded mass can be calculated by adding known masses of passengers, baggage, freight and fuel.

The mass of fuel can be calculated if its volume and specific gravity are known.

Actual mass of passengers and baggage can be used, alternatively standard masses given in Subpart J can be used.

Mass of freight must, on all occasions, be calculated from actual masses of individual items.

2.8. EQUIPMENT LIST

An equipment list is produced for most aeroplanes and details the mass and arm of all removable equipment fitted to the aircraft. This enables the operator to adjust the mass and balance documentation for any equipment that is removed for a particular flight.

2.9. CALCULATION OF FUEL MASS

Mass of fuel must be determined by using the actual density or specific gravity of the fuel. If the actual density is not known a standard fuel density specified in the Operations Manual may be used. Density is mass per unit volume, Specific Gravity (SG) is the mass compared to an equal volume of pure water.

Fuel is measured as it is being dispensed to the aircraft, it may be measured in various quantities ie. Imperial gallons, US gallons or litres. These quantities need to be converted into mass in Kilogrammes or Pounds to be entered in the load sheet.

The following chart is a handy method of converting volume to mass. You are advised to remember this chart as it will not be provided in the exams. When moving in the direction of the arrows multiply by the numbers above the line. When moving against the direction of the arrows divide by the numbers above the line.

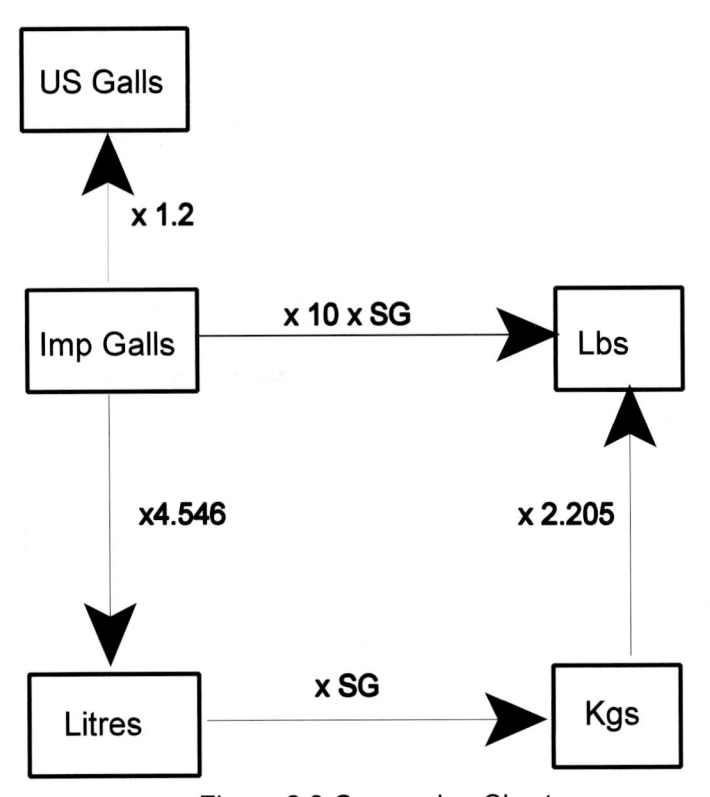

NB. Conversion factors have been rounded for simplicity hence small errors might occur.

Figure 2.3 Conversion Chart

MASS AND BALANCE — LOADING AND CENTRE OF GRAVITY

Example 1

a) Find the mass of 50 Imperial gallons of AVGAS with a specific gravity of 0.72.

Mass = 50 x 10 x 0.72 = 360 lb

b) For 50 US gallons this would be:

Mass = 50 ÷ 1.2 x 10 x 0.72 = 300 lb

To convert pounds to kilograms, divide the lb figure by 2.205.

Example 2

Find the mass of 2250 litres of fuel with a density of 0.82.

Mass = 2250 x 0.82 = 1845 kg.

To convert kg to lb multiply the kg figure by 2.205.

Try these on your own

1. You require 63,000 kg of fuel for your flight, the aircraft currently has 12,000 kg indicated on the gauges. How many US gallons of fuel do you request if the density is 0.81.

2. The refueller has metered 4596 Imperial Gallons, your fuel gauges indicated 5,600 lbs before refuelling, what should they indicate now. The density is 0.79.

3. If the mass of 6000 US gallons of fuel is 16780 kg, what is its S.G. given that 1 Imperial gallon of fuel equals 4.546 litres?

4. The bowser delivers 10,000 litres of fuel and this is incorrectly entered on the aircraft load sheet as 10,000 kgs. Given that the SG of the fuel is 0.75, determine if the aircraft is heavier or lighter than appears and by how much.

MASS AND BALANCE   **LOADING AND CENTRE OF GRAVITY**

2.10. CALCULATION OF CENTRE OF GRAVITY

The position of the Centre of Gravity is calculated using the principle of moments about a datum. The total mass of an aircraft acts through its Centre of Gravity and will cause a moment (turning force) about any chosen datum. The datum remember is a fixed point determined by the manufacturer.

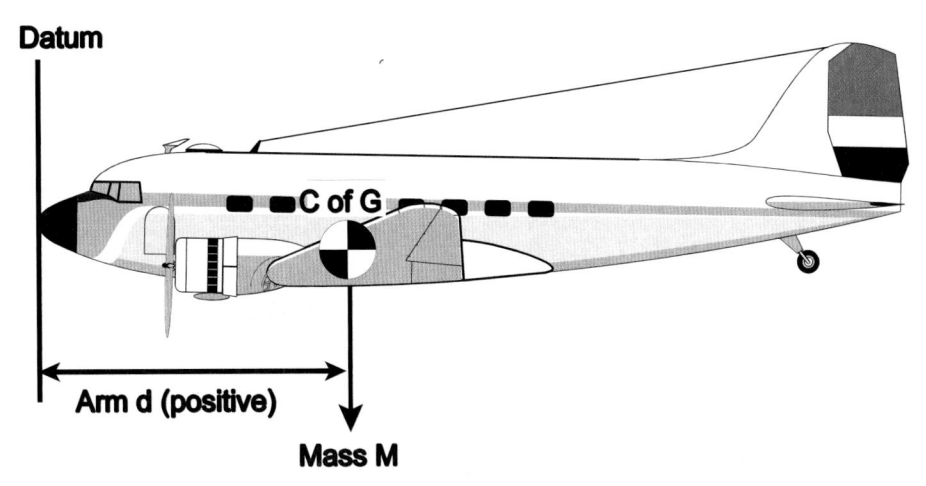

Figure 2.4 CG Datum, Arm and Mass

Total moment about datum = M x d

Equally the **total moment** about the datum can be considered to be made up of the component masses of the aircraft, each multiplied by its distance from the datum.

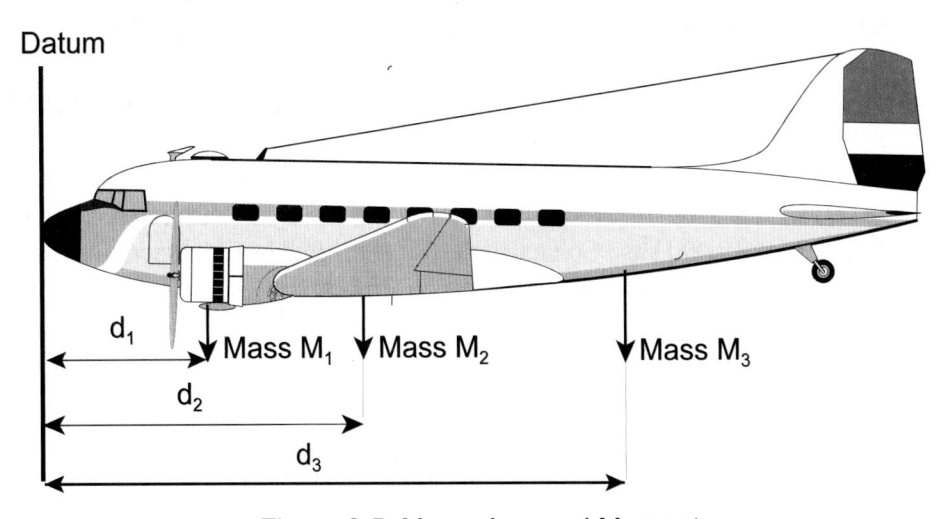

Figure 2.5 Mass, Arm and Moment

MASS AND BALANCE   **LOADING AND CENTRE OF GRAVITY**

$$\text{Total Moment} = (M_1 \times d_1) + (M_2 \times d_2) + (M_3 \times d_3)$$

Notice that in the above example the aircraft has the datum at the nose. This has the advantage that all the calculated moments are positive because all the masses are aft of the datum.

If the position of the datum is such that positive and negative moments are encountered they are added mathematically in the same way. **Don't forget that items forward of the datum have a negative arm and will give a negative moment.**

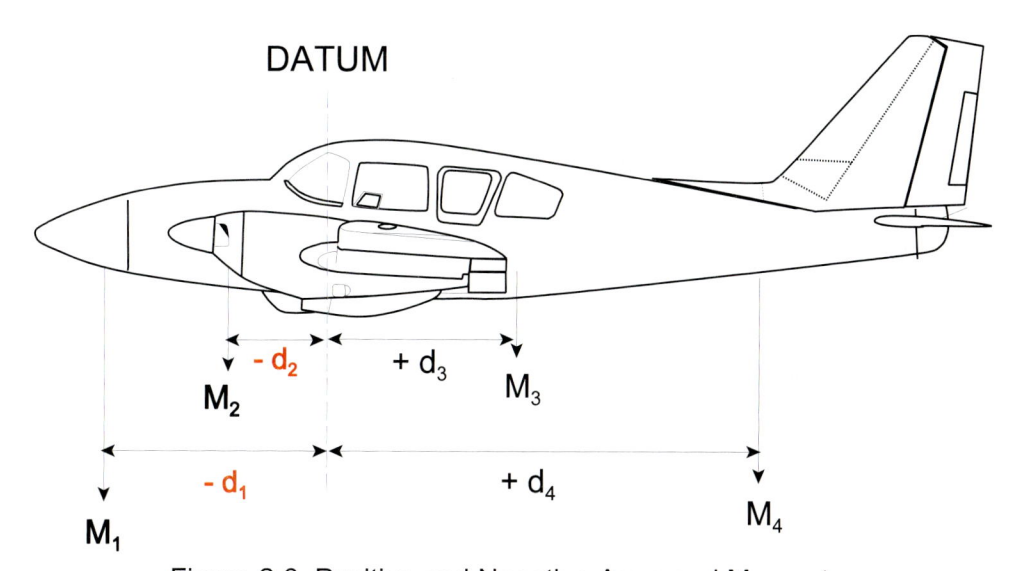

Figure 2.6 Positive and Negative Arms and Moments

In this case:

Total Moment about datum $= (M_1 \times -d_1) + (M_2 \times -d_2) + (M_3 \times +d_3) + (M_4 \times +d_4)$

and from Fig 2.6

Total Moment = Total Mass (M) x CG distance from Datum (d)

The C of G position from the datum can be found by using the formula:

CG position relative to datum   = $\dfrac{\textbf{Total Moment}}{\textbf{Total Mass}}$

*The negative balance arms forward of the datum are shown in red.

MASS AND BALANCE — LOADING AND CENTRE OF GRAVITY

2.11. CALCULATION OF CG FOR THE BASIC EMPTY MASS

The above procedure may be used to find the CG position for the Basic Empty Mass

The aircraft is weighed and the mass on each undercarriage leg is indicated by the weighing device.

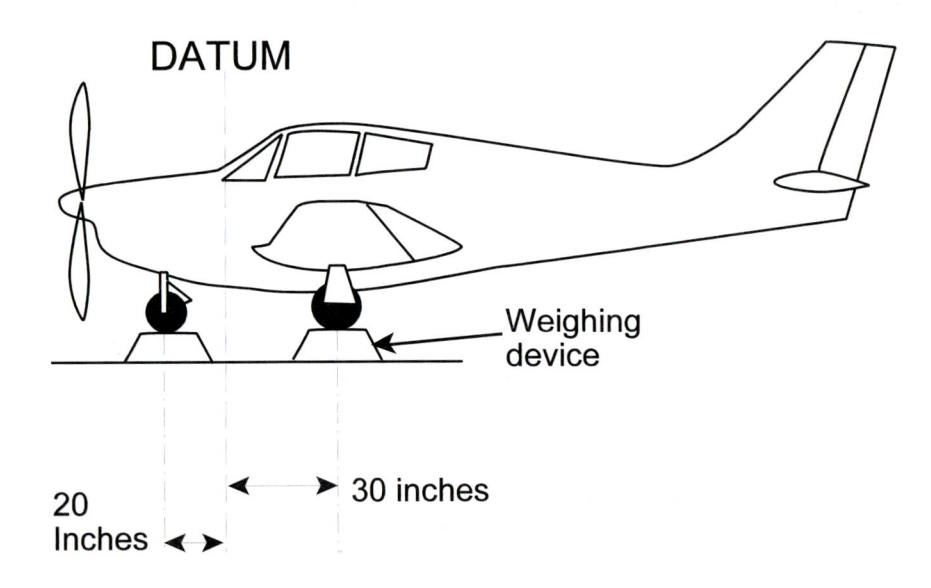

Figure 2.7 Basic Empty mass Weighing Calculations

Example:

ITEM	MASS (lb)	ARM	MOMENT
Nose wheel	500	-20	-10000
L. Main wheel	2000	30	+60,000
R. Main wheel	2000	30	+60,000
Total Mass =	4500	**Total Moment =**	+110,000

C of G = $\dfrac{\text{Total Moment}}{\text{Total Mass}}$ = $\dfrac{+110,000}{4,500}$

C of G position = +24.4   (i.e. 24.4 inches aft of the Datum)

Try this one yourself;
An aeroplane with a two wheel nose gear and four main wheels rests on the ground with a single nose wheel load of 725 kg and a single main wheel load of 6000 kg. The distance between nose wheels and the main wheels is 10 metres. How far is the centre of gravity in front of the main wheels?

MASS AND BALANCE   **LOADING AND CENTRE OF GRAVITY**

2.14. COMPILING A MASS AND BALANCE DOCUMENT (LOAD SHEET)

A load sheet in its simplest form is a list showing the DOM and CG position. Added in tabular form are elements of the traffic load and fuel by their individual masses, distribution and moments. From this list the take off mass and C of G position can be calculated.

A load sheet is individual to each type of aircraft and must be compiled before each flight.

A load sheet is required by JAR-OPS 1 Subpart J to contain some mandatory information:

1. The aeroplane registration and type
2. The Flight Number
3. The identity of the commander
4. The identity of the person who prepared the document.
5. The dry operating mass and CG position.
6. The mass of take off fuel and trip fuel.
7. The mass of consumables other than fuel.
8. The items of traffic load, passengers, baggage, freight.
9. The take of mass, landing mass and zero fuel mass.
10. The load distribution.
11. The aeroplane CG positions.
12. The limiting mass and CG values.

In the following pages we will concentrate on calculating the take off mass and CG position for the single engine piston aircraft SEP 1 using these values.

Basic empty mass	2415	lbs
Front seat occupants	340	lbs
3rd and 4th seat passengers	340	lbs
Baggage zone B	200	lbs
Fuel at engine start	60	US galls
Trip fuel (calculated fuel burn)	40	US galls

When completed the load sheet can be used to check that limiting values such as MZFM; RAMP MASS; MTOM and MLM have not been exceeded. The mass and CG limits are presented in graphical form (data sheet page 9, fig 2.5) against which the calculated values can be checked.

2.12. CALCULATION OF CG FOR LOADED MASS

When the CG of the Basic Empty Mass has been determined, the CG for the loaded mass, both for take-off and landing, may be calculated using the same principle of moments. For convenience the data may be tabulated on the Mass and Balance documentation (load sheet).

2.13. PROCEDURE FOR CALCULATING THE LOADED MASS

For light aeroplanes the Basic Empty Mass and C of G as recorded in the aircraft's Weight and Balance Record are used as the start position for the aircraft loaded calculations. For large aircraft in the Transport Category the Dry Operating Mass (DOM) and CG are generally used as the starting point. The DOM is obtained by adding the mass of the crew and specialist equipment to the Basic Empty Mass (Large fleet operators may use a fleet DOM and CG instead of calculating the DOM and CG for individual aircraft).

1. In the "Mass" column, enter the BEM (or DOM if applicable) of the aircraft followed by the masses of all the listed items.

2. Add all the masses listed to obtain the ZFM and TOM. Compare these to the limiting values to verify that they are within limits.

3. In the "Arm" columns, enter the arm of each item. Arms aft of the datum are positive (+), those forward of the datum are negative (-).

4. Calculate the moment of each item, that is, the mass multiplied by the arm. Remember that a positive mass and a negative arm when multiplied together give a negative moment.

5. Determine the sum of the moments.

6. If required divide the take-off Total Moment by the Take-Off Mass to obtain the position of the CG as a distance from the datum.

7. Compare the CG position with the CG limits given. The CG must lie between the limits.

8. After making adjustments for fuel usage (and any changes to the traffic load i.e. parachutists) to the take-off mass and moment determine the Landing Mass and C of G using moments as above. Again check against the limiting values.

Note: for large public transport aircraft it is normal practice to calculate the Take off and ZFM CG position. The reason being that if the C of G is in limits at both the Take-off Mass and the Zero Fuel Mass then it will remain within limits throughout the intended flight provided the fuel is used in the correct sequence from the fuel tanks. Some modern aircraft have transducers on the landing gear struts which will "weigh" the aircraft whilst on the ground and pass the information to the flight management system which can be read by the flight crew as a mass and CG position.

MASS AND BALANCE — LOADING AND CENTRE OF GRAVITY

ITEM	MASS (lbs)	ARM (in)	MOMENT/100
1. Basic Empty Mass	2415	77.7	1876.46
2. Front seat occupants	340	79	268.6
3. 3rd and 4th seat pax	340	117	397.8
4. Baggage zone A	nil	108	
5. 5th and 6th seat pax	nil	152	
6. Baggage zone B	200	150	300
7. Baggage zone C	nil	180	
SUB TOTAL = ZERO FUEL MASS	3295		
8. Fuel loading 60 US galls	360	75	270
SUB TOTAL = RAMP MASS	3655		
9. Subtract fuel for start taxi and run up. (see note)	-13		-10
SUB TOTAL = TAKE OFF MASS	3642		3102.9
10. Trip fuel	-240	75	-180
SUB TOTAL = LANDING MASS	3402		2922.9

Table 1. Completed Load Sheet Solution

NB. Fuel for start, taxi and run up is normally 13 lbs at an average entry of 10 in the column headed **moment / 100**

The mass and balance arm data is entered on the load sheet in the appropriate columns as shown above. The individual moments are calculated by multiplying the mass of an item by its balance arm from the datum and entering the figure in the moment column.

Moments are often large numbers containing more than six digits and this can be an extra source of difficulty. In the SEP example shown, the moments have been divided by one hundred to reduce the number of digits and make them more manageable. Take care if you use this procedure because you must remember at the end of any calculations to multiply the final answers by one hundred to arrive at the correct total moment.

The fuel load may be given as a quantity (Imperial or American gallons) rather than a mass and you must convert it to mass before you can complete the load sheet (see para 9). Fuel mass and distribution may also be given in tabular form as shown in the example above, where the fuel mass and moment have been taken from the S.E.P.1 fuel chart, fig. 2.3, of the JAR Loading Manual.

MASS AND BALANCE   **LOADING AND CENTRE OF GRAVITY**

The Take-off and Landing CG positions can now be determined by using the following procedure:

Take-off CG:

a. Sum (add up) the vertical 'MASS' column to determine in turn, the ZFM, the Ramp Mass and the TOM.
    (ZFM = 3295 lb; Ramp Mass = 3655 lb and TOM = Ramp Mass – Start/Taxi fuel = 3655 – 13 = 3642 lb)

b. Check the Operating Manual to ensure that limiting masses of MZFM, MSTM and Regulated TOM have not been exceeded.

c. Sum the vertical 'MOMENTS' column to determine the total moment for ZFM, Ramp Mass moment and TOM. (ZFM moment = 284286 lb in; Ramp Mass moment = 284286 + 27000 = 311286 lb in; TOM moment 311286 – 1000 = 310286 lb in). Note, the figures in the table are shown in the abbreviated form i.e. 310286/100 = 3102.9 which is less accurate but easier to cope with.

d. Divide the moment of the TOM by the TOM to determine the CG position at take-off. (TOM CG position = 310290 ÷ 3642 = 85.2 inches aft of the datum).

e. Check the Operating Manual to ensure that the CG is within limits at both the ZFM and TOM situations. If this is the case then the CG will remain within the limits throughout the flight and should not go out of limits during the journey provided the fuel is used in the correct sequence.

Landing CG

a. Determine the moment of the fuel used in flight (the trip fuel) by multiplying its mass by the fuel arm. In light aircraft the fuel arm will usually be the same as the one used previously to calculate the take-off CG position. However, caution is required because in some large aircraft the balance arm of the fuel may change with the quantity of fuel consumed.

THE FUEL CONSUMED WILL GIVE A NEGATIVE MASS IN THE MASS COLUMN AND THIS WILL CHANGE THE MOMENT SIGN I.E. IF THE FUEL ARM IS POSITIVE THE FUEL MOMENT WILL BECOME NEGATIVE.

b. In the 'MASS' column, subtract the fuel used during the flight from the TOM to determine the Landing Mass.
    (Landing Mass = 3642 – 240 = 3402 lbs).

c. Check the Operating Manual to ensure that the Regulated Landing Mass has not been exceeded.

MASS AND BALANCE   **LOADING AND CENTRE OF GRAVITY**

d.   In the 'MOMENTS' column, determine the sign of the fuel moment and add or subtract it as appropriate, to the TOM moment to determine the Landing moment. (Landing Moment = 310286 − [240 x 75] = 310286 − 18000 = 292286 lb in [or 2922.9 lb in the abbreviated form]).

e.   If the CG is within limits at the ZFM then, for large aircraft, it is not normally necessary to calculate the Landing CG position because, as stated previously, the CG will be within limits throughout the flight. However, for light aircraft it is usual to determine the landing CG position and this is simply achieved by divide the Landing moment by the Landing Mass. (Landing CG position = 292286 ÷ 3402 = 85.92 inches aft of the datum because it has a positive sign).

The C of G can also be found by using the Centre of Gravity Envelope for the SEP.1 shown below. This is a graphical representation of the mass and centre of gravity limits. The vertical axis is the mass in pounds, the horizontal axis is the CG position in inches aft of the datum and the slanted lines represent the moment/100.

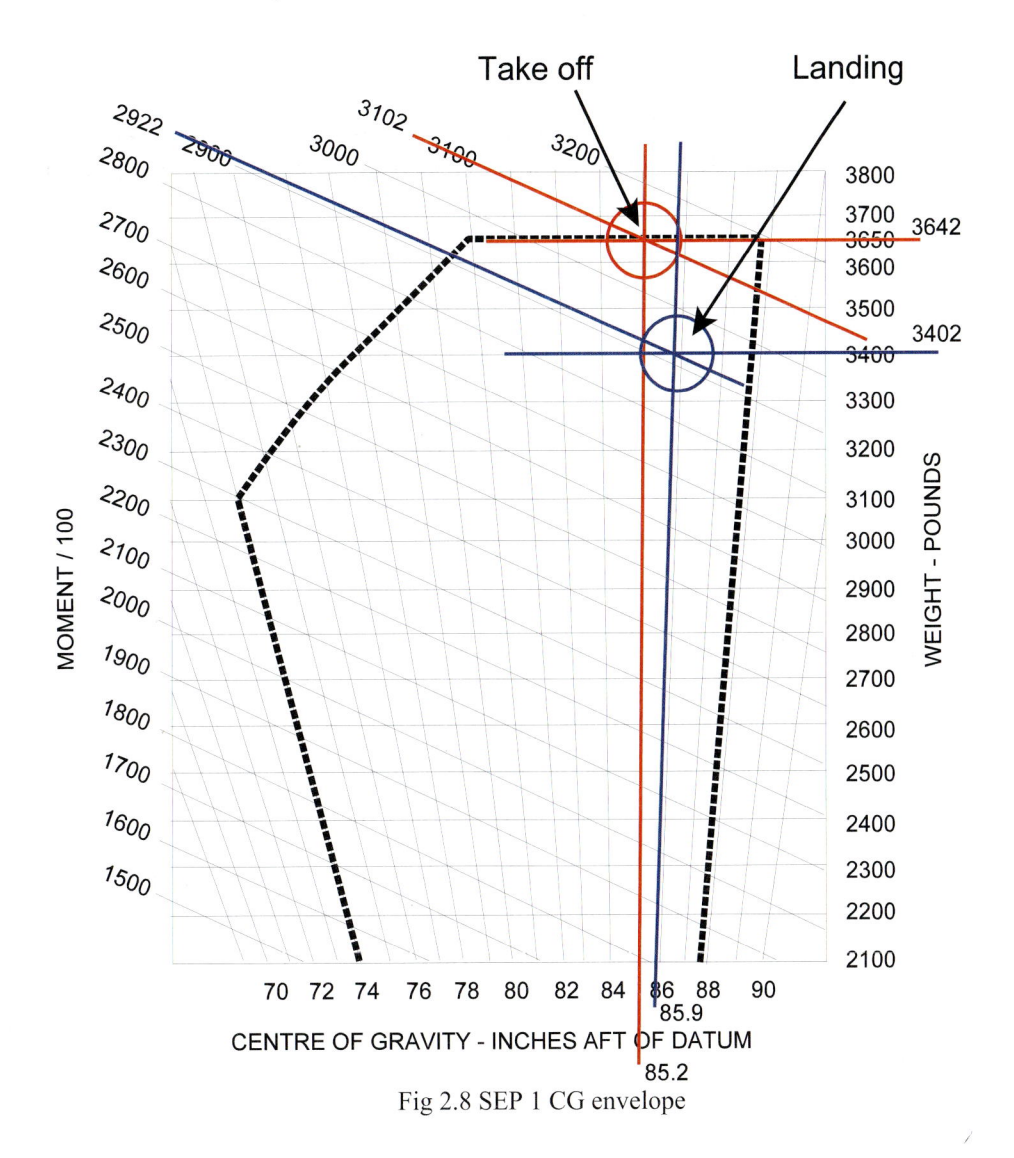

Fig 2.8 SEP 1 CG envelope

Example 3

Try this example of calculating the take off and landing mass and C of G position for a fictitious aircraft without using a formal load sheet.

For the data given below:

a) Find the CG for take off as loaded.

b) Find the CG for landing after a flight lasting for 1 hour 30 minutes.

Maximum Take-off Mass	2245 lb
Maximum Landing Mass	2100 lb
Centre of Gravity limits	2 in forward to 6 in aft of datum
Fuel Consumption	7.0 US Gallons per hour
Oil Consumption	1.0 US quart per hour

ITEM	MASS (lb)	ARM (in)
Basic Mass	1275	-5
Seats 1 and 2	340	-2
Seats 3 and 4	170	30
Fuel 35 US Gallons (SG 0.72)		2
Oil 8 US quarts (SG 0.9)		-48
Baggage	45	70

MASS AND BALANCE   **LOADING AND CENTRE OF GRAVITY**

Solution:

Lay out your solution in the same fashion as the load sheet;

Calculate fuel and oil mass using the density specified.

Fuel: $35 \div 1.2 \times .72 \times 10 = 210$ lb   (US galls to pounds)

Oil : $8 \div 4 \div 1.2 \times .9 \times 10 = 15$ lb   (US qts to pounds)

ITEM	MASS (lb)	ARM (in)	MOMENT
DOM	1275	-5	-6375
Seats 1 and 2	340	-2	-680
Seats 3 and 4	170	30	5100
Fuel 35 US Galls	210	2	420
Oil 8 US quarts 15	-48		-720
Baggage	45	70	3150
Take-off Mass	**2055 lb**	**Take-off Moment**	**895**

Take off CG = **0.435 inches aft of datum** (within limits)

To calculate landing CG

Fuel used in one and a half hours = $1.5 \times 7 \div 1.2 \times .72 \times 10 = 63$ lbs
Oil used = $1.5 \times .25 \div 1.2 \times .9 \times 10 = 2.8$ lbs

Landing mass	=	2055 - 63 - 2.8 = **1989.2 lbs**
Landing moment	=	Take off moment - Fuel moment - Oil moment
Landing moment	=	+895 - (63 x +2) - (2.8 x -48)
Landing moment	=	+895 - (126) - (-134) (minus and minus give plus)
Landing moment	=	+895 - 126 + 134
Landing moment	=	**+903**
Landing CG	=	$\dfrac{+903}{1989.2}$
	=	**+ 0.454 inches aft of datum** (within limits)

Note: Remember to check that the take-off mass and CG position and the landing mass and CG position are within the acceptable limits for the trip.

MASS AND BALANCE   **LOADING AND CENTRE OF GRAVITY**

2.15. CG POSITION AS A PERCENTAGE OF MEAN AERODYNAMIC CHORD (MAC)

In the previous examples the CG position and CG limits are given as distances from a datum. An alternative method is to state the C of G position and its limits as a percentage of the Mean Aerodynamic Chord (MAC). This is common practice with many swept wing airliners and it is so with the twin jet we shall be studying next.

The mean aerodynamic chord is one particular chord from the wing calculated from the aerodynamic characteristics of that particular wing. Because the CG affects many aerodynamic considerations, particularly stability, it is useful to know the CG position in relation to the aerodynamic forces.

The length of the MAC is a fixed distance for that aeroplane and it is at a fixed distance from the datum. The C of G is always calculated as a percentage of the MAC from the leading edge of the MAC. ie. A C of G position of 25% MAC would mean that the C of G was positioned at one quarter of the length of the chord measured backwards from the leading edge of the MAC.

The method of calculating the percentage of MAC is shown below.

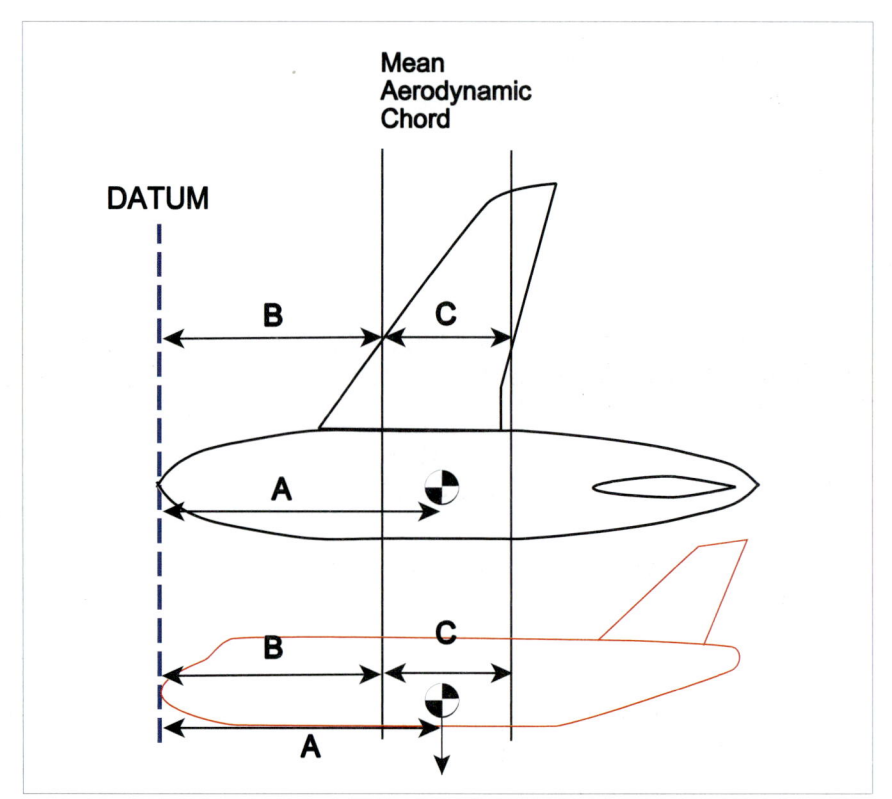

Figure 2 - 9

The % MAC is always a figure stated from the leading edge of the MAC which is specified from the datum.

- A = distance of CG from datum.
- B = distance of MAC leading edge from datum.
- C = length of MAC.

The CG as a percentage of MAC $= \dfrac{A - B}{C} \times 100$

Example 4

If the MAC is 152 inc and its leading edge is 40 in aft of the datum, and the CG is 66 in aft of the datum, what is the CG position as a percent of MAC.

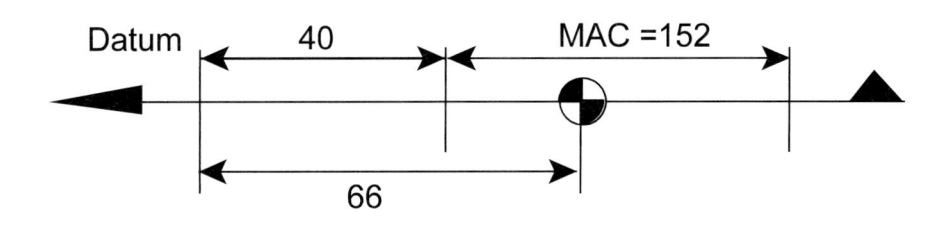

$\dfrac{66 - 40}{152} \times 100 =$ **17.1%**

Now try these:-

1. An aircraft has a MAC of 82 inches. The leading edge of the MAC is 103 inches aft of the datum. If the C.G. position is 14.7% MAC, what is the C.G. distance from the datum.

2. If the C.G. position is 21% MAC, the MAC is 73 inches, and the C.G. datum is 26 inches aft of the leading edge of the MAC, what is the C.G. position relative to the datum.

3. The C.G. limits are from 5 inches forward to 7 inches aft of the datum. If the MAC is 41 inches and its leading edge is 15 inches forward of the datum, what are the C.G. limits as % MAC.

4. The MAC is 58 inches. The C.G. limits are from 26% to 43% MAC. If the C.G. is found to be at 45.5% MAC, how many inches is it out of limits.

5. An aircraft of mass 62500 kg has the leading and trailing edges of the MAC at body stations +16 and +19.5 respectively (stations are measured in metres). What is the arm of the CG if the CG is at 30% MAC?

MASS AND BALANCE   **LOADING AND CENTRE OF GRAVITY**

2.16. RE-POSITIONING OF THE CENTRE OF GRAVITY

If the centre of gravity is found to be out of limits for any part of the flight, the aircraft must not take off until the load has been adjusted so as to bring the centre of gravity into limits.

This may be achieved in one of two ways:

a. By re-positioning mass which is already on board the aircraft. This will usually be baggage or passengers.
b. By adding or removing mass. Mass put on to the aircraft purely for the purpose of positioning or correcting the CG position is known as ballast.

The minimum mass to be moved, or the minimum amount of mass to be loaded or off-loaded, will be that which just brings the centre of gravity on to the nearest limit. It may be preferable of course to bring the CG further inside the limits. When the amount of mass adjustment has been calculated, it should be ascertained that this makes the aircraft safe for both take-off and landing.

2.17. RE-POSITIONING OF THE CENTRE OF GRAVITY BY REPOSITIONING MASS.

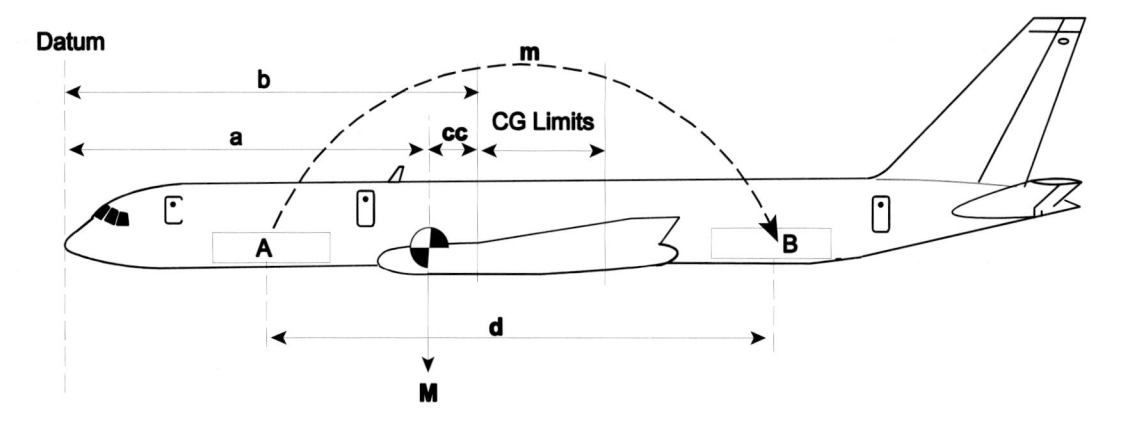

Figure 2 -10 Repositioning Mass

In figure 2.10 the centre of gravity has been found to be at a distance 'a' inches aft of the datum. The forward CG limit is 'b' inches aft of the datum. To bring the CG into limits, some baggage (**m**) will be moved from compartment A to compartment B.

If a mass of **m** lb is moved from A to B the change of moment will be

$$m \times d$$

That is, the change of moment is equal to the mass moved (**m**) multiplied by the distance through which it moves (**d**).

MASS AND BALANCE   **LOADING AND CENTRE OF GRAVITY**

If the total mass of the aircraft is **M**, with the CG at '**a**' inches aft of the datum, the total moment around the datum is **M x a.**

It is required to move the CG to (**b**) inches aft of the datum.
The new total moment will then be M x b.

> The change in moment required is therefore M x b. - M x a = M (b - a)
> And M (b - a) = m x d
> b - a is equal to the change to the C of G position '**cc**'.
> and so   **m x d = M x cc**

That is, the mass to move multiplied by the distance it moves is equal to the total aircraft mass multiplied by the change to the CG.

Example 5

The CG limits of an aircraft are from -4 to +3 inches from the datum. It is loaded as shown below:

ITEM	MASS (lb)	ARM (in)	MOMENT (lb in)
Basic Empty Mass	2800	2	5600
Crew	340	-20	-6800
Fuel	600	10	6000
Forward Hold	0	-70	0
Aft Hold	150	80	12000
Total Mass =	**3890**	**Total Moment =**	**16800**

Therefore C of G = $\frac{16800}{3890}$ = 4.32 in

The Aft Limit is +3 and so the CG is 1.32 in out of limits too far aft.

It can be corrected by moving some freight or baggage from the rear hold to the forward hold, a distance of 150 ins. (80 ins aft to 70 ins forward).

How much freight/baggage can be calculated by using our formula

       m x d   =   **M x cc**
       m x 150   =   3890 x 1.32

       m   = $\frac{3890 \times 1.32}{150}$

Mass of freight and/or   =   **34.232 lb**
baggage to move

MASS AND BALANCE **LOADING AND CENTRE OF GRAVITY**

To check that the aircraft is safe for all fuel states after take off, we will calculate the C of G at the Zero Fuel Mass with 35 lb of baggage moved to the forward hold.

ITEM	MASS (lb)	ARM (in)	MOMENT (lb in)
Basic Mass	2800	2	5600
Crew	340	-20	-6800
Fuel (Zero)	0		
Fwd. Hold	35	-70	-2450
Aft Hold	115	80	9200
ZFM =	**3290**	**ZMF moment =**	**5550**

$$\text{ZFM C of G} = \frac{5550}{3290}$$

= 1.69 in (in Limits)

Now try these

1. The CG limits of an aircraft are from 83 in. to 93 in. aft of the datum. The CG as loaded is found to be at 81 in. aft of the datum. The loaded mass is 3240 lb. How much mass must be moved from the forward hold, 25 in. aft of the datum, to the aft hold, 142 in. aft of the datum, to bring the CG onto the forward limit.

2. An aircraft has a loaded mass of 5500 lb. The CG is 22 in. aft of the datum. A passenger, mass 150 lb, moves aft from row 1 to row 3 a distance of 70 in. What will be the new position of the CG. (All dimension aft of the datum)

3. The loaded mass of an aircraft is 12 400 kg. The aft CG limit is 102 in. aft of the datum. If the CG as loaded is 104.5 in. aft of the datum, how many rows forward must two passengers move from the rear seat row (224 in. aft) to bring the CG on to the aft limit, if the seat pitch is 33 in. Assume a passenger mass of 75 kg each.

4. An aircraft of mass 17400 kg, has its CG at station 122.2. The CG limits are 118 to 122. How much cargo must be moved from the rear hold at station 162 to the forward hold at station -100 (forward of the datum) to bring the CG to the mid position of its range?

5. With reference to Fig. 1 how much load should be transferred from No. 2 hold to No. 1 hold in order to move the CG from the out-of-limits value of 5.5 m to the forward limit value of 4.8 m. The total mass of the aircraft is 13 600 kg.

MASS AND BALANCE **LOADING AND CENTRE OF GRAVITY**

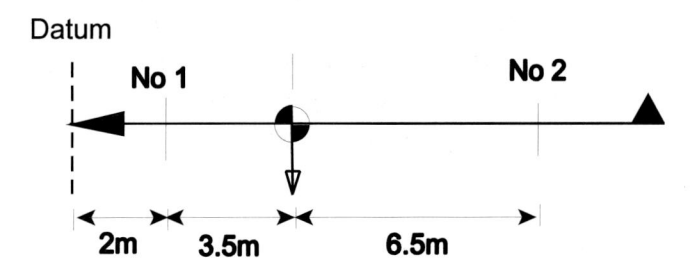

Datum
No 1   No 2
2m   3.5m   6.5m

Moving Mass Figure 1.

6. With reference to Fig. 2 the loaded mass of the aircraft is found to be 1850 lb and the CG moment 154 000 lb in. How much mass must be moved from the forward hold 40 in. aft of the datum, to the rear hold, 158 in. aft of the datum, to bring the CG on to the forward limit.

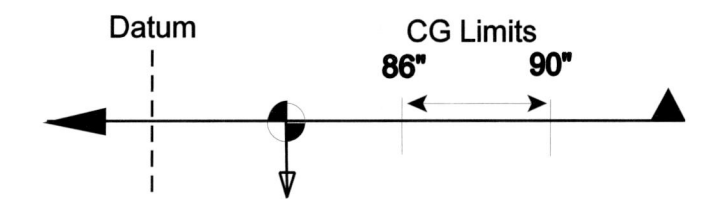

Datum   CG Limits
         86"   90"

Moving Mass Figure 2.

MASS AND BALANCE   **LOADING AND CENTRE OF GRAVITY**

2.18. RE-POSITIONING OF THE CENTRE OF GRAVITY BY ADDING OR SUBTRACTING MASS

The position of the C of G can also be adjusted by adding or subtracting mass. Mass added simply to re-position the C of G is called **ballast.**

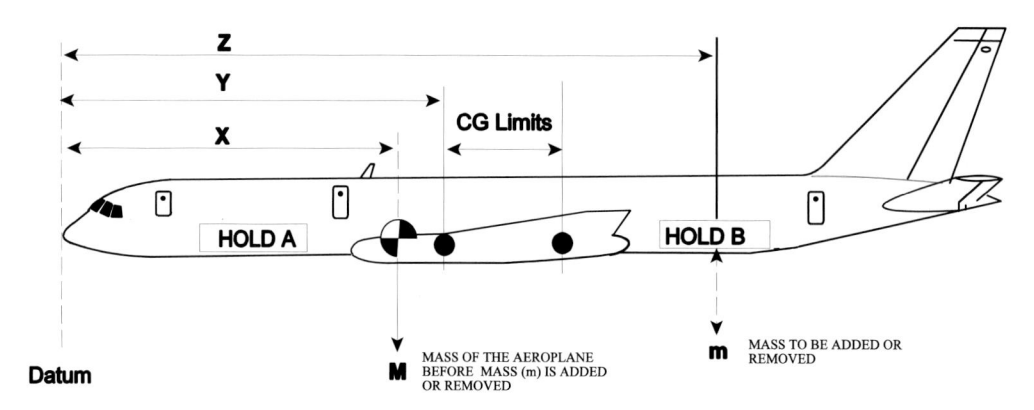

Figure 2.11 Adding or Removing Mass

To calculate the minimum amount of ballast required :

In fig 2.11 the CG has been found to be at a distance 'X' ins aft of the datum. The forward CG limit is at a distance 'Y' ins aft of the datum. To bring the CG into limits, ballast will be put in compartment B, a distance 'Z' ins aft of the datum.

If the total mass of the aircraft is M lb., the total moment will be M x a lb in.

If ballast of m lb is placed in compartment B in order to move the CG to its fwd limit, the mass will increase to M + m and the new total moment will be (M + m) x Y . Assuming equilibrium to be maintained the original total moment plus the moment of the added mass must equal the new total moment.

Algebraically using the above notation then :-

$$(M + m) \times Y = (M \times X) + (m \times Z)$$

New Total Moment = **Old Total Moment + Cargo Moment**

The same formula can be used for **removing** mass by changing the plus sign to a minus. So, for any calculation involving **adding or subtracting** mass remember the formula:-

New Total Moment = **Old total moment plus or minus the cargo moment**

Note that when calculating a change in CG position using the New Moment = Old Moment ± Change in Moment the distances (X, Y AND Z) are always measured from the datum itself.

MASS AND BALANCE   **LOADING AND CENTRE OF GRAVITY**

We will do an example calculation together.

Example 6

The CG limits of an aircraft are from 84 ins to 96 ins aft of datum at all masses. It is loaded as shown below:

ITEM	MASS (lb)	ARM (in)	MOMENT (lb in)
Basic Mass	1250	80	10000
Crew	340	82	27880
Fuel	300	72	22500
Baggage	0	140	0
	1890		**150380**

$$CG = \frac{150380}{1890} = 79.6 \text{ ins aft of datum}$$

The CG is out of limits by 4.4 ins too far forward. It will be brought into limits by putting ballast in the baggage compartment. The minimum ballast would be that required to bring the CG to 84 ins.

Using our formula

NTM = OTM + CM

(1890 + m) x 84 = (1890 x 79.6) + (m x 140)

158760 + 84 m = 150444 + 140m

158760 - 150444 = 140 m - 84 m

8316 = 56 m

$\frac{8316}{56}$ = m

m = 148.5

Therefore mass of ballast required = 148.5 lbs

It would be necessary to check that loading the ballast did not cause the total mass to exceed the Maximum Take-Off Mass and as before, that the aircraft was in limits for landing.

Although this may appear to be a long winded method it will always get you the correct answer, remember you may be asked to calculate a mass to add or remove, a change to the CG or the position to put ballast.

MASS AND BALANCE   **LOADING AND CENTRE OF GRAVITY**

Have a go at these adding or removing mass problems yourself:

Question 1.
An aircraft has three holds situated 10 in 100 in and 250 in aft of the datum, identified as holds A, B and C respectively. The total aircraft mass is 3500 kg and the CG is 70 in aft of the datum. The CG limits are from 40 in to 70 in aft of the datum. How much load must be removed from hold C to ensure that the CG is positioned on the forward limit.

Question 2.
An aircraft has a mass of 5000 lb and the CG is located at 80 in aft of the datum. The aft CG limit is at 80.5 in aft of the datum. What is the maximum mass that can be loaded into a hold situated 150 in aft of the datum without exceeding the limit.

Question 3.
The loaded mass of an aircraft is 108560 lb and the CG position is 86.3 ft aft of the datum. The aft CG limit is 85.6 ft. How much ballast must be placed in a hold which is located at 42 ft aft of the datum to bring the CG onto the aft limit.

Question 4.
The aft CG limit of an aircraft is 80 in aft of the datum. The loaded CG is found to be at 80.5 in aft of the datum. The mass is 6400 lb. How much mass must be removed from a hold situated 150 in. aft of the datum to bring the CG onto the aft limit.

Question 5.
An aircraft has a mass of 7900 kg and the CG is located at 81.2 in aft of the datum. If a package of mass 250 kg was loaded in a hold situated 32 in aft of the datum. What would the new C of G position be?

Question 6.
The CG limits of an aircraft are from 72 in. to 77 in. aft of the datum. If the mass is 3700 kg and the CG position is 76.5 in. aft of the datum, what will the change to the CG position be if 60 kgs is removed from the fwd hold located at 147 in fwd of the datum.

Question 7.
An aeroplane has a zero fuel mass of 47800 kg and a performance limited take-off mass of 62600 kg. The distances of the leading edge and trailing edge of the MAC from the datum are 16m and 19.5m respectively. What quantity of fuel in Imperial gallons, must be taken up to move the CG from 30% MAC to 23% MAC if the tank arm is 16m aft of the datum and the fuel SG is 0.72?

MASS AND BALANCE **LOADING AND CENTRE OF GRAVITY**

2.19. GRAPHICAL PRESENTATION

The aircraft mass and CG position are frequently calculated using one of a number of graphical methods. These notes give several examples of the graphs that may be used: see CAP 696 and these notes pages 2-17 and 2-50. There are two things in common on any of the graphs used:

    a.   mass and CG must be within the envelope or on the line of the envelope.
    b.   the mass of the aeroplane is always shown on the vertical scale.

The example of a mass and CG envelope for the SEP 1, shown in CAP 696 page 9, is unusual in that it uses both mass and moments on the vertical scale.

The horizontal scale may use the CG position (inches, metres or centimetres), the moment of the Cg (kg inches, kg metres or kg centimetres) or the percentage of the CG along the mean aerodynamic chord. The MEP 1, envelope on page 15 of CAP 696, is an example of using CG position and the MRJT envelope on page 31 of CAP 696, is an example of the use of the MAC percentage.

2.20. CARGO HANDLING

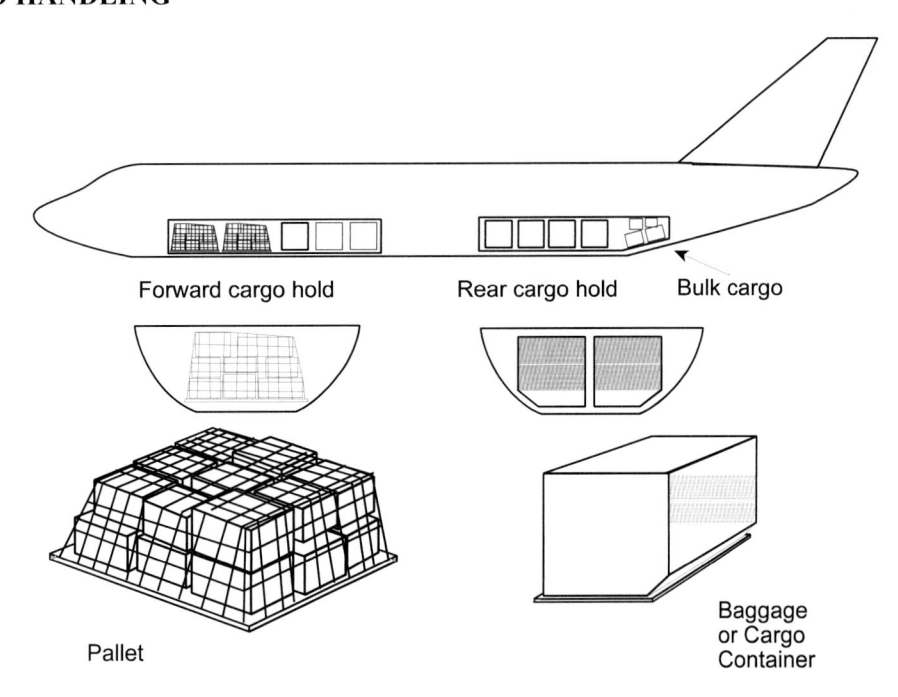

Figure 2.12 Cargo Handling

Cargo compartments

Compartments in the lower deck accommodate baggage and cargo. The compartments feature fire resistant sidewalls, ceilings and walkways. Cargo compartments are usually pressurised and heated and typically have fire detection and protection equipment. All cargo compartments have a maximum floor loading (kg/m^2) and maximum running load value (kg/m).

Containerised cargo

Baggage and cargo can be loaded into standard size containers designed to fit and lock into the cargo compartment. The containers have an individual maximum mass limit and an individual floor loading limit (mass per unit area).

Palletised cargo

Cargo can also be loaded onto standard sized pallets and restrained with cargo nets or strops. Typically the forward area of the forward cargo compartment is configured to take palletised freight.

Bulk cargo

Bulk cargo can be loosely loaded in the area at the aft of the rear cargo compartment and separated from the containers by a restraining net attached to the floor ceiling and sidewalls.

Cargo Handling Systems.

The forward and aft cargo compartments typically have separate cargo power drive systems to move containers and cargo pallets.

The power drive system is operated by a control panel at the door area of each cargo compartment and is capable of loading and unloading fully loaded containers or pallets in wet or dry conditions. A typical panel is shown below.

Power drive units (PDU) are mounted in the floor and are controlled by the control joystick to move containers laterally and longitudinally in the doorway area and longitudinally in the compartment forward and aft of the doorway.

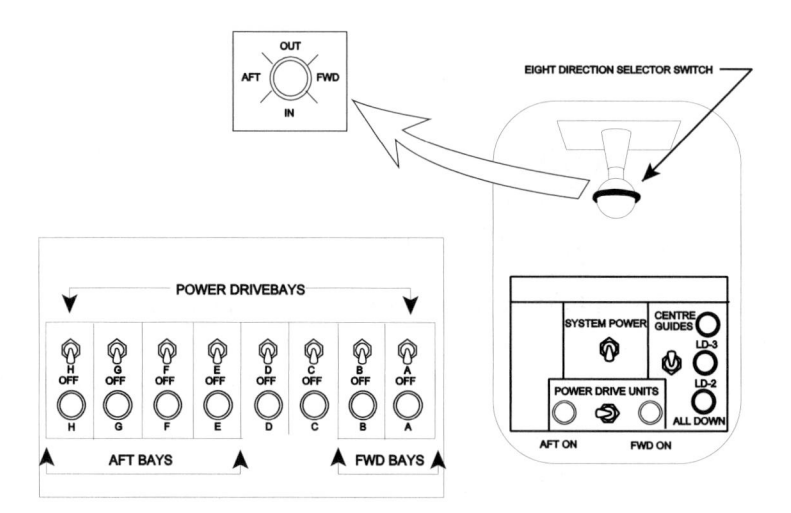

Figure 2.13 Cargo bay Control Panel

Guides, rollers, stops, locks and tie down fittings are included in the appropriate places to provide adequate 'G' restraint for the containers in flight.

MASS AND BALANCE   **LOADING AND CENTRE OF GRAVITY**

2.21. FLOOR LOADING

The floors of the passenger cabin and the freight areas of the aircraft are limited in the load that they can carry. Placing excessive loads on the structure may not only cause visible panel creasing and local indentations but is likely to significantly accelerate structural fatigue. Fatigue is cumulative and can lead to major structural collapse of the structure with little or no warning.

The floor loading is defined both by linear loading and by area loading intensity

2.22. LINEAR / RUNNING LOADS

The linear (or running) loading limitation (lb per linear foot or kg per linear inch) protects the aircraft under floor frames from excessive loads. Depending on the units used, it is the total load permitted in any one inch or one foot length of the aircraft (irrespective of load width).

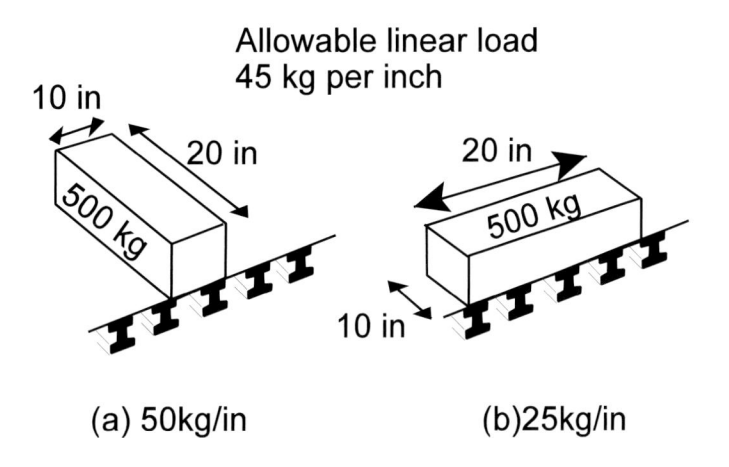

Figure 2.14   Linear Loading

Calculating the linear load distribution

As can be seen at Fig 2.14 (a), the linear load on the floor members is 500 kg divided by the 10 inch length i.e. 50 kg/inch; which is greater than the 45 kg/in allowed. However, the way in which an item of load is located in an aircraft can create an acceptable situation out of an unacceptable one.

In example (b), by simply rotating the load through the linear load on the floor members becomes 500 kg/20 inches i.e. 25 kg/in: which is well within the limit. It can be seen that the least linear loading occurs when the longest length is placed at right angles to the floor beams.

MASS AND BALANCE   **LOADING AND CENTRE OF GRAVITY**

2.23. AREA LOAD LIMITATIONS

Area load limitations (kg/sq metre or lb/sq ft) protect the aircraft floor panels. The two permissible intensities of pressure are "Uniformly Distributed" (UD) and "Concentrated" loads.

UD loads

The general floor loading limitations for cargo are referred to as UD loads and are given as allowable lb/sq ft. Providing that a load (or series of loads) is within the allowable UD limitations for the floor area on which it rests, the loading is subject only to :

i.   not exceeding the linear load limitations.

ii.   the individual and accumulated compartment load limitations.

Concentrated loads

Load intensities which exceed the UD load intensities "concentrated loads", can be carried providing approved load spreaders are used to distribute the load over sufficient area to ensure that loading limits are not exceeded.

Load spreaders

When concentrated loads or items of load with hard or sharp areas are carried on an aircraft, some form of floor protection is essential. The normal practice is to employ standard load spreaders of 2 inch thick timber.

An example of load intensity and running load for you to do:

An item of cargo has dimensions 3 ft x 4 ft x 12 ft and weighs 900 kg. Given that the maximum running load for the compartment is 20 kg/in and the maximum load intensity for the compartment is 70 kg/ft^2 what are the running load values and the floor intensity values for the cargo and are there any limitations in the way in which it may be carried? (1 ft = 12 inches)

Considering the RUNNING LOAD (underfloor protection) first:

     Max running load   =   Load / shortest length
     Mid running load   =   Load / mid length
     Min running load   =   Load / longest length

     Max running load   = $\dfrac{900 \text{ kg}}{3 \text{ft} \times 12 \text{inches/ft}}$ = 25 kg/in   ) Max = 20 kg/in
) Exceeds limit

MASS AND BALANCE   **LOADING AND CENTRE OF GRAVITY**

Mid running load = $\dfrac{900 \text{ kg}}{(4\text{ft} \times 12 \text{ inches/ft})}$ = 18.75 kg/in   ) OK
) Below limit

Min running load = $\dfrac{900 \text{ kg}}{(12\text{ft} \times 12 \text{ inches/ft})}$ = 6.25 kg/in   ) OK
) Below limit

Now considering the DISTRIBUTION LOAD INTENSITY (Floor protection)

Max distribution load on the floor   =   Load / smallest area
Mid distribution load on the floor   =   Load / medium area
Min distribution load on the floor   =   Load / largest area

Max dl   = $\dfrac{900 \text{ kg}}{3 \text{ ft} \times 4 \text{ ft}}$ = $\dfrac{900 \text{kg}}{12 \text{ ft}^2}$ = 75 kg/ft^2) Max = 70 kg/ft^2
) Exceeds limit

Mid dl   = $\dfrac{900 \text{ kg}}{3 \text{ ft} \times 12 \text{ ft}}$ = $\dfrac{900 \text{ kg}}{36 \text{ ft}^2}$ = 25 kg/ft^2) OK
) Below limit

Min dl   = $\dfrac{900 \text{ kg}}{4 \text{ ft} \times 12 \text{ ft}}$ = $\dfrac{900 \text{ kg}}{48 \text{ ft}^2}$ = 18.75 kg/ft^2) OK
) Below limit

The cargo can be carried providing it is placed on the cargo-bay floor on either its medium area or largest area to prevent floor damage and that either the 4 feet length or the 12 feet length is placed parallel to the longitudinal axis of the aeroplane to prevent under-floor damage (spreading the load across the under floor frames).

2.24. SINGLE ENGINE PISTON/ PROPELLER AIRCRAFT (SEP 1)

The procedure for compiling the mass and balance documentation for the single engine piston/propeller aircraft (SEP1) given in the data sheets is similar to the calculations previously specified in these notes, but notice that the load manifest and CG envelope are specific to that type of aircraft.

A number of self help examples are included at page 2 - 52

MASS AND BALANCE   **LOADING AND CENTRE OF GRAVITY**

2.25.   LIGHT TWIN PISTON/PROPELLER AIRCRAFT (MEP 1)

The procedure for compiling the mass and balance documentation for the twin engine aircraft (MEP1) given in the data sheets is also similar to the procedures defined in these note, but notice that the load manifest and CG envelope are specific to that type of aircraft.

A sample calculation based on compiling the load sheet for that aircraft is shown on pages 14 and 15 of the data sheets. Pages 12 and 13 of the data sheets contain specific information based on that particular aircraft.

These pages should be studied by the student after which the following self assessment questions should be attempted.

A number of self help questions are included at page 2 - 55

2.26.   MEDIUM RANGE JET TWIN (MRJT 1)

The MRJT (Medium Range Twin Jet) has a more complex loading and trim sheet which we shall deal with shortly.

Pages 20 - 25 of the data sheets contain information specific to the MRJT and should be studied in detail. Notes on particular items follow.

MRJT Fig 4.1 and 4.2 Body Station

Though it would be possible to locate components and parts of the airframe structure by actually measuring their distance from the CG datum, it would be difficult and impractical on anything other than a very small aircraft. Instead, aircraft are divided about their three axes by a system of station numbers, water lines and buttock lines. The system of structural identification is not part of this subject except to say that in the past, station numbers were used as balance arms about the datum for first-of-kind aircraft. However, new variants of original series aircraft are often made by inserting additional lengths of fuselage and consequently, the distances of components and structure from the original datum undergo a change. The station numbers could all be re-numbered to enable them to retain their use as balance arms but it is often more beneficial to retain the original station numbers as they are. This means that on some variant aircraft they can no longer be used as balance arms for CG purposes and in order to find out how far a particular station is from the datum a conversion chart is required.

Figure 4.1 of the JAR Loading Manual shows a fuselage side view which includes the balance arms about the datum. The table at Fig 4.2 shows how the station numbers for the aircraft can be converted into balance arms and vice versa. An examination of the table will show that this aircraft is a variant of a previous series aircraft in that two fuselage sections 500A to 500G and 727A to 727G have been inserted. This is the reason the balance arms and the station numbers are not coincidental and why the conversion chart is required. Notice from the chart that the centre section of the airframe (stations 540 to 727) which is the same as the original aircraft, has retained its original station numbers and they are coincidental with the balance arms.

MASS AND BALANCE   **LOADING AND CENTRE OF GRAVITY**

Students must know how to use this chart so here are two examples:

1. Convert body station 500E into a balance arm.

   Body station 500E = 348 + 110 = Balance arm 458 inches.

2. Convert balance arm 809 inches into a station number.

   Balance arm 809 – 82 = Station number 727

   Further examination of the chart shows that balance arm 809 is actually station number 727D.

Now try these:

1. What is the station number at the nose of the aircraft?

2. What is the station number 1365 inches from the datum?

3. What is the distance of station 500 from the datum?

4. What is the distance of station 727C from the datum?

MRJT Fig 4.3 and 4.4 Flap Position

The movement of the flaps on a large aircraft may have a considerable effect on the C of G position. table 4.3 shows the moment change to the aircraft when the flaps are extended or retracted,. for example retracting the flaps from 30° to 0° would cause a total moment change of minus 15,000 kg.inches. Conversely extension of the flaps from 0° to 40° would cause a total moment change of plus 16,000 kg. inches.

The stabiliser setting for take off is extracted from the graph at fig 4.4. The purpose of this is to allow the stabiliser trim to be set to allow the elevator sufficient authority to enable the aircraft to be rotated during the take off run and controlled during the first stages of flight. The position of the C of G will determine the stabiliser setting for take off.

MASS AND BALANCE   **LOADING AND CENTRE OF GRAVITY**

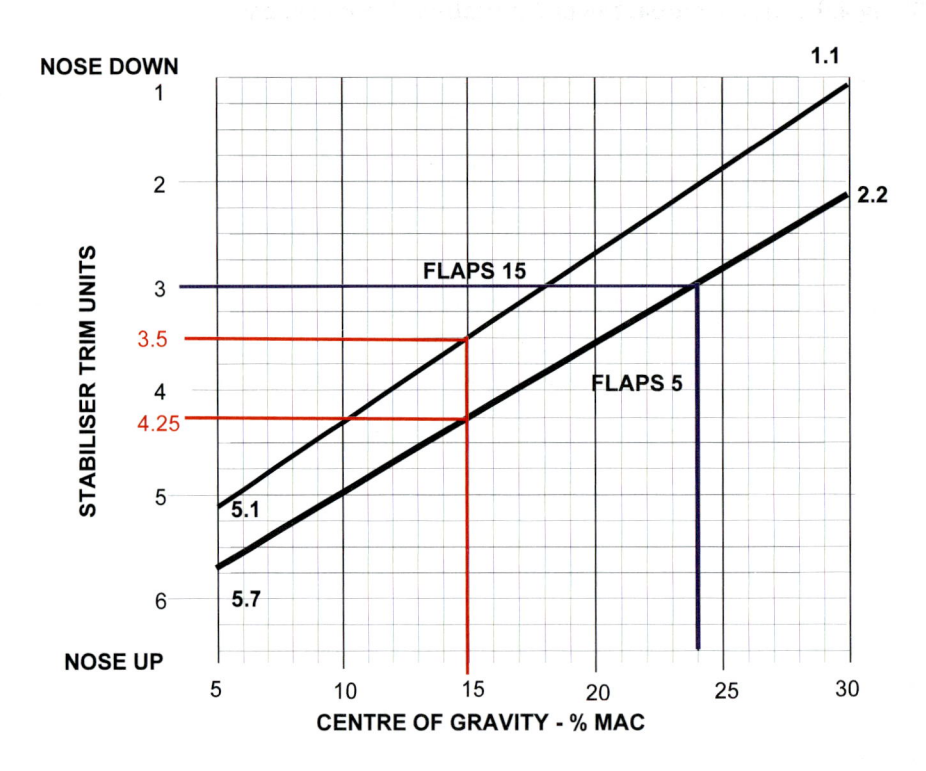

Fig 2.15 Stabiliser Trim Setting

Example 1.
What is the stabiliser trim setting if the CG is 15 % MAC and the flaps are moved from the 5 degree to the 15 degree position?

From the graph, if the C of G is 15% MAC and the flap setting is 5° then the stabiliser trim setting will be 4.25 units nose up.

If the C of G was the same 15% MAC with a flap setting of 15° then the stabiliser trim setting would be 3.5 units nose up.

Example 2.
If, as a result of the traffic and fuel load, the CG moved from 15% MAC to 24% MAC what would be the change in stabiliser trim?

From the graph, the stabiliser trim for 5 degrees take-off flap would change from 4.25 trim units at 15% MAC to 3 trim units at 24% MAC.

NB: To enable the pilot to correctly set the stabiliser trim for take off there will be a trim indicator on the instrument panel or as an integrated part of an electronic display unit.

MASS AND BALANCE   LOADING AND CENTRE OF GRAVITY

MRJT Fig 4.9 Cargo Compartment Limitation (CAP 696, page 24)

These tables detail the cargo compartment limitations which must be considered when items of cargo are loaded to ensure that the limitations are not exceeded.

a. Running Load

The running load is the fore/aft linear load. For example a box having dimensions of 3 feet by 3feet by 3feet and weighing 100 kg would have a running load of 100 ÷ 3 = 33.3 kg per **foot** which would be equal to 33.3 ÷ 12 = **2.78 kg per inch**. Be very careful to use the correct units (don't forget that there is a conversion chart on page 4 of the data sheets)

However if a box having the same weight was 3ft x 2ft x 3ft then it could be positioned so that the 2ft side was running fore/aft in which case the running load would be:

$$100 \div 2 \div 12 = 4.16 \text{ kg per inch}$$

The cargo may have to be orientated correctly to prevent exceeding the running load limitations.

b. Area Load

The first box would have a load intensity (mass per unit area) of 100 ÷ 9 = 11.1 kg per square foot as the area it would be standing on would be 3 x 3 = 9 square feet.

The second box however could be place on its side measuring 2ft x 3ft = 6 square feet area, it would therefore have a load intensity of 100 ÷ 6 = 16.66 kg per sq.ft

The cargo may have to be orientated correctly to prevent exceeding the distribution load limitations.

2.27. MASS AND BALANCE CALCULATIONS (MRJT)

a. Calculating the Underload using the formulae

The captain of the aircraft needs to know if he is able to accommodate any additional last minute changes to the load i.e. VIP's, emergency evacuation etc. Before the captain can allow any last minute additions to the load he must know if he has any spare load capacity or **underload** as it is referred to. Unfortunately, the underload is not simply the difference between the regulated take-off mass and the actual take-off mass, the MZFM and the Traffic Load have to be considered.

Clearly, if the aircraft is already at its MZFM it has no underload. Even if it is below the MZFM, any underload may have already been consumed by extra fuel uptake.

Any last minute load additions that are allowed will increase the size of the Traffic Load. The Traffic Load that can be carried is the lowest value of the Structural Traffic Load, the Take-off Traffic Load and the Landing Traffic Load.

The allowed traffic load is the lowest of the following (students are required to know these formulas well):

1. Structural Limited Traffic Load   =   MZFM – DOM

2. Take-off Limited Traffic Load   =   RTOM – DOM – Take-off fuel

3. Landing Limited Traffic Load   =   RLM – DOM – Fuel remaining

In the case of the MRJT aircraft, if the DOM = 34300 kg, the take off fuel = 12000 kg and the fuel remaining at landing = 4000 kg:

Structural Limited Traffic Load   =   51300 – 34300 = 17000 kg

Take-off Limited Traffic Load   =   62800 – 34300 – 12000 = 16500 kg

Landing Limited Traffic Load   =   54900 – 34300 – 4000 = 16600 kg

The allowed Traffic load is thus 16,500 kg. But if the actual traffic load was only 16,000 kg there would be a 500 kg underload.

Students are required to know how to determine the allowable traffic load using the formulae above.

Fuel Load Definitions
Students need to have a basic knowledge of the fuel load definitions before they attempt traffic load calculations.

i.   **Start and Taxi Fuel:**
The mass of fuel used in starting and operating the APU and the main engines and in taxying to the runway threshold for take off. It is assumed that at the point of releasing the brakes for take-off the aircraft is at or below the regulated take-off mass for the conditions prevailing. In operations where fuel is critical the start and taxi fuel must not be less than the amount expected to be consumed during the start and taxi procedures.

ii.   **Trip Fuel :**
This is the mass of fuel required to complete the take-off run, the climb, the cruise, the descent, the expected arrival procedures, the approach and landing at the designated airport.

iii. Contingency Fuel :

Fuel carried in addition to the trip fuel for unforeseen eventualities such as avoiding bad weather or having an extended hold duration at the destination airport. the contingency fuel in mass and balance calculations is usually given as a percentage of the trip fuel i.e. if the trip fuel is 1000 kg mass the contingency fuel at 5% of the trip fuel would be 50 kg. Don't forget that the contingency fuel is part of the landing mass if it is not actually used during the trip.

iv. Alternate (Diversion) Fuel :

That mass of fuel required to carry out a missed approach at the destination airfield, the subsequent climb out, transit to, expected arrival procedures, approach, descent and landing at an alternate airfield.

v. Final Reserve Fuel :

The minimum fuel that should be in the tanks on landing. Essentially it is a final reserve for unplanned eventualities and should allow a piston engine aircraft to fly for a further 45 minutes or a jet engine aeroplane to fly for a further 30 minutes at a given height and holding speed.

vi. Additional Fuel :

Only required if the sum of the trip, contingency, alternate and final reserve fuels are insufficient to cover the requirements of AMC OPS 1.255 (Instrument landings and power unit failures which not required for mass and balance calculations).

The take-off fuel for mass and balance calculations is simply the sum of ii to vi above. Note that the fuel state requirements vary with the intended flight plan and they are not always required.

the landing fuel mass is the actual amount of fuel remaining in the tanks at touch down. In a trip where no eventualities have occurred it will include the contingency, alternate, final reserve and additional fuel masses if they were included in the flight plan.

b. Calculating the Underload using the Load and Trim Sheet

The Load and Trim Sheet used for the MRJT automatically calculates the underload but does it using a slightly different method than that shown above. Instead of determining the lowest traffic load the Load and Trim sheet first determines the allowable Take-off Mass. It then calculates the allowable Traffic Load by deducting the Operating Mass from the allowable Take-off Mass. Finally, it calculates the underload by deducting the actual Traffic Load from the allowable Traffic Load.

The allowable Take-off Mass is the lowest of:

1. MZFM + Take-off Fuel

2. Regulated Take-off Mass

3. Regulated Landing Mass + Fuel used in flight

MASS AND BALANCE   **LOADING AND CENTRE OF GRAVITY**

For example. In the case of the MRJT, the MZFM is given as 51300 kg, the MTOM is 62800 kg and the MLM is 54900 kg. Let us assume that there are no Performance Limits so that the Regulated Take-off and Landing masses are equal to the Structural Limited Take-off Mass and the Structural Limited Land Mass respectively. Let us also assume that the DOM is 34,000 kg, that the actual traffic load is 12,400kg, the take off fuel load is 16,000 kg and 8,000 kg of fuel was used in flight.

Allowable Take-off mass is the lowest of:

1. MZFM + Take-off Fuel
      51300 + 16,000 kg   =   67,300 kg

2. Regulated Take-off Mass
      62800 kg   =   62,800 kg

3. Regulated Landing Mass + Estimated fuel consumption
      54900 + 8000   =   62,900 kg

The maximum allowable Traffic Load for the conditions prevailing can now be determined by subtracting the Operating Mass from the Regulated Take-off Mass i.e. :

Maximum Traffic Load   =   Regulated Take-off Mass – Operating Mass
                                        =   62800 – (34000 + 16000)
                                        =   12800 kg

The underload can now be determined by subtracting the actual traffic load from the maximum allowable traffic load.

        Underload   =   **maximum traffic load – actual traffic load**
                          =   12800 – 12400
                          =   400 kg

Fortunately, the MRJT Load and Trim Sheet makes easy work of the above calculations and students are advised to practice using the Load and Trim Sheet.

Example calculations.

Mass and balance calculations may be carried out using the Loading Manifest (data sheet fig 4.10) and CG limits envelope (fig 4.11) or the Load and Trim sheet (fig 4.12).

We will do a sample calculation using both methods.

Using the following values complete the Loading Manifest fig 4.10 and check the limiting values with the CG envelope.

MASS AND BALANCE   **LOADING AND CENTRE OF GRAVITY**

     DOM   34,300 kg and 15% MAC
     PAX   Total 116, standard weight 84 kg each
                     10 each in zone A and G
                     12 each in zone B and F
                     24 each in zone C, D and E
     CARGO   600 kg hold 1
                     1,500 kg hold 4 (includes checked baggage)
     FUEL   15,000kg at take off
                     260 kg start and taxi
                     10,000 kg trip fuel

Use the JAR data sheets where required to find balance arm from MAC and vice-versa. The moment/1000 is calculated from the mass and balance arm/1000. The balance arm is calculated by dividing the total moment by the total weight. The fuel balance arm and quantity in each tank is also found from the data sheets. Note that the centre fuel tank content is used before the wing tank fuel content and that the centre tank includes 24 kg of unusable fuel.

As a start, if the DOM CG position is 15% MAC, then that is 15% of 134.5 inches (LM page 21) or 20.175 inches. That makes the CG balance arm 20.175 + 625.6 = 645.8 inches aft of the datum

The fuel load balance arm can be extracted from fig 4.5 and 4.6 of the loading manual, for example the maximum contents of tanks one and two is 9084 kg with a balance arm of 650.7. Fill in this blank sheet using the above information then check the aircraft has not exceeded any of the limits in the CG envelope

See if your answer agrees with mine! (I have estimated changes to the fuel tank CG position and accounted for the unusable fuel in the centre tank. However, the fuel CG position will be fixed and there will be no unusable fuel to account for in JAR exams)

If you wish to check the mass values and CG positions using the load and trim sheet for the MRJT use a DOI of 40.5.

MASS AND BALANCE   **LOADING AND CENTRE OF GRAVITY**

Fig 2.16, LOADING MANIFEST - MRJT 1

Max Permissible Aeroplane Mass Values

TAXI MASS   ZERO FUEL MASS

TAKE-OFF MASS   LANDING MASS

ITEM	MASS (KG)	B.A. In	MOMENT KG-IN/1000	C.G. % MAC
1. DOM	34300	645.8	22150.9	15%
2. PAX Zone A		284		-
3. PAX Zone B		386		-
4. PAX Zone C		505		-
5. PAX Zone D		641		-
6. PAX Zone E		777		-
7. PAX Zone F		896		-
8. PAX Zone G		998		-
9. CARGO HOLD 1		367.9		-
10. CARGO HOLD 4		884.5		-
11. ADDITIONAL ITEMS				-
ZERO FUEL MASS				
12. FUEL TANKS 1&2				
13. CENTRE TANK				
TAXI MASS				
LESS TAXI FUEL				
TAKE OFF MASS				
LESS FLIGHT FUEL				
EST. LANDING MASS				

MASS AND BALANCE   LOADING AND CENTRE OF GRAVITY

Fig 2.17 LOADING MANIFEST - MRJT 1 (Table 3)

Max Permissible Aeroplane Mass Values

TAXI MASS   **63060kg**   ZERO FUEL MASS   **51300kg**

TAKE-OFF MASS   **62800kg**   LANDING MASS   **54900kg**

ITEM	MASS (KG)	B.A. In	MOMENT KG-IN/1000	C.G. % MAC
1. DOM	34300	645.8	22150.9	15%
2. PAX Zone A	840	284	238.5	-
3. PAX Zone B	1008	386	389	-
4. PAX Zone C	2016	505	1018	-
5. PAX Zone D	2016	641	1292	-
6. PAX Zone E	2016	777	1566	-
7. PAX Zone F	1008	896	903	-
8. PAX Zone G	840	998	838	-
9. CARGO HOLD 1	600	367.9	221	-
10. CARGO HOLD 4	1500	884.5	1327	-
11. ADDITIONAL ITEMS				-
ZERO FUEL MASS	**46144**	**649**	**29943.4**	**17.4%**
12. FUEL TANKS 1&2	9084	650.7	5911	-
13. CENTRE TANK	6176	600.4	3708	-
TAXI MASS	**61404**	**644.3**	**39562.4**	-
LESS TAXI FUEL (C/TANK)	-260	600.5	-156	-
TAKE OFF MASS	**61144**	**644.5**	**39406.4**	**14%**
LESS CENTRE TANK FUEL	-5892	600.9	-3540.5	-
LESS MAIN TANK FUEL	-4108	628	-2580	
EST. LANDING MASS	**51144**	**651**	**33286**	**18.9%**

Note: the fuel CG positions have been estimated for simplicity and allowance has been made for the 24 kg of unusable fuel in the centre tank. However, it is most unlikely that you will not be required to adjust for fuel tank CG changes or to account for unusable fuel quantities in JAR exams.

MASS AND BALANCE   **LOADING AND CENTRE OF GRAVITY**

2.28. LOAD AND TRIM SHEET MRJT

Figure 2.18 shows a combined Load and Trim sheet for a modern twin jet airliner designated JAA - Medium Range Jet Twin (MRJT). See page 29 of the Loading Manual. (Students may be required to complete a part of the load and trim sheet during the JAA exams).

The left hand side of the page (part A) is the loading document which itemises the mass and mass distribution within the aeroplane i.e. dry operating mass, traffic load and and fuel load. The right hand side (part B) indicates how each mass in turn subsequently affects the position of the C G in relation to the mean aerodynamic chord. The ZFM and the TOM must be within the relevant area of the graph envelop on completion of the mass calculations otherwise the aircraft is unsafe to fly.

Part A (loading summary) should be completed as follows:-

Section 1   is used to establish the limiting take off mass, maximum allowable traffic load and underload before last minute changes (LMC)

Section 2   Shows the distribution of the traffic load using the following abbreviations.

TR	Transit
B	Baggage
C	Cargo
M	Mail
Pax	Passengers
Pax F	First Class
Pax C	Club/Business
Pax Y	Economy

Section 3   Summarises the loading and is used to cross check that limiting values have not been exceeded.

Part B is the distribution and trim portion. The lower part of part B is the CG envelope graph. The vertical scale of which is given in terms of mass and the horizontal axis scale in terms of MAC.

1. Using data from the loading summary enter the Dry Operating Index (DOI) for the DOM.

2. Move the index vertically downwards into the centre of the first row of horizontal boxes. **Note** that each box within the row has a pitch which represents either a defined mass or a number of persons. There is also an arrow indicating the direction in which the pitch is to be read.

3. Move the index horizontally in the direction of the arrow to a pitch value corresponding to the value in the Mass/No box immediately to the left of the row.

MASS AND BALANCE   **LOADING AND CENTRE OF GRAVITY**

4. Repeat operations 1 to 3 above for each subsequent row of boxes down to and including row Og. After completing the index calculation for row Og drop a line vertically down until it bisects a mass on the vertical scale of the envelope corresponding to the Zero Fuel Mass. The point of bisection must occur within the MZFM envelope.

5. Go to the horizontal row of boxes marked 'Fuel index' and add the take-off fuel index.

6. After completing the fuel index calculation drop a line vertically down from it until it bisects a mass on the vertical scale of the envelope corresponding to the Take-Off Fuel Mass. The point of bisection must occur within the TOM envelope.

7. Providing the ZFM and the TOM are within the envelope as described above, the aircraft is safe for the intended flight - including any permitted diversions.

MASS AND BALANCE LOADING AND CENTRE OF GRAVITY

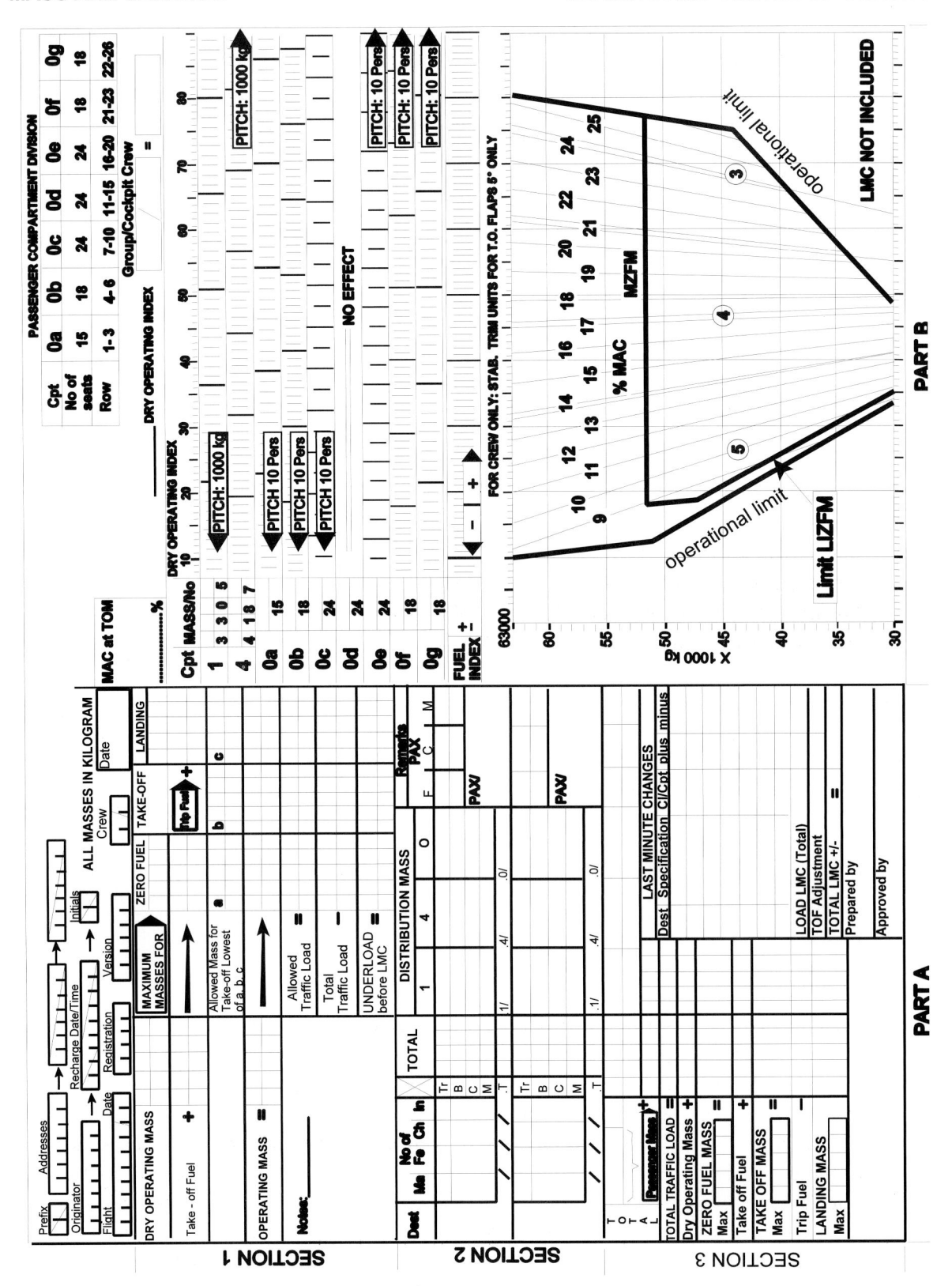

Figure 2.18. MRJT, Load and Trim Sheet

MASS AND BALANCE — LOADING AND CENTRE OF GRAVITY

Example:

The following example deals with part A and part B of the load sheet separately using the data shown.

DOM	=	34300 kg
DOI	=	45.0
MTOM	=	62800 kg
MZFM	=	51300 kg
MLM	=	54900 kg
Passengers	130	(Average mass 84 kg)
Baggage	130	(@14 kg per piece)
Cargo	630 kg	
Take off fuel	14500 kg	
Trip Fuel	8500 kg	

Limitations for JAA - Medium Range Twin Jet

Maximum Structural Taxi Mass	63050 kg
Maximum Structural Take Off Mass	62800 kg
Maximum Structural Landing Mass	54900 kg
Maximum Structural Zero Fuel Mass	51300 kg
Maximum Number of Passengers	141

Cargo

Hold 1	Max Volume	607 cu.ft
	Max Load	3305 kg
Hold 2	Max Volume	766 cu.ft
	Max Load	4187 kg

Standard Crew (Allowed for in DOM)

Flight Deck		2
Cabin Crew	forward	2
	aft	1

Section 1 of the load sheet is completed first in order to find the three potential take off masses (a, b and c). The value at (a) is the take-off mass you would achieve if you loaded the aeroplane to the MZFM and then added your intended fuel load. The value at (b) is the Regulated take-off mass for the takeoff airfield conditions as existing and the value at (c) is the take-off mass you would achieve if you were to land at the Regulated landing mass and then added back the mass of the trip fuel. The lowest of (a), (b) and (c) is the limiting take-off mass for traffic load calculations. The maximum allowable traffic load for the trip can be determined by subtracting the operating mass from the limiting take-off mass. Subsequently, any underload can be calculated by subtracting the actual traffic load from the allowable traffic load.. The underload sets the limiting mass for any last minute changes (LMC). Sections 2 and 3 of the load sheet detail the mass and distribution of the traffic load and give actual values of take-off and landing masses.

MASS AND BALANCE   **LOADING AND CENTRE OF GRAVITY**

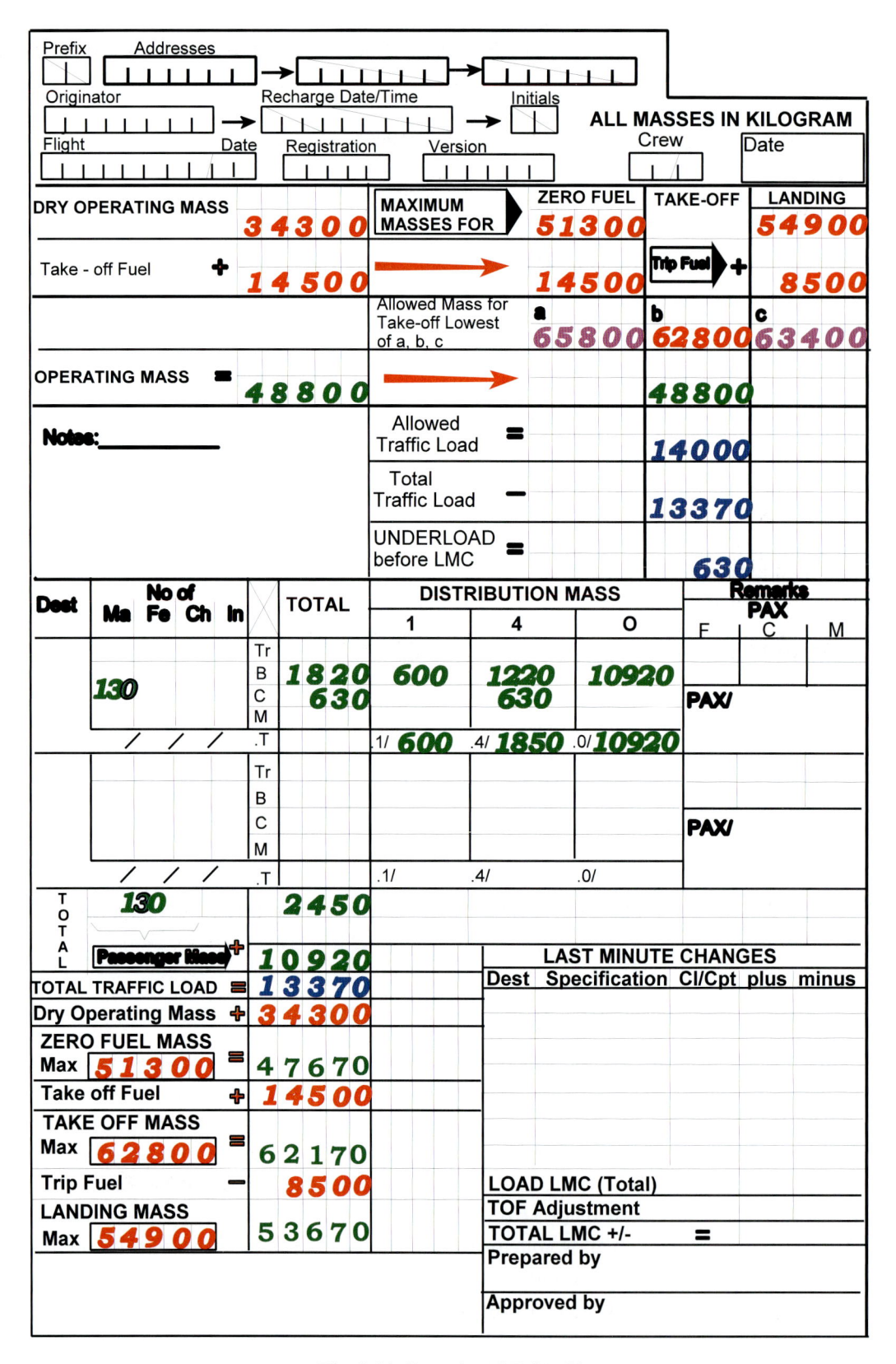

			ZERO FUEL	TAKE-OFF	LANDING
DRY OPERATING MASS	34300	MAXIMUM MASSES FOR	51300		54900
Take-off Fuel +	14500	→	14500	Trip Fuel +	8500
		Allowed Mass for Take-off Lowest of a, b, c	a 65800	b 62800	c 63400
OPERATING MASS =	48800	→		48800	
Notes:		Allowed Traffic Load =		14000	
		Total Traffic Load =		13370	
		UNDERLOAD before LMC =		630	

Dest	No of Ma Fe Ch In		TOTAL	DISTRIBUTION MASS 1 / 4 / 0	Remarks PAX F / C / M
	130	Tr B C M	1820 630	600 / 1220 / 10920 / 630 /	PAX/
	/ / /	.T		.1/ 600 .4/ 1850 .0/ 10920	
		Tr B C M			PAX/
	/ / /	.T		.1/ .4/ .0/	
TOTAL	130		2450		

Passenger Mass + 10920
TOTAL TRAFFIC LOAD = 13370
Dry Operating Mass + 34300
ZERO FUEL MASS
Max 51300 = 47670
Take off Fuel + 14500
TAKE OFF MASS
Max 62800 = 62170
Trip Fuel − 8500
LANDING MASS
Max 54900   53670

LAST MINUTE CHANGES
Dest   Specification   Cl/Cpt   plus   minus

LOAD LMC (Total)
TOF Adjustment
TOTAL LMC +/−   =
Prepared by
Approved by

Fig 2.19 Completed Trim Sheet

MASS AND BALANCE   **LOADING AND CENTRE OF GRAVITY**

In part B the graph is entered at the top by drawing a vertical lie from the DOM index of 45 into the row for cargo compartment 1. This row is split into sections by heavy lines representing 1000kg, each section is split again into 10, each line representing 100 kg. the arrow in the box represents the direction to move to adjust for that mass.

Follow the same procedure for each cargo compartment or seating compartment until have adjusted for all of the traffic load. Before adjusting for the fuel load draw the line down to intersect with the zero fuel mass to identify the ZFM CG position. Then adjust for the fuel index taken from the data sheets, page 30, and draw the line vertically down to identify the take-off CG position.

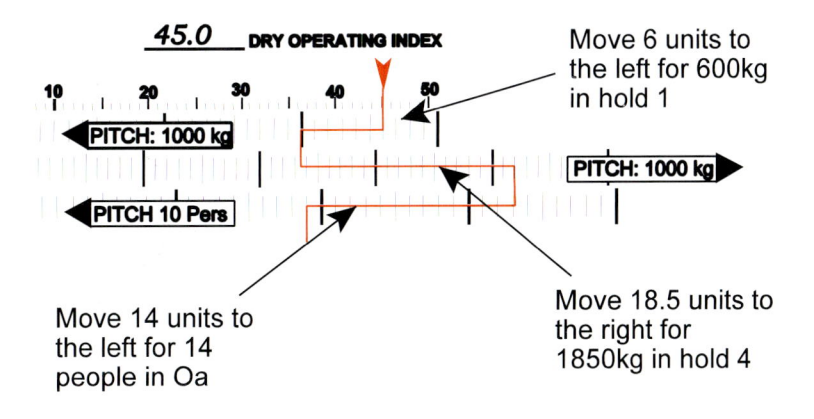

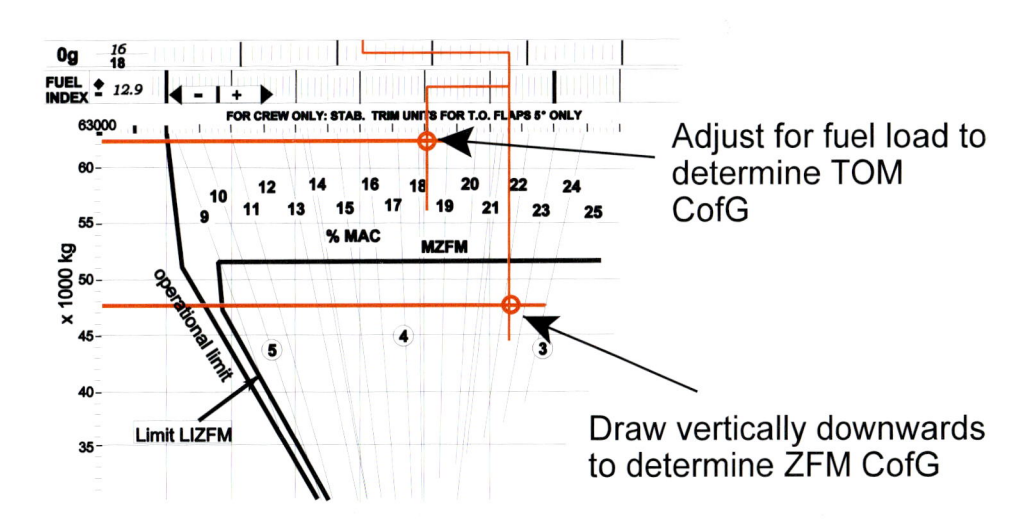

Fig 2.20 Load and trim Calculation Diagrams

MASS AND BALANCE

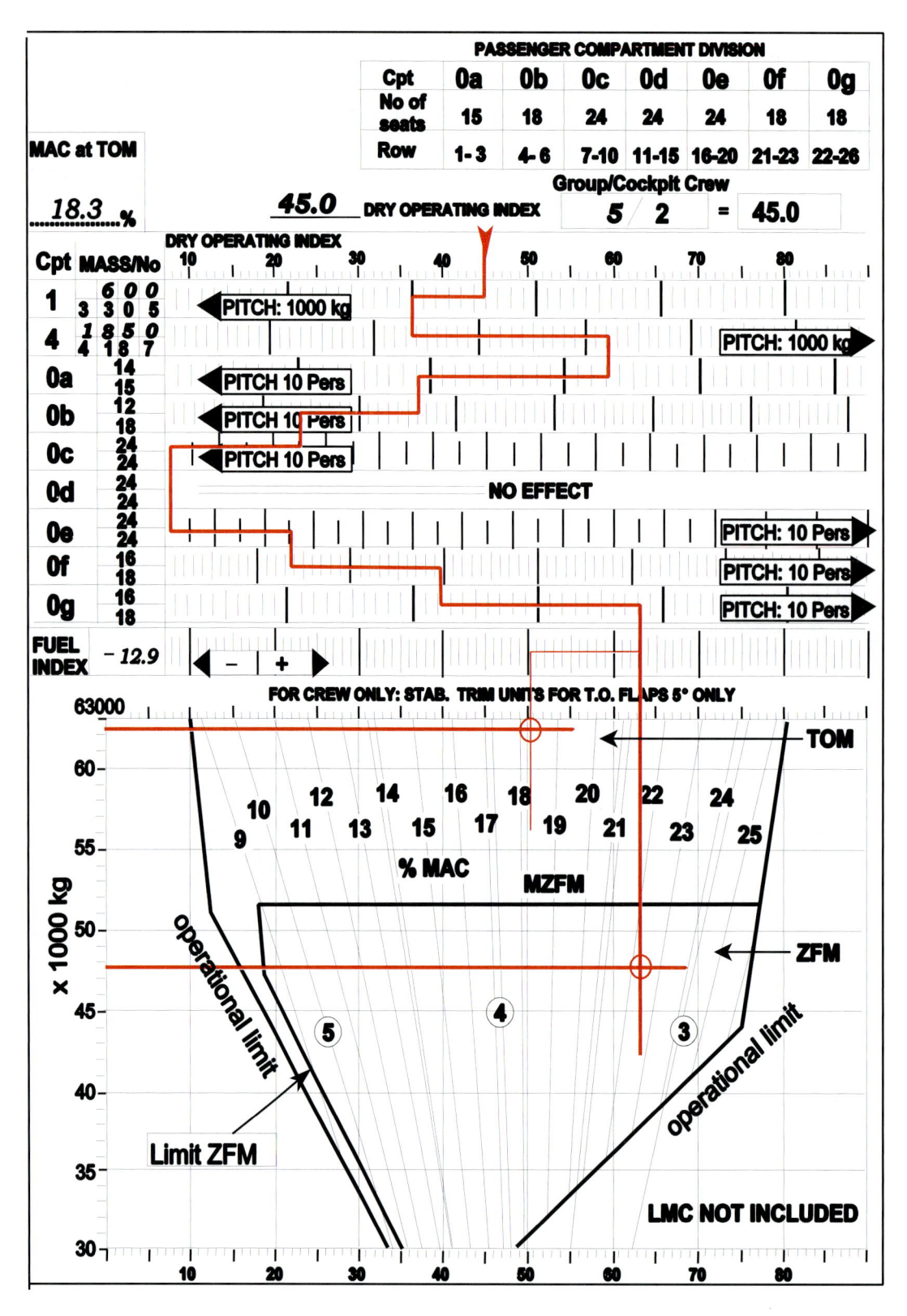

LOADING AND CENTRE OF GRAVITY

Figure 2.21

MASS AND BALANCE   **LOADING AND CENTRE OF GRAVITY**

Example.

With regard to Part 'A' above (Calculating the traffic/under load)~ A typical JAA question is given below. For practice work out the answer by using both the load sheet method described on page 2-47 and the calculation method described on page 2-39.

A scheduled flight of three hours estimated time, within Europe, is to be conducted. Using the data given calculate the maximum mass of freight that may be loaded in the following circumstances:

Performance limited take-off mass	67900 kg
Performance limited landing mass	56200 kg
MZFM	51300 kg
DOM	34960 kg
Fuel on board at ramp	15800 kg
Taxi fuel	450 kg
Trip fuel	10200 kg
Passengers (adults/each 84 kg)	115
(children/each 35 kg)	6
Flight crew (each 85 kg)	2
Cabin crew (each 75 kg)	5
Allow standard baggage for each passenger (13 kg)	

a.   1047 kg
b.   6147 kg
c.   4764 kg
d.   4647 kg

MASS AND BALANCE LOADING AND CENTRE OF GRAVITY

2.29. SELF ASSESSMENT QUESTIONS FOR S.E.P.1; M.E.P.1 AND MRJT

SELF ASSESSMENT QUESTIONS FOR SINGLE ENGINE PISTON/PROPELLER (SEP 1)

Unless told otherwise, assume that the maximum fuel capacity is 74 gallons. For all questions refer to CAP 696 (Loading Manual) in Book 15.

1. Where is the reference datum?

    a. 74 inches aft of the fwd CG position
    b. 80.4 inches aft of the rear CG position
    c. 87.7 inches aft of the rear CG position
    d. 39 inches forward of the firewall

2. What are the CG limits?

    a. fwd limit = 74 inches to 80.4 inches
    b. fwd limit = 74 inches, aft limit = 80.4 inches
    c. fwd limit = 74 inches, aft limit = 87.7 inches
    d. fwd limit = 74 inches to 80.4 inches and aft limit = 87.7 inches

3. What is the CG at the BEM?

    a. 77 inches
    b. 87 inches
    c. 77.7 metres
    d. 77.7 inches

4. What is the structural load limit for the floor at baggage zone 'C'?

    a. 50 lb per square foot
    b. 100 lb per cubic foot
    c. 100 lb per square foot
    d. 100 kg per square inch

5. What is the distance of the main undercarriage from the firewall?

    a. 97 inches
    b. 58 inches
    c. 87.7 inches
    d. 39 inches

MASS AND BALANCE **LOADING AND CENTRE OF GRAVITY**

6. The aircraft has six seats. Assuming no other cargo or baggage, what is the maximum fuel that can be carried if all six seats are occupied and the mass of each occupant is 180 lb?

    a. 50 lbs but the CG would be dangerously out of limits
    b. 155 lbs but the CG would be dangerously out of limits
    c. 50 lbs and the CG would be in limits
    d. 155lbs and the CG would be in limits

7. Where is the centroid of baggage zone B?

    a. 108 inches from the datum
    b. 120 inches from the datum
    c. 150 inches from the datum
    d. 180 inches from the datum

8. Assuming the weight and access is not a problem where can a box of mass 500 lb be positioned if the dimensions are 0.75 ft x 1.5 ft x 5 ft?

    a. in any of the baggage zones if placed on its smallest area
    b. in zones 'B' or 'C' if placed on its largest area
    c. in zone 'C' only if placed on its middle area
    d. in zone 'A' only if placed on its largest area

9. Assuming the weight and access is not a problem, where can a cubic box of mass 500 lb be positioned if the dimensions are 3.15 ft?

    a. in any of the baggage zones
    b. in zone 'B' or 'C' only
    c. in zone 'A' only
    d. in zone 'C' only

10. If the landing mass is 3155 lb and the trip fuel was 40 gallons, what was the ZFM if the fuel tanks held 60 gallons of fuel prior to take-off?

    a. 3001 lb
    b. 3035 lb
    c. 3098 lb
    d. 3111 lb

MASS AND BALANCE   **LOADING AND CENTRE OF GRAVITY**

11. What is the maximum ramp mass?

    a. 3650 lbs
    b. 3663 lbs
    c. 3780 lbs
    d. 3870 lbs

12. How far is the main wheel from the aft CG limit?

    a. 0.7 inches behind the rear datum
    b. 0.7 inches forward of the rear datum
    c. 6.6 inches forward of the rear datum
    d. 9.3 inches aft of the rear datum

13. How far is the firewall from the fuel tank centroid?

    a. 36 inches
    b. 37 inches
    c. 38 inches
    d. 39 inches

14. If the total moment is less than the minimum moment allowed:

    a. useful load items must be shifted aft
    b. useful load items must be shifted forward
    c. forward load items must be increased
    d. aft load items must be reduced

15. The CG is on the lower of the fwd CG limits:

    a. at a mass of 2500 lb and moment of 185000 lb in
    b. at a moment of 175,000 lb in and a mass of 2350 lb
    c. at a moment of 192,000 lb in and a mass of 2600 lb
    d. all the above

MASS AND BALANCE   LOADING AND CENTRE OF GRAVITY

SELF ASSESSMENT QUESTIONS FOR MEP 1

1. What performance class does the aircraft belong to?

    a. Performance class 'A'
    b. Performance class 'B'
    c. Performance class 'C'
    d. Performance class 'D'

2. Where is the reference datum?

    a. 78.4 inches forward of the wing leading edge at the inboard edge of the inboard fuel tank
    b. 25.3 inches forward of the nose wheel
    c. 109.8 inches forward of the main wheel
    d. all the above

3. The main wheel is :

    a. 19 inches forward of the fwd CG limit at the maximum take-off mass
    b. 27.8 inches behind the fwd CG limit at a take-off mass of 3400 lbs
    c. 15.2 inches forward of the rear CG limit at the maximum take-off mass
    d. all the above

4. The nose wheel is :

    a. 56.7 inches forward of the fwd CG limit at maximum take-off mass
    b. 65.5 inches forward of the fwd CG limit at maximum take-off mass
    c. 69.3 inches aft of the rear CG limit at maximum take-off mass
    d. all the above

5. What is the minimum fuel mass that must be consumed if the aircraft, having become airborne at maximum weight, decides to abort the flight.

    a. 1260 lb
    b. 280 lb
    c. 237 lb
    d. 202 lb

6. If the pilot has a mass of 200 lb, what is the maximum traffic load?

    a. 1060 lb
    b. 1600 lb
    c. 1006 lb
    d. 6001 lb

MASS AND BALANCE
LOADING AND CENTRE OF GRAVITY

7. Assuming the maximum zero fuel mass and maximum take-off mass, what fuel load can be carried?

    a. 38.9 Imperial gallons
    b. 46.6 US gallons
    c. 176.8 litres
    d. any one of the above

8. A box of mass 100 lb is to be transported. The box dimensions are 9 x 9 x 12 inches. Which zones can it be carried in?

    a. all zones, both the mass and structural loading are within limits
    b. zones 2 and 3 only
    c. no zones, both the mass and structural loading would be exceeded.
    d. no zones, the structural loading would be exceeded.

9. A box of mass 360 lb is to be transported. The dimensions of the box are 1.7ft x 1.7ft x 1.8ft. Which zones can it be carried in?

    a. zones 2 and 3only but placed on the 1.7 x 1.7 face
    b. zones 2 and 3 only but placed on the 1.7 x 1.8 face
    c. no zones, both the mass and structural loading would be exceeded
    d. no zones, the structural loading would be exceeded.

10. Assuming floor loading limits are acceptable, how much freight and fuel load can be carried for MTOM if the pilot's mass was 200 lb?

    a. A full load in each zone plus 380 lb of fuel
    b. 50 lb in zones 1 or 4 but full loads in each of the other zones, plus 280 lbs of fuel.
    c. 350 lbs load in zone 4 but full loads in all the other zones, plus 280 lbs of fuel.
    d A full freight load in each zone plus 280 lb of fuel

11. What is the maximum fuel tank capacity?

    a. not given.
    b. 123 US gallons
    c. 46.6 US gallons
    d. TOM minus ZFM

12. If the aircraft is at MTOM with full fuel tanks and a pilot of mass 200 lb, what traffic load can be carried?

    a. nil
    b. 579 lbs providing at least 20.5 gallons of fuel are consumed in start, taxi and flight
    c. 625 lbs providing at least 43.3 gallons of fuel are consumed in start, taxi and flight
    d. 759 lbs providing at least 59.5 gallons of fuel are consumed in start, taxi and flight

MASS AND BALANCE **LOADING AND CENTRE OF GRAVITY**

13. The CG when the TOM is 4300 lb and the corresponding moment is 408500 lb in is :

    a. 95 inches
    b. 59 inches
    c. 0.4 inches tail heavy
    d. 0.4 inches rear of the aft limit

14. If the CG is 86 inches and the TOM is 4100 lb the aircraft is :

    a. just on the forward CG limit
    b. just outside the forward CG limit
    c. just inside the aft CG limit
    d. within the two forward limits

MASS AND BALANCE   **LOADING AND CENTRE OF GRAVITY**

SELF ASSESSMENT QUESTIONS FOR MRJT 1

1. What is the total length of the fuselage?

    a. 1365 inches
    b. 1375 inches
    c. 1387 inches
    d. 1395 inches

2. How far is the front spar from the datum?

    a. 562 inches
    b. 540 inches
    c. 500 inches
    d. 458 inches

3. What is the distance between the two main access doors?

    a. 940 inches
    b. 947 inches
    c. 974 inches
    d. 984 inches

4. How far is the leading edge of the mean aerodynamic chord from the datum?

    a. 540 inches forward of the datum
    b. 589.5 inches forward of the datum
    c. 625.6 inches aft of the datum
    d. 627.5 inches aft of the datum

5. What is the length of the mean aerodynamic chord?

    a. 104.5 inches
    b. 114.5 inches
    c. 124.5 inches
    d. 134.5 inches

6. What moment change occurs when the flaps are fully retracted from the 15 degree position?

    a. a reduction of 14 kg in
    b. an increase of 14 kg in
    c. a reduction of 14000 kg in
    d. an increase of 14000 kg in

MASS AND BALANCE   LOADING AND CENTRE OF GRAVITY

7. What change in moment occurs when the flaps are retracted from 40 degrees to 5 degrees?

   a. a negative moment of 5 kg in
   b. a negative moment of 11 kg in
   c. a negative moment of 16 kg in
   d. a negative moment of 5000 kg in

8. What stabiliser trim setting is required for take-off when the CG is 19% MAC for 5 degrees of take-off flap?

   a. 2.75
   b. 3.75
   c. 4.75
   d. 5.75

9. What is the maximum structural take-off mass?

   a. 63060 kg
   b. 62800 kg
   c. 54900 kg
   d. 51300 kg

10. What is the CG range for maximum zero fuel mass?

    a. 8% MAC to 27% MAC
    b. 12% MAC to 20% MAC
    c. 7.5% MAC to 27.5% MAC
    d. 8.5% MAC to 26% MAC

11. Assuming the MZFM, what is the maximum allowable fuel mass for take-off?

    a. 10015 kg
    b. 10150 kg
    c. 11500 kg
    d. 15000 kg

12. Assuming the standard masses have been used for both passengers and baggage, what is the mass of a full passenger and baggage load?

    a. 13027 kg
    b. 13677 kg
    c. 14127 kg
    d. 15127 kg

MASS AND BALANCE **LOADING AND CENTRE OF GRAVITY**

13. What is the allowable hold baggage load for an aircraft with a full passenger complement?

    a. 1533 kg
    b. 1633 kg
    c. 1733 kg
    d. 1833 kg

14. What is the underload if only maximum passenger hold baggage is carried?

    a. 3305 kg - 1833 kg = 1472 kg
    b. 4187 kg - 1833 kg = 2354 kg
    c. 7492 kg - 1833 kg = 5659 kg
    d. 9247 kg - 1833 kg = 7414 kg

15. If the crew mass is 450 kg and the Zero Fuel Mass is 51300 kg, what is the Basic Empty Mass if a full traffic load is carried?

    a. 31514 kg
    b. 31773 kg
    c. 37713 kg
    d. 33177 kg

16. Using the values for the data given in the Loading Manual, would the aircraft be able to carry both a full fuel load and a full Traffic load at take-off?

    a. No
    b. Yes, providing the BEM was not more than 31145 kg
    c. Yes, providing the BEM was not less than 31451 kg
    d. Yes, providing the BEM was not more than 31514 kg

17. If the DOM is given as 34300 kg and the aircraft has a full load of passengers and baggage, what additional cargo mass could it carry i.e. what is the underload?

    a. none
    b. 3123 kg
    c. 3223 kg
    d. 3323 kg

18. What is the maximum usable fuel quantity?

    a. 5311 US gallons
    b. 5294 US gallons
    c. 5123 US gallons
    d. 5032 US gallons

MASS AND BALANCE   **LOADING AND CENTRE OF GRAVITY**

19. What is the maximum usable fuel mass?

    a. 16092 kg
    b. 16078 kg
    c. 16064 kg
    d. 16040 kg

20. What is the allowable start and taxi fuel?

    a. 160 kg
    b. 260 kg
    c. 360 kg
    d. 460 kg

21. What are the preferred zones for passenger loads if the pax load is low?

    a. zones E, F and G
    b. zones C, D and E
    c. zones B, C and D
    d. A, B and C

22. How many seats are there in zone B?

    a. 15
    b. 18
    c. 21
    d. 24

23. The leading edge of the MAC is given as 625.6 inches aft of the datum. What is the distance of the CG from the datum if it is found to be 16% of the MAC?

    a. 547 inches
    b. 647 inches
    c. 747 inches
    d. 674 inches

24. The CG is found to be 652.5 inches aft of the datum. What percentage is the CG to the MAC?

    a. 10%
    b. 15%
    c. 20%
    d. 25%

MASS AND BALANCE   **LOADING AND CENTRE OF GRAVITY**

25. If a passenger moves from a seat position corresponding to the balance arm at zone D to a position corresponding to the balance arm at zone F, what distance will the passenger have travelled and how many seat rows will he have passes?

    a. 255 inches and 8 seat rows
    b. 260 inches and 7 seat rows
    c. 265 inches and 6 seat rows
    d. 270 inches and 5 seat rows

26. The balance arm for each of the seat zones is measured from the datum to:

    a. the front border line of the zone
    b. the centre line of the zone
    c. the rear border line of the zone
    d. the front border line of the next zone in sequence

27. What is the maximum and minimum running load of a box of mass 500 kg and dimensions of 1m x 1.2m x 1.2m?

    a. 12.7 kg/in and 10.6 kg/in
    b. 10 kg/in and 12.4 kg/in
    c. 11 kg/in and 9.5 kg/in
    d. 15 kg/in and 13.1 kg/in

28. What is the maximum and minimum distribution load intensity for a box of mass 500 kg and dimensions of 1m x 1.2m x 1.2m?

    a. 50.5 kg/sq ft and 40.6 kg/sq ft
    b. 47.3 kg/sq ft and 37.7 kg/sq ft
    c. 45.1 kg/sq ft and 35.8 kg/sq ft
    d. 38.7 kg/sq ft and 32.3 kg/sq ft

29. All other parameters being acceptable, a box with a maximum and minimum running load of 12 kg/in and 7 kg/in and a mass of 800 kg can be fitted into:

    a. any compartment of either the forward or aft cargo compartment
    b. the front section of the aft cargo compartment or the rear section of the forward cargo compartment
    c. the rear section of the forward cargo compartment or the rear section of the aft cargo compartment
    d. the centre section of forward cargo compartment only

MASS AND BALANCE   **LOADING AND CENTRE OF GRAVITY**

30. A box with a mass of 500 kg and dimensions 0.8m and 0.9m x 1.3m has a maximum and minimum distribution load intensity of:

    a. 64.6 kg/sq ft max and 39.7 kg/sq ft min
    b. 39.7 kg/sq ft max and 64.6 kg/sq ft min
    c. 44.7 kg/sq ft max and 39.7 kg/sq ft min
    d. 64.6 kg/sq ft max and 44.7 kg/sq ft min

31. The maximum freight mass allowed is:

    a. 17017 lb
    b. 16517 lb
    c. 16017 lb
    d. 15517 lb

32. Assuming all other parameters are acceptable, a box with a mass of 500 kg and with equal sides of 8.5 ft would fit into:

    a. either the front or rear cargo compartment
    b. the forward cargo compartment only
    c. neither cargo compartment
    d. the aft cargo compartment only

33. The front compartment of the front cargo hold is situated below:

    a. passenger zone A
    b. passenger zone B
    c. passenger zone C
    d. passenger zone D

34. The balance arm of the centroid of the forward hold compartment is:

    a. 228 inches
    b. midway between 228 inches and 286 inches
    c. midway between 286 inches and 343 inches
    d. 367.9 inches

35. The maximum distribution load intensity for the cargo compartments is:

    a. 68 lb per sq ft
    b. 68 kg per sq metre
    c. 68 kg per sq in
    d. 68 kg per sq ft

MASS AND BALANCE — LOADING AND CENTRE OF GRAVITY

36. Between 44,000 kg and 63,000 kg the rear CG limit as a percentage of the MAC:

    a. is constant at 28%
    b. increases from 28% to 29.5%
    c. decreases from 28% to 26%
    d. decreases from 28% to 9%

Referring to the Load and Trim Sheet on page 29, answer questions 37 to 49 inclusive:

37. The Traffic Load is:

    a. 39800 kg obtained from ZFM, 51300 kg less fuel mass 11500 kg
    b. obtained from the sum of pax mass plus baggage mass plus total cargo compartment mass
    c. 13370 kg obtained from 10920 kg pax mass plus 2450 kg baggage mass plus 630 kg cargo mass
    d. 13370 kg obtained from 10920 kg pax mass, 1820 baggage mass and 630 kg cargo mass

38. The cargo distribution in section 4 is:

    a. 1220 kg
    b. 630 kg
    c. 1850 kg
    d. 1820 kg plus 630 kg

39. The actual take-off mass is:

    a. 51300 kg ZFM plus 14500 kg take-off fuel
    b. 62800 kg less 8500 kg trip fuel
    c. 53670 kg less 14500 kg take-off fuel
    d. 47670 kg ZFM plus 14500 kg take-off fuel mass

40. The landing mass is:

    a. 62800 kg take-off mass less 8500 kg trip fuel
    b. 62170 kg take-off mass less 8500 kg trip fuel
    c. 62170 kg take-off mass plus 8500 kg trip fuel
    d. 62800 kg take-off mass plus 8500 kg trip fuel

41. In order to determine the underload the pilot starts by selecting the lowest mass from the three key masses given. The key masses are:

    a. Dry Operating Mass, Max'm Zero Fuel Mass and Take Off Mass
    b. Max'm Zero Fuel Mass, Take Off Mass and Landing Mass
    c. Dry Operating Mass, Max'm Zero Fuel Mass and Landing Mass
    d. Traffic Load, Take-off Mass and Landing Mass

MASS AND BALANCE **LOADING AND CENTRE OF GRAVITY**

42. From the figures given, if the actual take-off fuel mass (14500 kg) was added to the Maximum Zero Fuel Mass the aircraft would be:

    a. Below the maximum Take-off mass by 350 kg
    b. Over the maximum Take-off mass by 530 kg
    c. Over the maximum Take-off mass by 3000 kg
    d. Below the maximum Take-off mass by 630 kg

43. The actual underload for the aircraft after the traffic load and fuel load have been accounted for is:

    a. zero
    b. 720 kg
    c. 630 kg
    d. 960 kg

44. What is the Dry Operating Index?

    a. 45
    b. 12
    c. 54
    d. 10

45. What are the seat row numbers in pax zone 'Oc'?

    a. 4 - 6
    b. 6 - 8
    c. 7 - 10
    d. 8 - 13

46. What is the Take-off Mass as a percentage of the MAC?

    a. 18.3%
    b. 19.3%
    c. 20.3%
    d. 21.3%

47. Prior to take-off there is a change in destination and so the pilot decides to take 2000 kg of fuel less. Using the Load and Trim Sheet, calculate the new Take-off mass and CG position.

    a. Can't be calculated because the landing mass will be too high
    b. 60800 kg Take-off mass and CG 17.5% MAC
    c. 60170 kg Take-off mass and CG 18.8% MAC
    d. 60170 kg Take-off mass and 19.3% MAC

MASS AND BALANCE   **LOADING AND CENTRE OF GRAVITY**

48. When adjusting the CG index for the fuel load, why is the line moved to the left as a minus index?

    a. Because the fuel will be consumed in flight
    b. Because the fuel is given a minus index in the fuel index correction table
    c. Because the centroid of the tanks is behind the CG position
    d. Because the graph would run out of range

49. For a fuel mass of 11800 kg the index is:

    a. minus 4.5
    b. minus 5.7
    c. minus 6.3
    d. none of the above

50. A scheduled flight of three hours estimated flight time, within Europe, is being planned. Calculate the maximum mass of freight that may be loaded in the following circumstances:

Structural limited take-off mass	62800 kg
Structural limited landing mass	54900 kg
MZFM	51300 kg
Dry Operating Mass	34960 kg
Fuel on board at ramp	15800 kg
Taxi fuel	450 kg
Trip fuel	10200 kg
Passengers (adults each 84 kg)	115
Passengers (children each 35 kg)	6
Flight crew (each 85 kg)	2
Cabin crew (each 75 kg)	3
Standard baggage for each passenger	13 kg

    a. 4647 kg
    b. 4102 kg
    c. 1047 kg
    d. 5545 kg

MASS AND BALANCE   **LOADING AND CENTRE OF GRAVITY**

Answers to 'definition' example questions
1. = a, 2 = c, 3 = d, 4 = a

Answers to Fuel Mass Conversions
1. = 16660 US gallons
2. = 41908 lb
3. = 0.74
4. = Lighter by 2500 kg

Answer to Basic empty mass and CG position
The CG is 57 cm in front of the main wheels.

Answers to Percentage Mean Aerodynamic Chord Problems
1 = 115.054 inches
2. = 10.67 inches fwd of datum
3. = Fwd limit 24.3%, Aft limit 53.6%
4. = 1.45 inches out of limits
5 = 17.05m

Answers to Moving Mass Problems
1 = 55.3816 lb
2. = 23.9 inches
3. = 7 rows
4. = 146.1 kg
5. = 952 kg
6. = 43.2 lb

Answers to Adding or Removing Mass Problems
1 = 500 kg
2. = 35.97 lb
3. = 1742.911 lb
4. = 45.7 lb
5. = 79.7 inches
6. = 3.68 inches
7. = 45.71 lb
8 = 4455 Imperial gallons

Answer to intensity and running load example
The boxes of cargo must go into hold B and the baggage into hold C

Answers to Station Numbers
1. = Station 130
2. = Station 1217
3. = 348 inches
4. = 787 inches

MASS AND BALANCE — LOADING AND CENTRE OF GRAVITY

Answers to SEP1 Self Assessment Questions

1	d	11	b
2	d	12	d
3	d	13	a
4	c	14	a
5	b	15	d
6	b	16	b
7	c	17	
8	b	18	
9	b	19	
10	b	20	

Answer to example Traffic Load Calculation

Maximum allowable freight mass = 4647 kg

Answers to MEP1. Self Assessment Questions

1	b	6	a	11	b
2	d	7	d	12	c
3	b	8	d	13	d
4	b	9	b	14	a
5	c	10	a		

Answers to MRJT Self Assessment Questions

1	c	21	c	41	b
2	b	22	b	42	c
3	b	23	b	43	c
4	c	24	c	44	a
5	d	25	a	45	c
6	c	26	b	46	a
7	d	27	a	47	d
8	b	28	d	48	b
9	b	29	b	49	c
10	a	30	a	50	c
11	c	31	b	51	
12	b	32	a	52	
13	d	33	a	53	
14	c	34	d	54	
15	a	35	d	55	
16	a	36	c	56	
17	d	37	d	57	
18	b	38	b	58	
19	d	39	d	59	
20	b	40	b		

CHAPTER THREE – SPECIMEN QUESTIONS

Contents

                                                                                                                   Page

SPECIMEN QUESTIONS……………………...……….….……..……………….3 – 1

ANSWERS TO SPECIMEN QUESTIONS...………….……………...……………3 – 9

SPECIMEN EXAM PAPER...……………………………….…..……………...3 – 11

ANSWERS TO SPECIMEN EXAM PAPER...……….…………….…..…………3 – 15

EXPLANATIONS TO SPECIMEN EXAM PAPER ...…………………….…...……3 – 17

SPECIMEN QUESTIONS

1. Define the useful load:

    a. traffic load plus dry operating mass
    b. traffic load plus usable fuel mass
    c. dry operating mass plus usable fuel load
    d. that part of the traffic load which generates revenue

2. Determine the position of the CG as a percentage of the MAC given that the balance arm of the CG is 724 and the MAC balance arms are 517 to 1706.

    a. 14.2 %
    b. 15.3 %
    c. 16.3 %
    d. 17.4 %

3. The distance from the datum to the CG is:

    a. the index
    b. the moment
    c. the balance arm
    d. the station

4. Using CAP696, MRJT, fig 4.9. What is the balance arm, the maximum compartment load and the running load for the most aft compartment of the fwd cargo hold?

    a. 421.5 cm   3305 kg   13.12 kg per inch
    b. 1046.5 inches   711 kg   7.18 kg per kg
    c. 421.5 inches   2059 kg   13.12 kg per inch
    d. 1046.5 m   711 kg   7.18 kg per in

5. If the maximum structural landing mass is exceeded:

    a. The aircraft will be unable to get airborne
    b. The undercarriage could collapse on landing
    c. No damage will occur providing the aircraft is within the regulated landing mass.
    d. No damage will occur providing the aircraft is within the performance limited landing mass.

6. Use CAP 696, MRJT as appropriate. Prior to departure a MRJT is loaded with maximum fuel of 20100 ltr at an SG of 0.78. Calculate the maximum allowable traffic load that can be carried given the following data:

    PLTOM = 67200 kg
    PLLM = 54200 kg
    DOM = 34930 kg
    Taxi fuel = 250 kg
    Trip fuel = 9250 kg
    Contingency and holding fuel = 850 kg
    Alternate fuel = 700 kg

    a. 13092 kg
    b. 12442 kg
    c. 16370 kg
    d. 16842 kg

7. Using CAP 696, fig 4.-12. Assuming the fuel index moves minus 5.7 from the ZFM index, what is the take-off CG as a percentage of the MAC?

   a. 20.1 %
   b. 19.1 %
   c. 23.0 %
   d. 18.2 %

8. For a conventional light aeroplane with a tricycle undercarriage configuration, the higher the take-off mass:

   1. Stick forces at rotation will increase
   2. Range will decrease but endurance will increase
   3. Gliding range will reduce
   4. Stalling speed will increase

   a. all statements are correct
   b. Statement 3 only is correct
   c. Statements 1 and 4 only are correct
   d. Statement 4 only is correct

9. Due to a mistake in the load sheet the aeroplane is 1000 kg heavier than you believe it to be. As a consequence:

   a. V_1 will be later
   b. V_{MU} will be later
   c. V_R will be later
   d. V_1, V_{MU}, V_R will all occur earlier

10. If the aeroplane was neutrally stable this would suggest that:

    a. the CG is forward
    b. the CG is in mid range
    c. the CG is on the rear limit
    d. the CG is behind the rear limit

11. The CG position is:

    a. set by the pilot
    b. set by the manufacturer
    c. able to exist within a range
    d. fixed

12. Which of the following would not affect the CG?

    a. Cabin crew members performing their normal duties.
    b. Fuel usage
    c. Stabilator trim setting
    d. Mass added or removed at the neutral point

13. Using the data for the MRJT in CAP 696, what is the CG as a percentage of the MAC if the CG is 650 inches from the datum.

    a. 17.03%
    b. 18.14%
    c. 19.25%
    d. 20.36%

14. The CG datum has to be along the longitudinal axis:

    a. between the nose and the tail.
    b. between the leading and trailing edge of the MAC.
    c. but does not have to be between the nose and the tail.
    d. at the fire

15. The CG is :

    a. the point on the aircraft where the datum is located.
    b. the point on the aircraft at which gravity appears to act.
    c. the point on the aircraft from where the dihedral angle is measured.
    d. the point on the aircraft where the lift acts through.

16. The aircraft basic mass and CG position is found on :

    a. the weighing schedule and the aeroplane must be re-weighed if equipment change causes a change in mass or balance
    b. On the loading manifest and is DOM – traffic load.
    c. On the loading manifest and is ZFM – useful load
    d. On the weighing schedule and is adjusted to take account of any mass changes.

17. When determining the mass of fuel/oil and the value of the SG is not known, the value to use is:

    a. determined by the operator (and laid down in the aeroplane OPS Manual. A pilot simply has to look it up)
    b. set out in JAR OPS Section 1
    c. determined by the aviation authority
    d. determined by the pilot

18. In mass and balance terms, what is an index?

    a. A cut down version of a force
    b. A moment divided by a constant
    c. A moment divided by a mass
    d. A mass divided by a moment

19. Standard masses for baggage can be used when:

    a. 9 seats or more
    b. 20 seats or more
    c. 30 seats or more
    d. less than 30 seats

MASS AND BALANCE   SPECIMEN QUESTIONS

20. What is the zero fuel mass?

   a. MTOM minus fuel to destination minus fuel to alternative airfield.
   b. Maximum allowable mass of the aircraft with no usable fuel on board.
   c. Operating mass minus the fuel load.
   d. Actual loaded mass of the aircraft with no usable fuel on board.

21. If an aeroplane comes into lands below its MSLM but above the PLLM for the arrival airfield:

   1. Airframe structural damage will occur.
   2. Tyre temperature limits could be exceeded.
   3. It might not have sufficient runway length in which to stop safely.
   4 A go-around might not be achievable.
   5. Brake fade could occur..

   a. All the answers are correct
   b. 3 and 4 only are correct
   c. 2, 3, 4 and 5 only are correct
   d. 1, 3, 4 and 5 only are correct

22. A twin engine aeroplane of mass 2500 kg is in balanced level flight. The CG limits are 82 in to 95 in from the nose position of the aeroplane and the CG is approximately mid range. A passenger of mass 85 kg, moves from the front seat 85.5 inches aft of the nose to the rear seat 157.6 inches from the nose. What is the new CG position approximately?

   a. 2.5 inches
   b. 87.5 inches
   c. 91 inches
   d. 92.5 inches

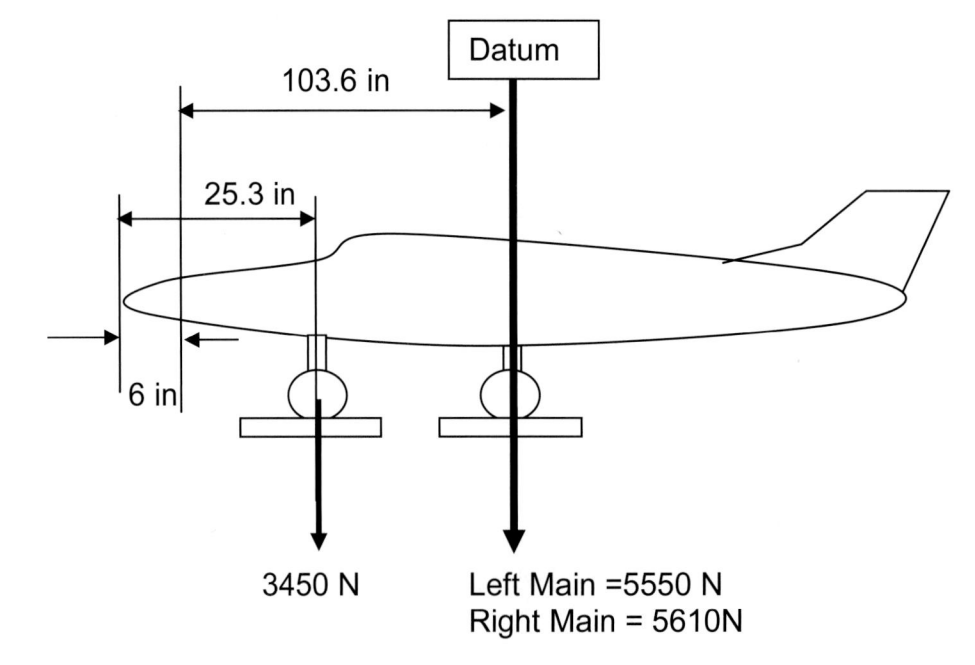

23. [Diagram: Aeroplane with Datum line. Distances shown: 103.6 in from nose gear to datum, 25.3 in from main gear to datum, 6 in. Nose gear reaction: 3450 N. Left Main = 5550 N, Right Main = 5610 N.]

Calculate the Basic Empty mass and CG position for the MEP 1 shown above.

   a. BEM = 1489 kg and CG is 20 inches forward of datum
   b. BEM = 1456 kg and CG is 20 inches aft of the nose
   c. BEM = 1489 kg and CG is 20 inches aft of datum
   d. BEM = 1456 kg and CG is 89.6 inches aft of the nose

24. A twin engine aeroplane is certified for a MSTOM and a MLM of 58000 kg and 55000 kg respectfully. What is the limiting take-off mass for the aeroplane?

PLTOM	61000 kg
PLLM	54000 KG
MZFM	36000 kg
Operating mass	55000 kg
Trip fuel	30000 kg
Contingency fuel 5% of trip fuel	
Alternative fuel	500 kg
Final reserve	500 kg
Flight duration	3 hours
Fuel consumption	500 kg per hour per engine
Useful load	41500 kg

a. 58000 kg
b. 61000 kg
c. 56145 kg
d. 56545 kg

Refer to CAP 696 for answers to 25, 26 and 27

25. With reference to CAP 696 figure 4.9, the centroid of the forward hold is :

a. half way between stations 228 and station 500
b. 314.5 inches forward of the aft cargo bay centroid
c. 367.9 inches from the datum
d. 367.9 inches from the nose of the aeroplane.

26. The distance of the leading edge of the wing MAC from the datum is:

a. undefined
b. 525.6 m
c. 625.6 in
d. 525.6 in

27. What is the CG as a percentage of the MAC of the fully loaded aircraft?

BEM	=	12000 kg
Arm	=	3 m
CG	=	25% MAC
MAC	=	2 m

Item	Balance arm
Front seats	2.5 m
Rear seats	3 m
Fuel @ 0.74 410 Ltr	
Fuel arm	2.5 m
Rear seats	Empty
Pilot	80 kg
Front seat Pax	80 kg

a. 16%
b. 19%
c. 21%
d. 24%

28. The maximum aircraft mass excluding all usable fuel is:

    a. fixed and listed in the aircraft's Operations Manual
    b. variable and is set by the payload for the trip.
    c. fixed by the physical size of the fuselage and cargo holds.
    d. variable and depends on the actual fuel load for the trip.

29. Just prior to take-off, a baggage handler put an extra box of significant mass into the hold without recording it in the LMC's. What are the effects of this action? The aeroplane has a normal, tricycle undercarriage.

    1. V_{MC} will increase if the extra load is forward of the datum.
    2. Stick forces in flight will decrease if the extra load is behind the datum.
    3. Stick forces at V_R will increase if the box is forward of the main wheels
    4. V_{MU} will occur later
    5. The safe stopping distance will increase.

    a. 3, 4 and 5 only
    b. 2, 3 and 4 only
    c. 1 and 5 only
    d. all the above

30. What is the maximum take-off mass given:

MSTOM	43,000 kg
MSLM	35,000 kg
PLLM	33,000 kg
MZFM	31,000 kg
DOM	19,000 kg
Total Fuel capacity	12,500 kg
Maximum Trip Fuel	9,000 kg
Contingency fuel	1000 kg
Alternate fuel	500 kg
Final reserve fuel	400 kg

    a. 43,000 kg
    b. 42,000 kg
    c. 41,000 kg
    d. 40,000 kg

31. What is the maximum mass an aeroplane can be loaded to before it moves under its own power?

    a. Maximum Structural Ramp mass
    b. Maximum Structural take-off mass
    c. Maximum Regulated Ramp Mass
    d. Maximum Regulated Take-off mass

32. The weight of an aircraft in all flight conditions acts:

    a. parallel to the CG
    b. at right angles to the aeroplane's flight path
    c. always through the MAC
    d. vertically downwards

33. With reference to MRJT Load and trim sheet (CAP696 Pg 31). If the DOM is 35000 kg and the CG is 14%, what is the D.O.I?

    a. 41.5
    b. 33
    c. 40
    d. 30

34. If the CG moves rearwards during flight:

    a. range will decrease
    b. range will increase
    c. stability will increase
    d. range will remain the same but stalling speed will decrease

35. The CG of an aeroplane is situated at 115.8 arm and the mass is 4750 kg. A weight of 160 kg is moved from a hold situated at 80 arm to a hold at 120 arm. What would be the new CG arm?.

    a. 117.14
    b. 118.33
    c. 118.50
    d. 120.01

36. What is the effect of moving the CG from the front to the rear limit at constant altitude, CAS and temperature?

    a. Reduced optimum cruise range
    b. Reduced cruise range
    c. Increased cruise range
    d. Increased stall speed.

37. The baggage compartment floor-loading limit is 650 kg/m^2. What is the maximum mass of baggage that can be placed in the baggage compartment on a pallet of dimensions 0.8m by 0.8m. The pallet has a mass of 6 kg?

    a. 416 kg
    b. 1015 kg
    c. 650 kg
    d. 410 kg

38. An aeroplane of 110,000kg has its CG at 22.6m aft of the datum. The CG limits are 18m to 22m aft of the datum. How much mass must be removed from a hold 30m aft of the datum to bring the CG to its mid point?

    a. 26800 kg
    b. 28600 kg
    c. 86200 kg
    d. 62800 kg

39. Where does the mass act through when the aircraft is stationary on the ground?

    a. The centre of gravity
    b. The main wheels
    c. It doesn't act through anywhere.
    d. The aerodynamic centre

MASS AND BALANCE   SPECIMEN QUESTIONS

40. If an aircraft is weighed prior to entry into service who is responsible for doing the re-weigh to prepare the plane for operations?

   a.   The manufacturer.
   b.   The operator
   c.   The pilot
   d.   The flight engineer.

41. An aeroplane has a tank capacity of 50000 Imperial gallons. It is loaded with fuel to a quantity of 165000 kg (790 kg/m^3). What is the specific gravity of the fuel and approximately how much more fuel could be taken up given that mass limits would not be exceeded?

   a.   0.73   46053 gallons
   b.   0.81   3940 gallons
   c.   0.72   46000 gallons
   d.   0.79   3946 gallons

42. Define Balance Arm :

   a.   BA = Mass / Moment
   b.   BA = Moment / Mass
   c.   BA = Mass / Distance
   d.   BA = Moment / Distance

43. You have been given 16500 litres of fuel at SG 0.78 but written down is 16500 kg. As a result you will experience :

   a.   heavier stick forces at rotation and improved climb performance.
   b.   heavier stick forces on rotation and distance to take-off increases.
   c.   lighter stick forces on rotation and calculated V1 will be too high.
   d.   lighter stick forces on rotation and V2 will be too low.

MASS AND BALANCE

Answers

1	A	16	A	31	A
2	D	17	A	32	D
3	C	18	B	33	C
4	C	19	B	34	B
5	B	20	D	35	A
6	B	21	C	36	C
7	A	22	C	37	D
8	C	23	A	38	B
9	B	24	A	39	A
10	D	25	C	40	B
11	C	26	C	41	D
12	C	27	D	42	C
13	B	28	A	43	C
14	C	29	A		
15	B	30	B		

MASS AND BALANCE — SPECIMEN QUESTIONS

SPECIMEN EXAMINATION PAPER

1. Define the useful load:

    a. traffic load plus dry operating mass
    b. traffic load plus usable fuel mass
    c. dry operating mass plus usable fuel load
    d. that part of the traffic load which generates revenue

2. Determine the position of the CG as a percentage of the MAC given that the balance arm of the CG is 724 and the MAC balance arms are 517 to 1706.
(2 marks)

    a. 14.2 %
    b. 15.3 %
    c. 16.3 %
    d. 17.4 %

3. The distance from the datum to the CG is:

    a. the index
    b. the moment
    c. the balance arm
    d. the station

4. Using CAP696, MRJT, fig 4.9. What is the balance arm, the maximum compartment load and the running load for the most aft compartment of the fwd cargo hold?

    a. 421.5 cm   3305 kg   13.12 kg per inch
    b. 1046.5 inches   711 kg   7.18 kg per kg
    c. 421.5 inches   2059 kg   13.12 kg per inch
    d. 1046.5 m   711 kg   7.18 kg per in

5. Individual aircraft should be weighed in an air conditioned hangar :

    a. on entry into service and subsequently every 4 years
    b. when the effects of modifications or repairs are not known
    c. with the hangar doors closed and the air conditioning off.
    d. all the above.

6. If a compartment takes a maximum load of 500 kg, with a running load limit of of 350 kg/m and a distribution load limit of 300 kg/m^2 max. Which of the following boxes, each of 500 kg, can be carried?
(2 marks)

    1. 100 cm x 110 cm x 145 cm
    2. 125 cm x 135 cm x 142 cm
    3. 120 cm x 140 cm x 143 cm
    4. 125 cm x 135 cm x 144 cm

    a. any one of the boxes if loaded with due care as to its positioning
    b. either of boxes 2, 3 and 4 in any configuration
    c. box 2 with its longest length perpendicular to the floor cross beam or box 3 in any configuration
    d. either of boxes 3 and 4 with their longest length parallel to the aircraft longitudinal axis.

MASS AND BALANCE   SPECIMEN QUESTIONS

7. Use CAP 696, MRJT as appropriate. Prior to departure a MRJT is loaded with maximum fuel of 20100 ltr at an SG of 0.78. Calculate the maximum allowable traffic load that can be carried given the following data:

(3 Marks)

    PLTOM = 67200 kg
    PLLM = 54200 kg
    DOM = 34930 kg
    Taxi fuel = 250 kg
    Trip fuel = 9250 kg
    Contingency and holding fuel = 850 kg
    Alternate fuel = 700 kg

    a.   13092 kg
    b.   12442 kg
    c.   16370 kg
    d.   16842 kg

8. If the maximum structural landing mass is exceeded:

    a.   The aircraft will be unable to get airborne
    b.   The undercarriage could collapse on landing
    c.   No damage will occur providing the aircraft is within the regulated landing mass.
    d.   No damage will occur providing the aircraft is within the performance limited landing mass.

9. For a conventional light aeroplane with a tricycle undercarriage configuration, the higher the take-off mass (assume a stab trim system is not fitted):

    1.   Stick forces at rotation will increase
    2.   Range will decrease but endurance will increase
    3.   Gliding range will reduce
    4.   Stalling speed will increase

    a.   all statements are correct
    b.   Statement 3 only is correct
    c.   Statements 1 and 4 only are correct
    d.   Statement 4 only is correct

10. Due to a mistake in the load sheet the aeroplane is 100 kg heavier than you believe it to be. As a consequence:

    a. V_1 will be later
    b. V_{MU} will be
    c. V_R will be later
    d. V_1, V_{MU}, V_R will all occur earlier

11. If the aeroplane was neutrally stable this would suggest that:

    a.   the CG is forward
    b.   the CG is in mid range
    c.   the CG is on the rear limit
    d.   the CG is behind the rear limit

12. The CG position is:

    a. set by the pilot
    b. set by the manufacturer
    c. able to exist within a range
    d. fixed

13. Which of the following would not affect the CG position?

    a. Cabin crew members performing their normal duties.
    b. Fuel consumption during flight.
    c. Horizontal stabilator trim setting
    d. Mass added or removed at the neutral point

14. An aircraft is about to depart on an oceanic sector from a high elevation airfield with an exceptionally long runway in the tropics at 1400 local time. The regulated take-off mass is likely to be limited by :

    a. MZFM
    b. Obstacle clearance
    c. Maximum certified Take-off mass
    d. Climb gradient

15. An aircraft is flying at 1.3 V_{S1g} in order to provide an adequate margin above the low speed buffet and transonic speeds. If the 1.3 V_{S1g} speed is 180 kts CAS the mass increases from 285000 kg to 320000 kg, What is the new 1g stalling speed?
(2 Marks)

    a. 146.7 kts, drag will increase and nautical mile per kg fuel burn will decrease.
    b. 191 kts, drag will increase and range nm/kg will increase.
    c. 191 kts, drag will increase and nm/kg fuel burn will decrease.
    d. 147 kts, drag will remain the same and nm/kg fuel burn will increase

16. The CG datum has to be along the longitudinal axis:

    a. between the nose and the tail.
    b. between the leading and trailing edge of the MAC.
    c. but does not have to be between the nose and the tail.
    d. at the fire wall.

17. The useful load is:

    a. TOM – fuel mass
    b. BEM plus fuel load
    c. TOM minus the DOM
    d. TOM minus the operating mass

18. In mass and balance terms, what is an index?

    a. A cut down version of a force
    b. A moment divided by a constant
    c. A moment divided by a mass
    d. A mass divided by a moment

MASS AND BALANCE   SPECIMEN QUESTIONS

19.   Standard masses for baggage can be used for aircraft with:

        a.   9 seats or more
        b.   20 seats or more
        c.   30 seats or more
        d.   less than 30 seats

20.   If an aeroplane comes into lands below its MSLM but above the PLLM for the arrival airfield:

                1.   Airframe structural damage will occur.
                2.   Tyre temperature limits could be exceeded.
                3.   The runway length might be inadequate.
                4   A go-around might not be achievable.
                5.   Brake fade could occur.

        a.   1 and 5 only
        b.   3 and 4 only
        c.   2, 3, 4 and 5 only
        d.   1, 3, 4 and 5 only

21.   What is the zero fuel mass?

        a.   MTOM minus fuel to destination minus fuel to alternative airfield.
        b.   Maximum allowable mass of the aircraft with no usable fuel on board.
        c.   Operating mass minus the fuel load.
        d.   Actual loaded mass of the aircraft with no usable fuel on board.

22.   An aeroplane develops a serious maintenance problem shortly after take-off and has to return to its departure airfield. In order to land safely the aircraft must jettison fuel. How much fuel must be jettisoned.

        a.   Sufficient to reduce the mass to the zero fuel mass
        b.   The pilot calculates the amount of fuel to jettison to reduce the mass to a safe level at or below the RLM.
        c.   The fuel system automatically stops the jettison at the RLM.
        d.   As much as the pilot feels is just insufficient to land safely

23.   Calculate the amount of cargo that could be loaded into the aircraft given the following information and using the CAP MRJT 4.14

Dry Operating Mass	34900 kg
Performance Limited Landing Mass	55000 kg
Trip Fuel	9700 kg
Contingency Fuel	1200 kg
Alternate Fuel	1400 kg
130 passengers at 84kg each	10920 kg
130 bags at 14kg each	1820 kg

        a.   2860 kg
        b.   3660 kg
        c.   4660 kg
        d.   5423 kg

MASS AND BALANCE SPECIMEN QUESTIONS

ANSWERS TO SPECIMEN EXAMINATION PAPER

1	B	13	D
2	D	14	D
3	C	15	A
4	C	16	C
5	D	17	C
6	D	18	B
7	B	19	B
8	B	20	C
9	C	21	D
10	B	22	B
11	D	23	A
12	C		

MASS AND BALANCE   SPECIMEN QUESTIONS

DEBRIEF TO SPECIMEN EXAMINATION PAPER

1. b. (See CAP 696 page 3)

2. d.

   $$\% \text{ MAC} = \frac{A - B}{C} \times 100$$

   $$\% \text{ MAC} = \frac{724 - 517}{706 - 517} \times 100$$

3. c. (See CAP 696 page 3)

4. c. See CAP 696 page 24. Note the values have been placed in reverse order

5. d. See JAR Ops, sub part J 1.605. Page 1-7 in your notes

6. d.

   Max Running load $= \dfrac{\text{Load}}{\text{Min Length}}$

   Therefore Min length = 500 kg / 350kg/m = 1.428 m
   Thus, anything shorter than 1.428m will exceed the maximum running load.

   Max Distribution load $= \dfrac{\text{Load}}{\text{Min Area}}$

   Therefore Min Area = 500kg/300kg/m^2 = 1.66 m^2
   Thus, any area less than 1.66m^2 will exceed the max floor intensity.
   Using the above, only boxes 3 and 4 meet the distribution load requirements but the load must be placed with its longest length parallel to the longitudinal axis to meet the running load requirement.

7. b.

   Ramp Fuel mass   =   20100 lt x 0.78   =   15678 kg
   Take-off fuel   =   15678 – 250 {start/taxi} = 15428 kg

   Fuel remaining   =   15428 – 9250   =   6178 kg

   From CAP 696, MZFM = 51300 kg
   1.   MZFM –DOM =   51300 – 34930   = 16370 kg

   2.   MSTOM is lower than PLTOM thus
        MSTOM – DOM – TO Fuel   =   62800 – 34930 – 15428 = 12442 kg

   3.   PLLM is lower than SLLM, thus
        PLLM - DOM - Fuel Remaining =   54200 – 34930 – 6178 = 13092 kg

        Allowable TL = lowest of 1, 2 or 3 above =   12442 kg

MASS AND BALANCE   SPECIMEN QUESTIONS

8.   b.

     The Maximum Structural Landing Mass is set by the manufacturer to meet with the Design
     Limit Loads (DLL) of the structure. If exceeded, the structure will be subject to excessive
     fatigue and could even be permanently damaged.

9.   c.

     1.   Stick forces at rotation will increase [weight is fwd of wheel rotation]
     2.   Range will decrease but endurance will increase [both will decrease]
     3.   Gliding range will reduce [gliding range is not affected by weight]
     4.   Stalling speed will increase [stalling speed increases with weight]

10. a.   Wrong. You will still believe it to be the speed you calculated
     b.   **Correct.** V_{MU} will be later (the extra mass will prolong the point of minimum lift-off)
     c.   Wrong. You will pull the stick back to rotate at the speed you originally calculated
     d.   Wrong. They will all occur later

11   d.   the CG is behind the rear limit (review the Principles of Flight notes on Static
          Margin. If the CG were positioned on the neutral point and the aeroplane was
          disturbed in pitch by a gust of wind, it would retain the new attitude because the
          moments about the CG would all equal one another.)

12. b.   Wrong. The manufacturer sets the limits not the position
     c.   Correct. within the range set by the manufacturer

13. c.   alters the moments about the CG but not the CG position

14. d.

     Oceanic means there are no obstacles to consider. Though we have an unlimited runway the
     high elevation of the airfield will result in a low air density. Also, the time, being at the hottest
     part of the day will further reduce the air density. The reduced density will seriously reduce the
     engine performance limits. Weight would be limited in order to achieve a suitable climb
     gradient.

15. a.

     $1.3 V_S$ = 180 kts, therefore V_S = $\frac{180}{1.3}$ = 138.46 kts

     New V_S = Old V_S × √(New weight/Old Weight)

              = 138.46 × √(320000/285000) = 146.7 kts

     Assuming the aircraft continues to fly at $1.3V_S$ its new speed will be 146.7 × 1.3 = 190.7 kts

     Naturally, drag will increase and range (nautical miles per kg of fuel) will decrease).

16. c

     But does not have to be between the nose and the tail. (The datum an be anywhere in front of,
     on or behind the aircraft so long as it is on the longitudinal axis of the aeroplane).

MASS AND BALANCE SPECIMEN QUESTIONS

17. c.

   TOM = DOM + Traffic Load + Fuel Load. But, Traffic Load + Fuel Load = Useful Load. Thus TOM = DOM + Useful Load, Rearrange formula, UL = TOM – DOM

18. b.   To simplify M&B calculations. See Cap 696 page 4

19. b.   See JAR OPS Sub-part J 1.620, para f

20. c.   2, 3, 4 and 5 only

   In this example the performance limitation is not stated and could be anything from a runway length restriction, a sloping runway, an obstruction limitation and/or altitude/temperature limitations. The aeroplane might sustain a burst tyre, brake fade, and/or brake fire as a result of heavy braking. Tyre temperatures might exceed limits and delay the take-off time even if they don't burst. The climb slope for obstacle clearance during a go-around might be reduced. As the landing is below the MSLM, the structure itself should not suffer direct damage providing the aeroplane comes to a stop without hitting anything.

21. d.   See CAP 696 page 3

22. b.   The pilot calculates the amount of fuel to jettison to reduce the mass to a safe level at or below the RLM. (Before jettisoning the fuel the pilot should attempt to declare an emergency if time permits and advise Air Traffic Control of his intensions).

23. a.

When attempting this sort of question the golden rule is to work out the fuel states first. Once the fuel states are known you can simply use the three formulas to determine the answer.

```
TOF   =   9700 + 1200 + 1400 = 12300
FR    =   1200 + 1400 = 2600
DOM   =   34900
MZFM  =   53000
RTOM  =   62800
RLM   =   54900
```

Formulas
1. MZFM – DOM   =   53000 – 34900   =   18100
2. RTOM – DOM – TOF = 62800 -34900 – 12300   =   15600
3. RLM – DOM – FR   =   54900 -34900 -2600   =   17480

ALLOWABLE TL   =   LOWEST OF 1, 2, 3 ABOVE =   15600 kg
ACTUAL TL   =   PAX + BAGGAGE   =   (130 x 84) + (130 x 14)
ACTUAL TL   =   12740 kg

DIFFERENCE BETWEEN ALLOWED TL AND ACTUAL TL = UNDERLOAD.
CARGO THAT CAN BE TAKEN   = UNDERLOAD

CARGO   =   15600 – 12740   =   2860 kg

JAAatpl
JOINT AVIATION AUTHORITIES

Theoretical Training Manuals

Revised Edition

FLIGHT PERFORMANCE & PLANNING 1

PERFORMANCE

This learning material has been approved as JAA compliant by the United Kingdom Civil Aviation Authority

CHAPTER ONE - DEFINITIONS

Contents

Page

1.1	DEFINITIONS	1 - 1
1.2	ABBREVIATIONS	1 - 4
1.3	JAR PERFORMANCE CLASSIFICATION	1 - 7
	PAPER 1	1 - 9

PERFORMANCE   DEFINITIONS

1.1 DEFINITIONS

Alternate airport means an airport at which an aircraft may land if a landing at the intended airport becomes inadvisable.

Accelerate-Stop Distance Available means Take-off Run Available + Stopway

Atmosphere, International Standard means the atmosphere defined in ICAO Document 7488/2.
For the purposes of JAR the following is acceptable:

a) The air is a perfect dry gas;

b) The temperature at sea-level is 15°C ,

c) The pressure at sea-level is 1.013250×10^5 Pa (29.92 in Hg) (1013.2 m bar);

d) The temperature gradient from sea-level to the altitude at which the temperature becomes -56.5°C is 3.25°C per 500 m (1.98°C/1000 ft);

e) The density at sea level ρ_o, under the above conditions is 1.2250 kg/m³

   NOTE: ρ is the density appropriate to the altitude and ρ/ρ_o the relative density is indicated by σ.

Balanced Field means that TODA = ASDA

Calibrated airspeed means indicated airspeed of an aircraft, corrected for position and instrument error. Calibrated airspeed is equal to true airspeed in standard atmosphere at sea level.

Ceiling

a) Absolute ceiling means the pressure altitude at which the rate of climb is zero

b) Service ceiling means the pressure altitude at which the rate of climb is a defined value.

Climb gradient means the ratio, in the same units, and expressed as a percentage,

of: $\dfrac{\text{Change in Height}}{\text{Horizontal Distance Travelled}}$

Clearway means, for turbine engine powered aeroplanes certificated after August 29, 1959, an area beyond the runway, not less than 152m (500 ft) wide, centrally located about the extended centreline of the runway, and under the control of the airport authorities. The clearway is expressed in terms of a clearway plane, extending from the end of the runway with an upward slope not exceeding 1.25%, above which no object or terrain protrudes. However threshold lights may protrude above the plane if their height above the end of the runway is 0.66 m (25 ins) or less and if they are located to each side of the runway.

Critical engine means the engine whose failure would most adversely affect the performance or handling qualities of an aircraft.

Density altitude means the height in the International Standard Atmosphere at which the prevailing density occurs.

Equivalent airspeed means the calibrated airspeed of an aircraft corrected for adiabatic compressible flow for the particular altitude. Equivalent airspeed is equal to calibrated airspeed in standard atmosphere at sea level.

Final take-off speed means the speed of the aeroplane that exists at the end of the take-off path in the en-route configuration with one engine inoperative.

Flap extended speed means the highest speed permissible with wing-flaps in a prescribed extended position.

Gross performance means the average performance which a fleet of aeroplanes should achieve if satisfactorily maintained and flown in accordance with the techniques described in the manual.

Indicated airspeed means the speed of an aircraft as shown on its pitot static airspeed indicator calibrated to reflect standard atmosphere adiabatic compressible flow at sea level uncorrected for airspeed system errors.

Landing distance available means the length of runway which is declared available by the appropriate Authority and suitable for the ground run of an aeroplane landing.

Landing gear extended speed means the maximum speed at which an aircraft can be safely flown with the landing gear extended.

Landing gear operating speed means the maximum speed at which the landing gear can be safely extended or retracted.

Load factor means the ratio of a specified load to the total weight of the aircraft. The specified load is expressed in terms of any of the following: aerodynamic forces, inertia forces, or ground or water reactions.

Mach number means the ratio of true air speed to the speed of sound.

Net performance means the gross performance diminished by a margin laid down by the appropriate Authority

Pressure altitude means the height in the International Standard Atmosphere at which the prevailing pressure occurs; obtained by setting the subscale of a pressure altimeter to 1013.2 mb (29.92 in/760 mm mercury).

Reference landing speed means the speed of the aeroplane, in a specified landing configuration, at the point where it descends through the landing screen height in the determination of the landing distance for manual landings.

Stopway means an area beyond the take-off runway, no less wide than the runway and centred upon the extended centreline of the runway, able to support the aeroplane during an abortive take-off, without causing structural damage to the aeroplane, and designated by the airport authorities for use in decelerating the aeroplane during an abortive take-off.

Runway slope means the difference in height between two points on the aerodrome surface divided by the distance between those points, in the same units, and expressed as a percentage.

Take-off Distance Available means Take-off Run Available + Clearway or 1.5 x Take-off Run Available. Whichever is the lesser.

Take-off Run Available means the length of runway which is declared available by the appropriate Authority and suitable for the ground run of an aeroplane taking off.

Take-off safety speed means a referenced airspeed obtained after lift-off at which the required one-engine-inoperative climb performance can be achieved.

True airspeed means the airspeed of an aircraft relative to undisturbed air. True airspeed is equal to equivalent airspeed multiplied by $(\rho_o/\rho)^{1/2}$.

Unaccelerated flight means flight at a steady speed with no longitudinal, lateral or normal accelerations.

PERFORMANCE   DEFINITIONS

1.2 ABBREVIATIONS

ACN   means Aircraft classification number

ASD   means Accelerate stop distance

BFL   means Balanced field length :   TOD = ASD

CAS   means Calibrated airspeed

EAS   means Equivalent airspeed

EPR   means Engine pressure ratio

ETOPS means Extended range twin operations

IAS   means Indicated airspeed

JAR   means Joint Aviation Requirements

M   means Mach number

MAP   means Manifold air pressure

PCN   means Pavement classification number

SR   means Specific range

SFC   means Specific fuel consumption

TAS   means True air speed

TOD   means Take-off distance

TOR   means Take-off run

V_A means Design manoeuvring speed

V_{EF} means The **CAS** at which the critical engine is assumed to fail

V_F means Design flap speed

V_{FE} means Maximum flap extended speed

V_{FTO} means Final take-off speed

PERFORMANCE — DEFINITIONS

V_{LE} means Maximum landing gear extended speed

V_{LO} means Maximum landing gear operating speed

V_{LOF} means Lift-off speed

V_{MBE} means Maximum brake energy speed

V_{MC} means Minimum control speed with the critical engine inoperative

V_{MCA} means The minimum control speed, take-off climb

V_{MCG} means The minimum control speed, on or near ground

V_{MCL} means The minimum control speed, approach and landing

V_{MO}/M_{MO} means Maximum operating limit speed

V_{MU} means minimum unstick speed

V_{NE} means Never-exceed speed

V_{NO} means Normal operating speed

V_R means Rotation speed

V_{RA} means Rough airspeed

V_{REF} means Reference landing speed

V_S means The stall speed or the minimum steady flight speed at which the aeroplane is controllable

V_{SO} means The stall speed or the minimum steady flight speed in the landing configuration

V_{S1} means The stall speed or the minimum steady flight speed obtained in a specified configuration

V_{SR} means the speed at which the stick pusher would activate, if fitted.

V_T means Threshold speed

V_{TMAX} means Maximum threshold speed

PERFORMANCE **DEFINITIONS**

V_X means Speed for best angle of climb

V_Y means Speed for best rate of climb

V_1 means Take-off decision speed

V_2 means Take-off safety speed, with the critical engine inoperative.

V_{2min} means Minimum take-off safety speed, with the critical engine inoperative.

V_3 means Steady initial climb speed with all engines operating

1.3 JAR PERFORMANCE CLASSIFICATION   JAR - OPS 1.470

Performance Class A

Multi-engined aeroplanes powered by turbo-propeller engines with a maximum approved passenger seating configuration of more than 9 or a maximum take-off mass exceeding 5700 kg., and all multi-engined turbo-jet powered aeroplanes.

Performance Class B

Propeller driven aeroplanes with a maximum approved passenger seating configuration of 9 or less, and a maximum take-off mass of 5700 kg. or less

Performance Class C

Aeroplanes powered by reciprocating engines with a maximum approved passenger seating configuration of more than 9 or a maximum take-off mass exceeding 5700 kg.

	Multi-engined Jet	Propeller driven	
		Multi-engined Turbo-prop	Piston
Weight : Greater than 5700 kg. Passengers : More than 9	A	A	C
Weight : 5700 kg. or less Passengers : 9 or less	A	B	B

PERFORMANCE   DEFINITIONS

PAPER 1

1. The Accelerate-Stop Distance Available is :

   a) TORA + Clearway + Stopway
   b) TORA + Stopway
   c) TORA + Clearway
   d) TODA + Stopway

2. The International Standard Atmosphere defines an atmosphere where Sea level temperature (i) Sea level pressure (ii) Sea level density (iii) temperature lapse rate (iv) are :

	(i)	(ii)	(iii)	(iv)
a)	15°C	1013 mb	1.225 kg/m^3	1.98°C/1000m
b)	15°C	1013 mb	1.225 kg/m^3	6.5°C/1000 m
c)	0°C	1.013 Bar	1225 g/m^3	1.98°C/1000 ft
d)	15°C	29.92 in.Hg	1013 kg/m^3	1.98°C/1000 ft.

3. The Service Ceiling is the pressure altitude where :

   a) the rate of climb is zero
   b) the low speed and high speed buffet are coincident
   c) the lift becomes less than the weight
   d) the rate of climb reaches a specified value

4. The Gross performance of an aircraft is :

   a) the average performance achieved by a number of aircraft of the type.
   b) the minimum performance achieved by a number of aircraft of the type
   c) the average performance achieved by a number of aircraft of the type reduced by a specified margin
   d) the minimum performance achieved by the individual aircraft reduced by a specified margin.

5. The Mach number is the ratio of :

   a) True Air speed : Speed of sound at sea level
   b) Indicated Air speed : local speed of sound
   c) True Air speed : Local speed of sound
   d) True Air speed : Speed of sound in ISA conditions

6. In relation to an aerodrome, a Balanced Field is when

   a) TODA = TORA
   b) TODA = ASDA
   c) TORA = ASDA
   d) The runway is usable in both directions.

PERFORMANCE   DEFINITIONS

7. The Clearway at an aerodrome is an area beginning :

   a) at the end of the stopway, with a width equal to the runway width, and clear of obstacles.
   b) at the end of the runway, having a minimum required width, disposed equally about the extended centre line, with no obstacles protruding above a plane sloping upwards with a slope of 1.25%
   c) at the end of the runway, with a minimum width of 60 m each side of the centre line and clear of obstacles.
   d) at the end of the runway, clear of obstacles and capable of supporting the weight of the aircraft during an emergency stop.

8. Which of the following statements is correct :

   a) Gross gradient is less than net gradient
   b) Gross take-off distance is less than net take-off distance
   c) Gross landing distance is greater than net landing distance
   d) Gross acceleration is less than net acceleration

9. The load factor is the ratio of :

   a) Lift : Drag at the optimum angle of attack
   b) Weight : Maximum Authorised Weight
   c) Thrust : Weight
   d) Total lift : Weight

10. An aerodrome has a pressure of 1013 mb and a temperature of 25°C, the pressure altitude (i) and the density altitude (ii) would be :

	(i)	(ii)
a)	Sea level	Sea level
b)	above sea level	below sea level
c)	sea level	above sea level
d)	below sea level	above sea level

ANSWERS

1	2	3	4	5	6	7	8	9	10
b	b	d	a	c	b	b	b	d	c

CHAPTER TWO - GENERAL PRINCIPLES: TAKE OFF

Contents

		Page
2.1	TAKE OFF.	2 - 1
2.2	AVAILABLE DISTANCES	2 - 1
2.3	REQUIRED DISTANCES	2 - 2
2.4	DISTANCE TO ACCELERATE.	2 - 2
2.5	FORCES ON THE AIRCRAFT.	2 - 3
2.6	TAKE OFF SPEED	2 - 7
2.7	EFFECT OF VARIABLE FACTORS ON TAKE OFF DISTANCE	2 - 7
	TAKE OFF PAPER 1	2 - 11

© Oxford Aviation Services Limited

PERFORMANCE   GENERAL PRINCIPLES: TAKE OFF

2.1 TAKE OFF.

The take off part of the flight is the distance from the brake release point to the point at which the aircraft reaches a 'screen' of defined height.

For any particular take off it must be shown that the distance required for take off in the prevailing conditions does not exceed the distance available at the take off aerodrome

2.2 AVAILABLE DISTANCES

The available distances at an aerodrome are defined in chapter 1. They are :

a)   The Take off Run Available. (TORA)
     The length of runway suitable for normal operations

b)   The Take off Distance Available. (TODA)
     The length of runway plus any clearway. Clearway is an area beyond the runway, having a minimum required width, which is clear of obstacles. The purpose of the clearway is to ensure that the aircraft will not hit any obstacles after it leaves the runway

c)   The Accelerate -Stop Distance Available (ASDA)
     If a take off is aborted the aircraft can be brought to a stop either on the runway or on a stopway. The stopway must be able to support the weight of the aircraft without structural damage, but is not suitable for normal operation.

Figure 2.1 illustrates the available distances at an aerodrome

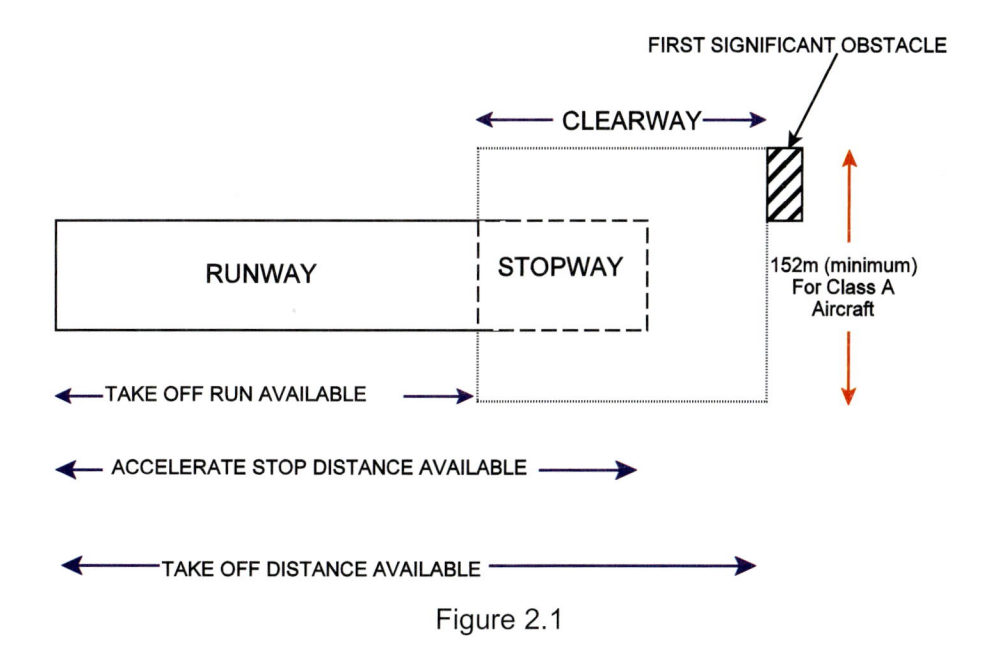

Figure 2.1

PERFORMANCE   **GENERAL PRINCIPLES: TAKE OFF**

2.3 REQUIRED DISTANCES

The distance required for take off may be considered as two segments:

a) The take off roll or ground run.

b) The airborne distance to a "screen" of defined height. This will consist of a transition to climbing flight, followed by a steady climb.

The total distance from the start of the run to the screen is called the Take off Distance.

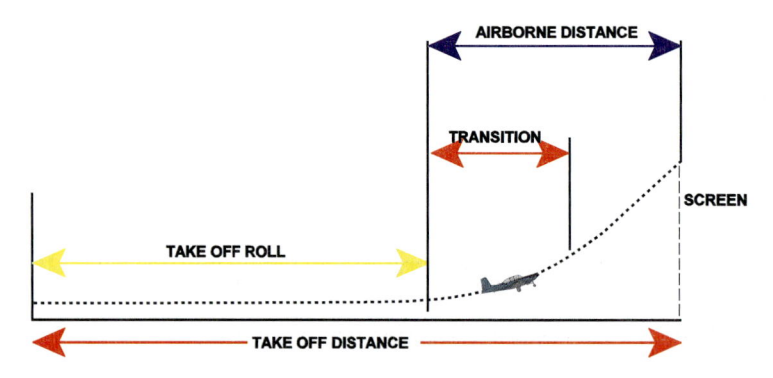

Figure 2.2.

The Performance Regulations require factors to be applied to the estimated distances to give the scheduled Take off Run required and Take off Distance required, so that an adequate margin of safety is incorporated.

For any take off, the take off run required and the take off distance required must not exceed the distances available at the take off aerodrome, specified by the Regulations as being suitable for that manoeuvre.

2.4 DISTANCE TO ACCELERATE.

From Newton's laws of motion, the distance required for a body to reach a speed 'V' from rest with a constant acceleration 'a' is $\frac{V^2}{2a}$ ....(Equation 2.1)

The acceleration depends on the net accelerating force 'F' and the mass M

$$a = \frac{F}{M}$$ ....(Equation 2.2)

PERFORMANCE   **GENERAL PRINCIPLES: TAKE OFF**

For an aircraft taking off the net accelerating force is thrust minus drag. Both of these forces will change as the speed changes, and so the acceleration will not be constant during the take off..

Also, during the airborne part of the take off the laws of motion will be somewhat different, but the distance required will still depend on the speed to be achieved and the acceleration.

During the airborne part of the take off, there will be a transition to climbing flight, followed by a steady climb.

2.5   FORCES ON THE AIRCRAFT.

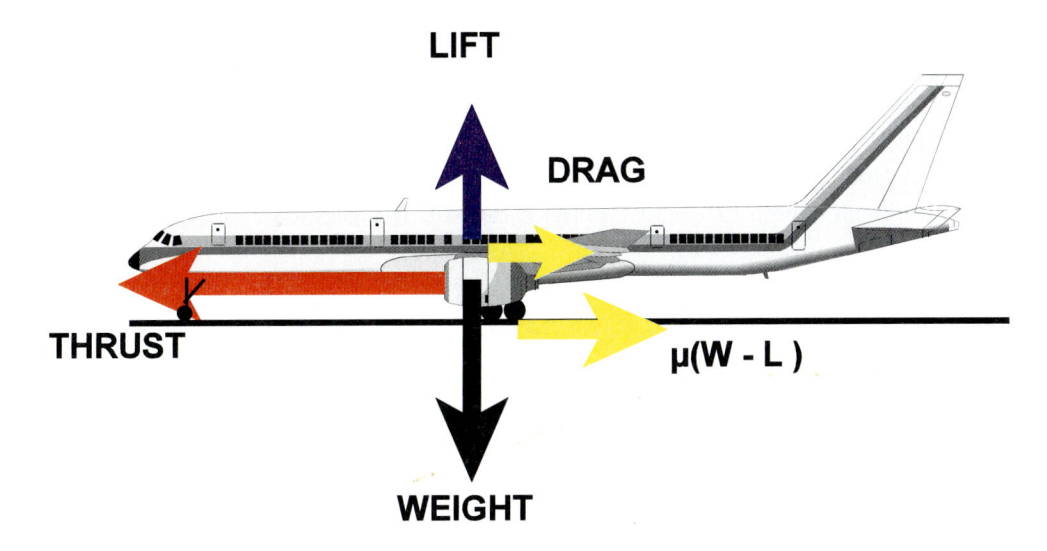

Figure 2.3.

Figure 2.3 illustrates the forces acting on an aircraft during the take-off run on a level runway

Thrust

The engine thrust will vary during take off, and the variation of thrust with speed will be different for jet and propeller engines.

a)   **Jet engine**. For a jet engine the net thrust is the difference between the gross thrust and the intake momentum drag. Increasing speed increases the intake momentum drag, which reduces the thrust. However the increased intake pressure due to ram effect offsets this loss of thrust, and eventually cause the net thrust to increase again. Initially the net thrust will decrease with speed, but at higher speeds it will increase again. In general there will be a decrease of thrust during take off.

PERFORMANCE   GENERAL PRINCIPLES: TAKE OFF

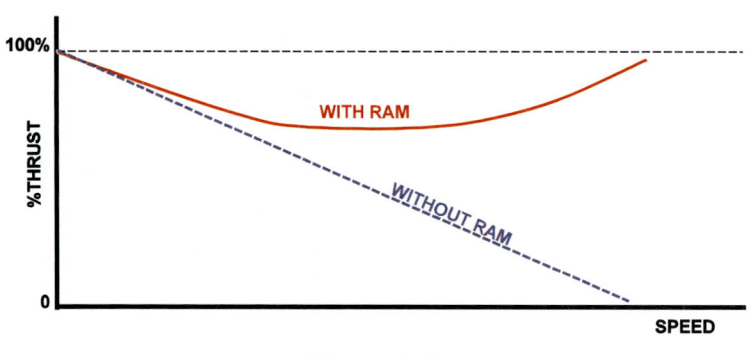

Figure 2.4.

Flat rated engines. The thrust produced by an engine at a given rpm. will depend on the air density, and hence on air pressure and temperature. At a given pressure altitude, decreasing temperature will give increasing thrust. However, many jet engines are "flat rated", that is, they are restricted to a maximum thrust even though the engine is capable of producing higher thrust. Consequently at temperatures below the flat rating cut off, (typically about ISA + 15°C) engine thrust is not affected by temperature.

Figure 2.5

b) **Propeller** For a propeller driven aircraft, thrust is produced by a propeller converting the shaft torque into propulsive force, and depends on the propeller efficiency. Propeller efficiency depends on the propeller angle of attack. For a fixed pitch propeller, angle of attack decreases as forward speed increases, giving initially an increase in efficiency, but also bringing the propeller nearer to its "experimental pitch" where it will give zero thrust. Thrust therefore decreases with increasing speed. For a variable pitch propeller, the propeller will initially be held in the fine pitch position during take off and the propeller angle of attack will decrease with increasing speed Above the selected rpm. the propeller governor will come into operation, increasing the propeller pitch, and reducing the rate at which the thrust decreases.

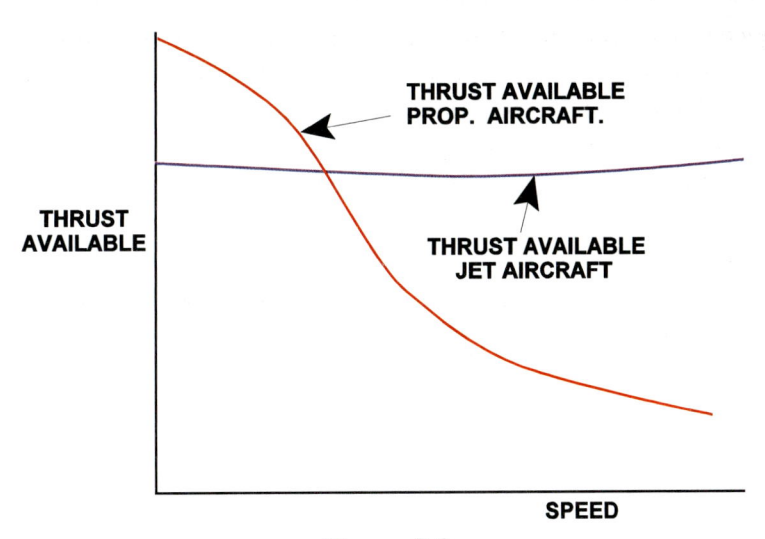

Figure 2.6.

Supercharged engines. If the engine is unsupercharged, the power produced will decrease with decreasing density, (higher temperature or lower pressure). For a supercharged engine, power may be maintained with increasing altitude, although increasing temperature will cause a decrease in density and a loss of power.

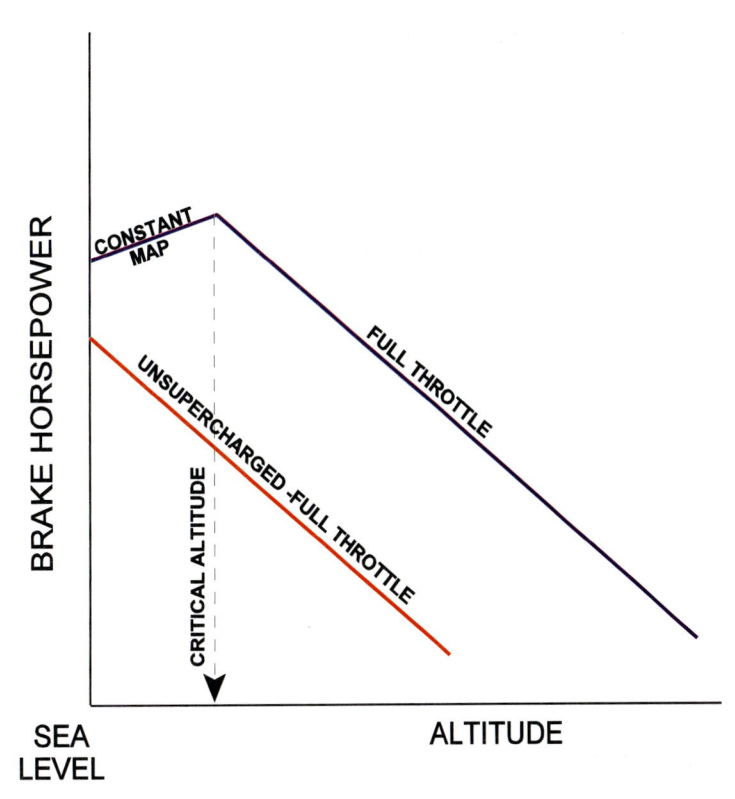

Figure 2.7.

PERFORMANCE — GENERAL PRINCIPLES: TAKE OFF

DRAG. The drag of an aircraft during take off results from aerodynamic drag and wheel drag. The aerodynamic drag depends on the IAS, configuration (flap position) and angle of attack. For a nose-wheel aircraft, the angle of attack will be approximately constant until rotation, and so for a given flap position the aerodynamic drag will increase with IAS squared. At rotation the drag will increase due to the higher angle of attack. It is therefore important that the aircraft is not rotated before V_R, as this will increase the take off run required.

The wheel drag depends on the load on the wheel and the runway surface resistance. The load on the wheels is initially the weight of the aircraft, but as the wings begin to develop lift the load on the wheels decreases, and so the wheel drag decreases with increasing speed.

The drag of the aircraft during take off is therefore:

$$D = D_A + \mu (W - L) \quad \text{...............(Equation 2.3)}$$

Where D_A = aerodynamic drag

μ = Coefficient of rolling friction

W = Total aircraft weight

L = Total lift

and the acceleration $a = \dfrac{F}{M} = \dfrac{g}{W}\left[T - \{D_A + \mu(W-L)\}\right]$ ....(Equation 2.4)

Figure 2.8 shows the variation of the forces on the aircraft as speed increases during the take off run.

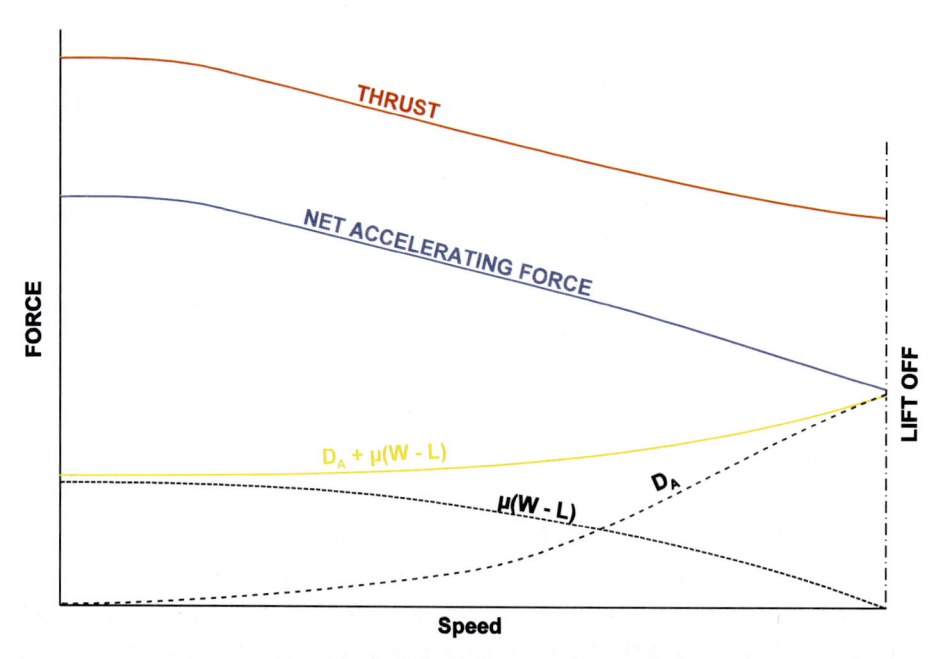

Figure 2.8. Variation of Forces with Speed.

PERFORMANCE GENERAL PRINCIPLES: TAKE OFF

2.6 TAKE OFF SPEED

The speed in equation (2.1) is True Ground Speed, whereas the lift off speed is an IAS. When calculating the take-off run required, account must therefore be taken of the effect of density on TAS for a given IAS, and of the effect of wind on TGS for a given TAS.

The speed to be reached at the screen is determined by the Regulations, and is required to be a safe margin above the stall speed and the minimum control speed, a speed that gives adequate climb performance, and that takes account of the acceleration that will occur after lift off.

2.7 EFFECT OF VARIABLE FACTORS ON TAKE OFF DISTANCE

Aeroplane mass. The mass of the aeroplane affects:

a) The acceleration for a given accelerating force. As the mass increases the acceleration decreases.

b) The wheel drag. Increased mass increases the wheel drag, and so reduces the accelerating force

c) The take off safety speed

Increased mass therefore reduces acceleration and increases the take off distance.

Air density. Density is determined by pressure (altitude) temperature and humidity. Density affects:

a) The power or thrust of the engine

b) The TAS for a given IAS

Low density gives reduced power or thrust, and increased TAS for a given IAS, both effects increase the take off distance.

Wind. The lift and drag depend on air speed, but distance depends on ground speed. A headwind will therefore reduce the ground speed at the required take-off air speed and reduce the take off distance. A tailwind will increase the ground speed and increase the take off distance.
The Regulations for all classes of aircraft require that in calculating the take off distance, only 50% of the headwind component is assumed, and that 150% of a tailwind component is assumed This is to allow for variations in the reported winds during take off.
Note that for any headwind the distance required to take-off will be less than the calculated distance, as only half the headwind is allowed for. Equally for any tailwind the distance required will be less, as a stronger tailwind is allowed for. If the wind is a 90° crosswind the distance required to take off will be the same as the distance calculated for zero wind component.

PERFORMANCE   **GENERAL PRINCIPLES: TAKE OFF**

Runway slope. If the runway is sloping a component of the weight will act along the runway, and increase or decrease the accelerating force.

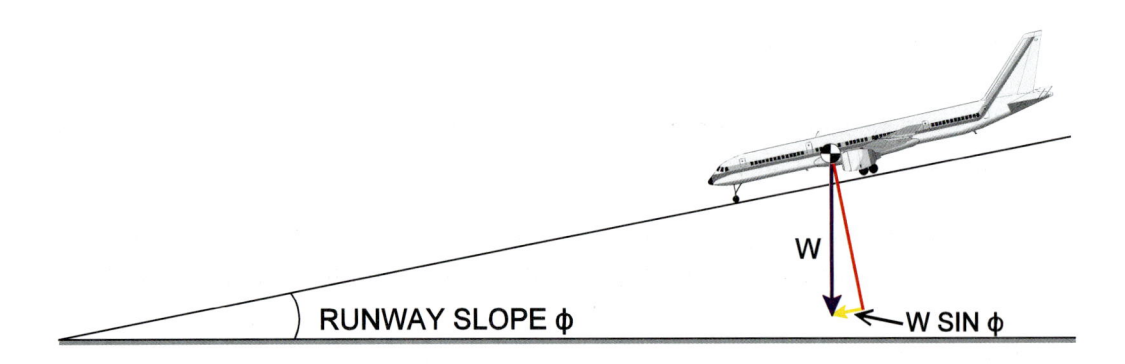

Figure 2.9

A downhill slope will increase the accelerating force, and reduce the take off distance, whereas an uphill slope will reduce the accelerating force and increase the take off distance.

Runway surface. Even on a smooth runway there will be rolling resistance due to the bearing friction and tyre distortion. If the runway is contaminated by snow, slush or standing water, there will be additional drag due to fluid resistance and impingement. This drag will increase with speed, until a critical speed is reached, the hydroplaning speed, above which the drag decreases again.

If the take off is abandoned and braking is required, the coefficient of braking friction is reduced on a runway which is wet, icy or contaminated by snow or slush and the stopping distance is greatly increased.

Figure 2.10.

Airframe contamination. The performance data given assumes that the aircraft is not contaminated by frost, ice or snow during take off. If any of these contaminants is present the performance of the aircraft will be reduced, and the take off distance will be increased.

PERFORMANCE   GENERAL PRINCIPLES: TAKE OFF

Flap setting. Flaps affect:

a)   The C_{Lmax} of the wing

b)   The drag

Increasing flap angle increases C_{Lmax}, which reduces stalling speed and take off speed. This reduces the take off distance.

Increasing flap angle increases drag, reducing acceleration, and increasing the take off distance. The net effect is that take off distance will decrease with increase of flap angle initially but above a certain flap angle the take off distance will increase again. An optimum setting can be determined for each type of aircraft, and any deviation from this setting will give an increase in the take off distance.

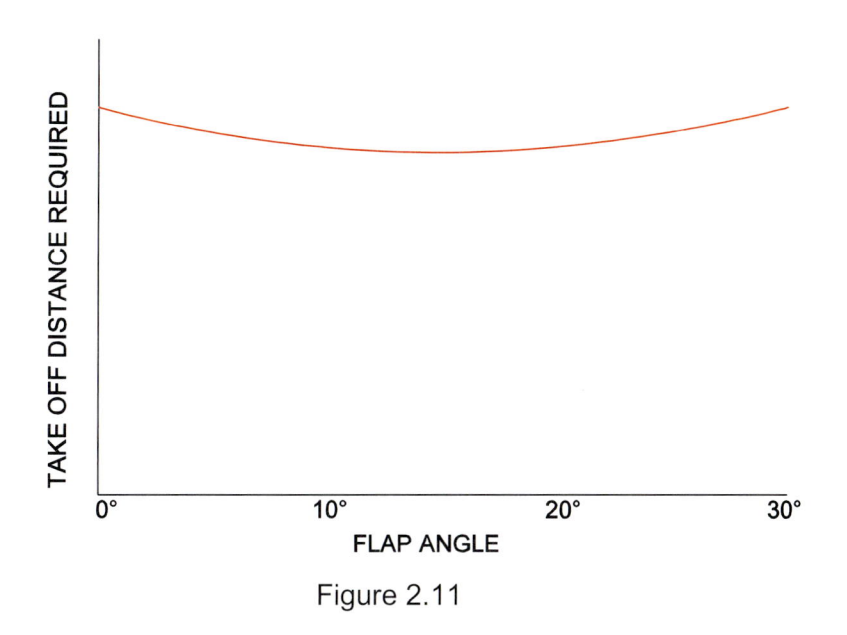

Figure 2.11

The flap setting will also affect the climb gradient, and this will affect the Maximum Mass for Altitude and Temperature, which is determined by a climb gradient requirement, and the clearance of obstacles in the take-off flight path.

Increasing the flap angle increases the drag, and so reduces the climb gradient for a given aircraft mass. The maximum permissible mass for the required gradient will therefore be reduced. In hot and high conditions this could make the Mass-Altitude-Temperature requirement more limiting than the field length requirement if the flap setting for the shortest take-off distance is used. A greater take-off mass may be obtained in these conditions by using a lower flap angle.

PERFORMANCE   GENERAL PRINCIPLES: TAKE OFF

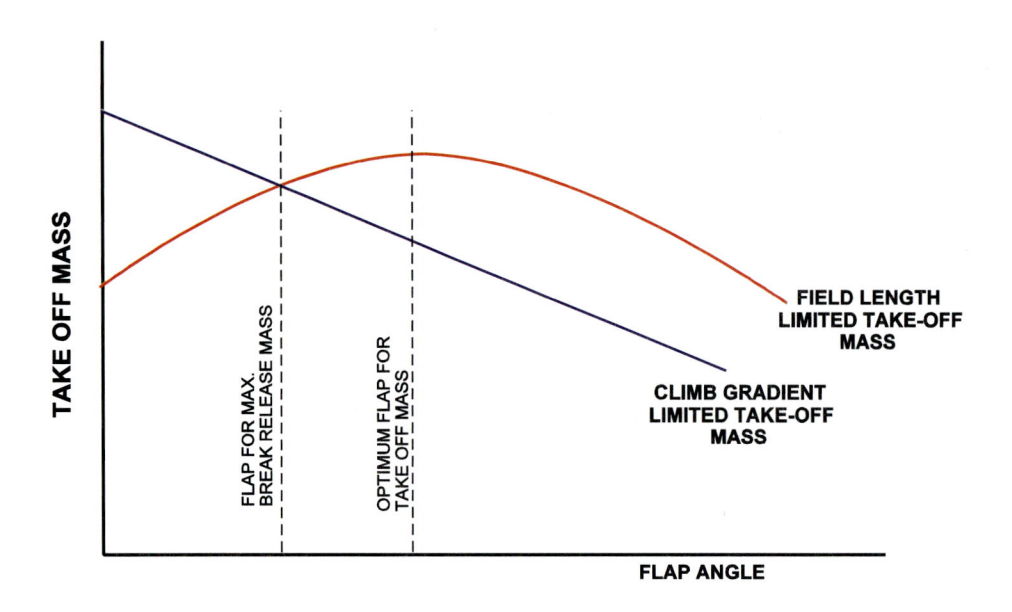

Figure 2.12 Optimum flap setting

If there are obstacles to be considered in the take-off flight path, the flap setting that gives the shortest take-off distance may not give the maximum possible take-off mass if the Take-off Distance Available is greater than the Take-off Distance Required. If close-in obstacles are not cleared, using a lower flap angle will use a greater proportion of the Take-off Distance Available but may give a sufficiently improved gradient to clear the obstacles.

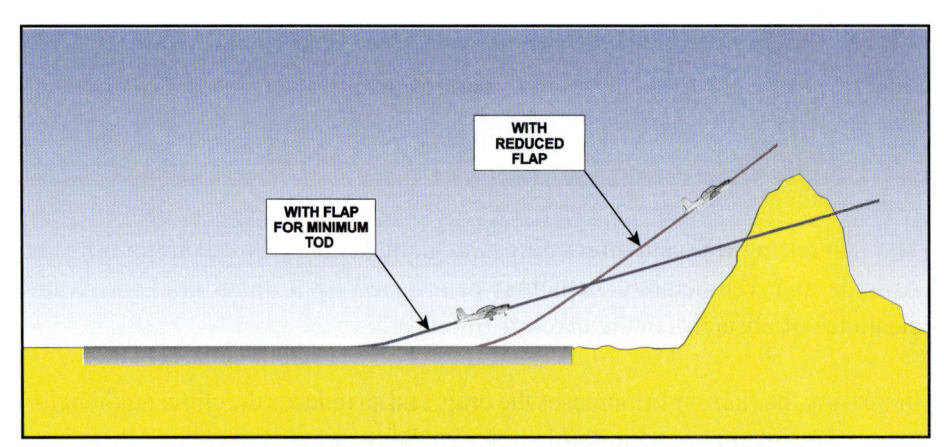

Figure 2.13 Optimum flap for obstacle clearance

PERFORMANCE   GENERAL PRINCIPLES: TAKE OFF

PAPER 1

1. Assuming that the acceleration is constant during the take-off, if the take-off speed is increased by 3%, the Take-off distance will increase by :

   a) 3%
   b) 6%
   c) 9%
   d) 12%

2. The forces acting on an aircraft during the take-off run are :

   a) Lift, thrust and drag
   b) Lift, weight, aerodynamic drag, thrust
   c) Lift, weight, aerodynamic drag, wheel drag, thrust
   d) Weight, thrust, drag.

3. During the take-off run the thrust of a jet engine :

   a) Is decreased due to ram effect
   b) Is increased due to intake momentum drag
   c) Is decreased due to reducing difference between jet velocity and aircraft velocity
   d) Is increased due to increasing intake ram temperature rise.

4. A "flat rated" jet engine will give :

   a) a constant thrust for temperatures below a cut-off value
   b) a constant thrust for temperatures above a cut-off value
   c) decreasing thrust as temperature decreases below a cut-off value
   d) increasing thrust as temperature increases above a cut-off value

5. For a jet engine without limiters, thrust will increase as a result of :

   a) Increased pressure altitude
   b) Increased ambient temperature
   c) Decreased pressure altitude
   d) Increased atmospheric humidity

6. As speed increases the thrust of a fixed pitch propeller will :

   a) Decrease to a constant value
   b) Increase to a constant value
   c) Decrease initially and then increase
   d) Eventually decrease to zero

PERFORMANCE   GENERAL PRINCIPLES: TAKE OFF

7. The rolling friction drag of an aircraft's wheels during take-off:

   a) Depends on the aircraft weight and is constant during take-off
   b) Depends on the total load on the wheels and decreases during take-off
   c) Depends on the wheel bearing friction and increases with speed
   d) Depends on tyre distortion and increases with speed

8. For a given wind speed, the regulations on wind factor give the least margin of safety on take-off:

   a) if the wind is at 45° to the runway
   b) if the wind is at 90° to the runway
   c) if the wind is a pure headwind
   d) if the wind is a pure tailwind

9. The take-off distance required will increase as a result of:

   a) increasing mass, reducing flap below the optimum setting, increasing density
   b) decreasing mass, increasing flap above the optimum setting, increasing density
   c) decreasing mass, increasing flap above the optimum setting, decreasing density
   d) increasing mass, reducing flap below the optimum setting, decreasing density

10. If the flap angle is reduced below the optimum take-off setting, the field limited take-off mass (i) and the climb gradient limited mass (ii) will:

	(i)	(ii)
a)	increase	decrease
b)	decrease	increase
c)	decrease	decrease
d)	increase	increase

PERFORMANCE **GENERAL PRINCIPLES: TAKE OFF**

ANSWERS

Questions	Answers
1	B
2	C
3	C
4	A
5	C
6	D
7	B
8	B
9	D
10	B

CHAPTER THREE - GENERAL PRINCIPLES: CLIMB AND DESCENT

Contents

		Page
3.1	CLIMB	3 - 1
3.2	EFFECT OF VARIABLES ON CLIMB GRADIENT.	3 - 2
3.3	RATE OF CLIMB	3 - 4
3.4	EFFECT OF VARIABLES ON RATE OF CLIMB.	3 - 6
3.5	CEILING	3 - 7
3.6	DESCENT	3 - 7
3.7	GLIDE	3 - 9
3.8	RATE OF DESCENT IN A GLIDE	3 - 10
	CLIMB AND DESCENT PAPER 1	3 - 11
	CLIMB AND DESCENT PAPER 2	3 - 13

© Oxford Aviation Services Limited

© Oxford Aviation Services Limited

PERFORMANCE GENERAL PRINCIPLES: CLIMB AND DESCENT

3.1 CLIMB

Balance of forces in the climb. In a steady climb, the weight has a component along the flight path, which adds to the drag force

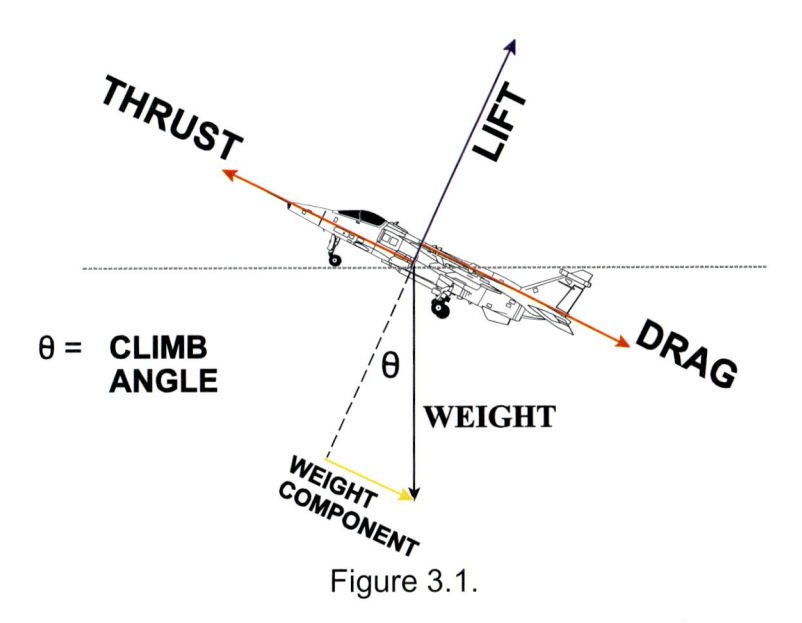

Figure 3.1.

To maintain a steady speed along the flight path, the opposite forces along the flight path must be equal.

$$T = D + W \sin \theta \quad \text{......................................(Equation 3.1)}$$

$$\text{hence } \sin \theta = \frac{T-D}{W} \quad \text{......................................(Equation 3.2)}$$

It is usually more convenient to know the climb gradient rather than the angle of climb. Climb gradient is the ratio of height gained to distance travelled, and is therefore the tangent of the climb angle.

For small angles, $\sin \theta \approx \tan \theta = \gamma = $ climb gradient.
Therefore climb gradient $\approx \dfrac{T - D}{W}$

This shows that the climb gradient depends on:

a) the excess thrust (T-D)

b) the mass of the aircraft

PERFORMANCE **GENERAL PRINCIPLES: CLIMB AND DESCENT**

For a given mass the maximum climb gradient will occur when the excess thrust is greatest. As both drag and thrust vary with speed, there will be a particular speed at which this occurs. This speed is called V_X.

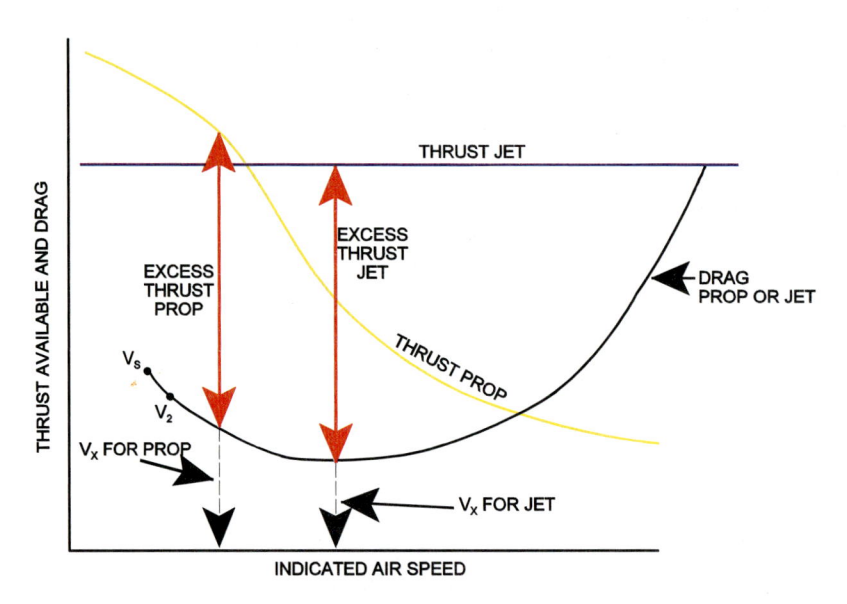

Figure 3.2. V_X for Prop and Jet Aircraft

It can be seen from Fig. 3.2 that because of the different variation of thrust with speed for jet and propeller aircraft, V_X will be different for the two types. For the jet, since thrust is approximately constant with speed, V_X will be close to minimum drag speed. (V_{md} corresponds to maximum $\frac{L}{D}$ or maximum $\frac{C_L}{C_D}$)

For the propeller engined aircraft, V_X will be much lower, and nearer to the Take off safety speed V_2.

3.2 EFFECT OF VARIABLES ON CLIMB GRADIENT.

Mass

Aeroplane mass affects:

a) climb gradient for a given excess thrust
b) the drag

Increased mass gives higher drag which reduces the excess thrust, and secondly reduces the gradient for a given excess thrust. Increased mass therefore reduces the climb gradient for a given thrust. The speed V_X will increase with increasing mass.

PERFORMANCE — GENERAL PRINCIPLES: CLIMB AND DESCENT

Configuration.

The drag of an aircraft will depend on its configuration. The highest drag will occur with gear down and landing flap. The balked landing climb requirement has to be met in this configuration. After take-off, the configuration for the first segment of climb is with gear down and take-off flap, and for the second segment with gear up and take-off flap. If higher flap angles are used to reduce he take-off distance required, the first and second segment climb gradients will be reduced.

Climb gradient decreases with gear or flap extended, and V_X will decrease, since V_{md} decreases.

Density

The drag of an aircraft at any IAS is not affected by changes of density, but the thrust will decrease with decreasing density.

Increasing altitude or increasing temperature (decreasing density) will reduce thrust and decrease climb gradient because the excess thrust has decreased. The speed for best climb angle V_X for the jet aircraft will be essentially a constant IAS, but will increase as a TAS. For the piston engined aircraft the IAS for best climb angle may increase with altitude.

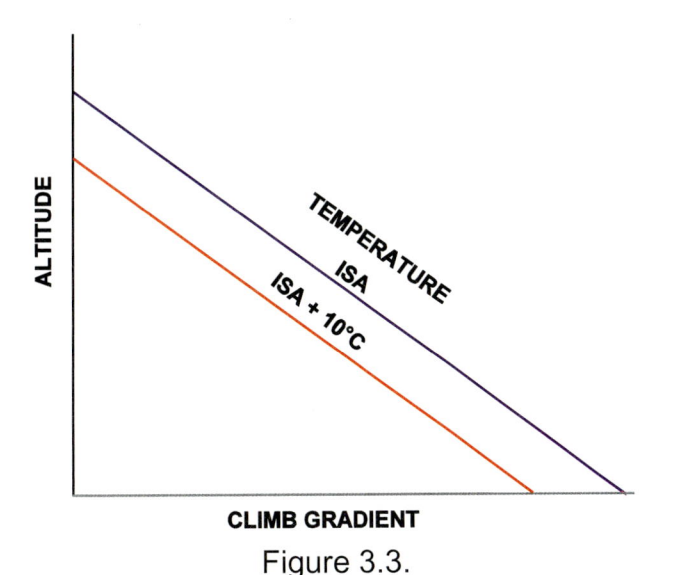

Figure 3.3.

Acceleration

Equation (3.2) assumes that all of the excess thrust (T - D) is used to climb, and that the speed remains constant. If the aircraft is allowed to accelerate during the climb some proportion of the excess thrust is required to accelerate, and the climb gradient will be reduced.

PERFORMANCE — GENERAL PRINCIPLES: CLIMB AND DESCENT

Wind

The climb angle given by equation 3.2 is relative to the air mass. If there is a wind, the climb angle relative to the ground will be different. A headwind will give a steeper angle and a tailwind a flatter angle. In the consideration of clearance of ground obstacles after take off the climb angle relative to the ground must be taken into account.

3.3 RATE OF CLIMB.

Although the angle of climb is important for clearance of obstacles after take off, an aircraft normally requires to climb at the maximum rate of climb so as to reach the required altitude in the least possible time.

The rate of climb is the vertical component of the aircraft's velocity and depends on the aircraft's velocity and the climb angle.

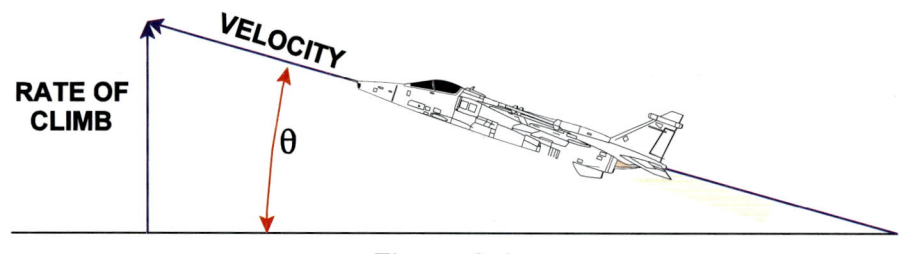

Figure 3.4.

$$\text{Rate of climb} = V \sin \theta \quad \quad \text{(Equation 3.3)}$$

From equation (3.2)
$$\text{Rate of climb} = V \times \frac{T - D}{W} \quad \text{(Equation 3.4)}$$

$$= \frac{VT - VD}{W} \quad \text{(Equation 3.5)}$$

$$= \frac{\text{Power available - Power required}}{W} \quad \text{(Equation 3.6)}$$

This shows that the rate of climb depends on:

a) the excess power

b) the mass

PERFORMANCE — GENERAL PRINCIPLES: CLIMB AND DESCENT

For a given mass the rate of climb is a maximum when the excess power is greatest. The speed at which this occurs is called V_Y.

The maximum rate of climb does not occur at the speed which gives the maximum angle of climb. If the speed is increased above that for the best angle of climb although the climb angle will decrease, the rate of climb will initially increase.

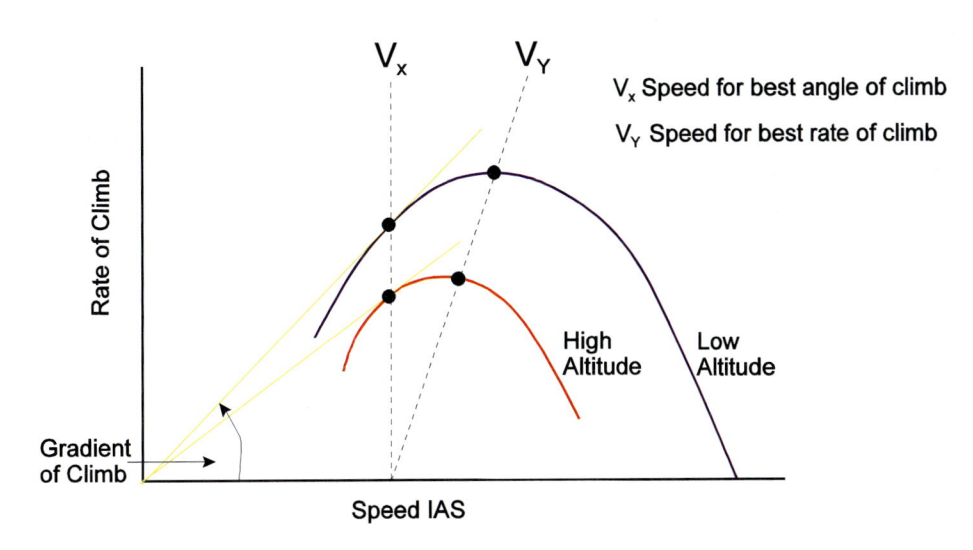

Figure 3.5 Variation of rate of climb with speed

The speed for maximum rate of climb (maximum excess power) is shown in Figure 3.6 for jet and propeller aircraft.

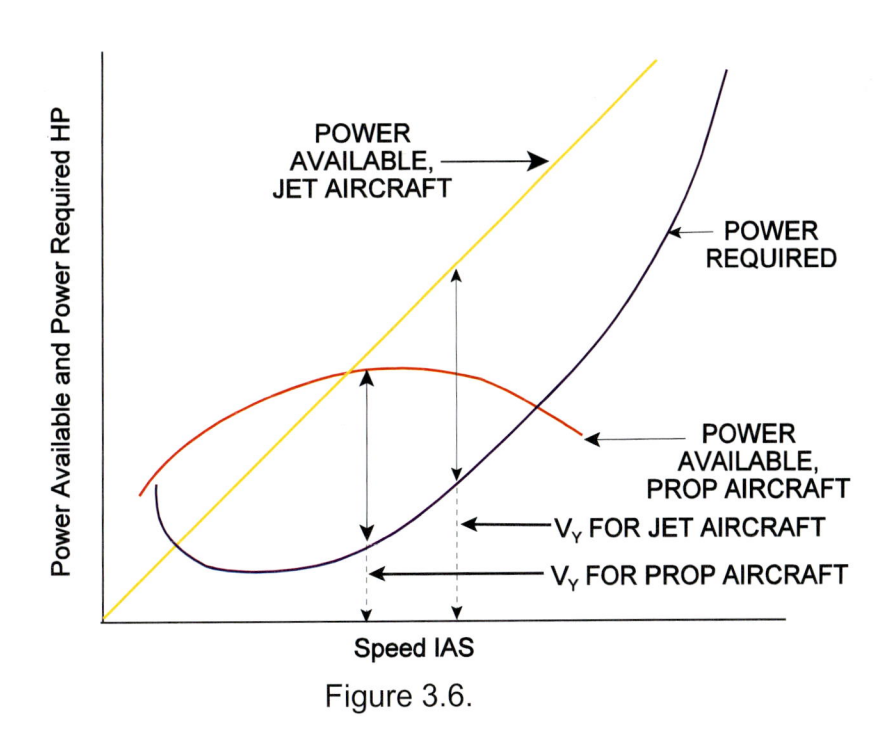

Figure 3.6.

PERFORMANCE   **GENERAL PRINCIPLES: CLIMB AND DESCENT**

3.4 EFFECT OF VARIABLES ON RATE OF CLIMB.

Mass

Aeroplane mass affects:

a) rate of climb for a given excess power
b) the drag and hence the power required

Increased mass gives increased power required which reduces the excess power, and secondly reduces the rate of climb for a given excess power. The speed V_Y will increase with increasing mass.

Configuration

With gear or flap extended the rate of climb will decrease due to the increased drag increasing the power required, and the speed for best rate, V_Y, will decrease.

Density

The air density affects:

a) the power available
b) the power required at a given IAS

Power available will decrease with decreasing density (increasing altitude or temperature).

Power required will increase with decreasing density, since the TAS increases at a given IAS, and power required is drag x TAS. hence the excess power, (power available - power required) will reduce and the rate of climb will decrease as density decreases.

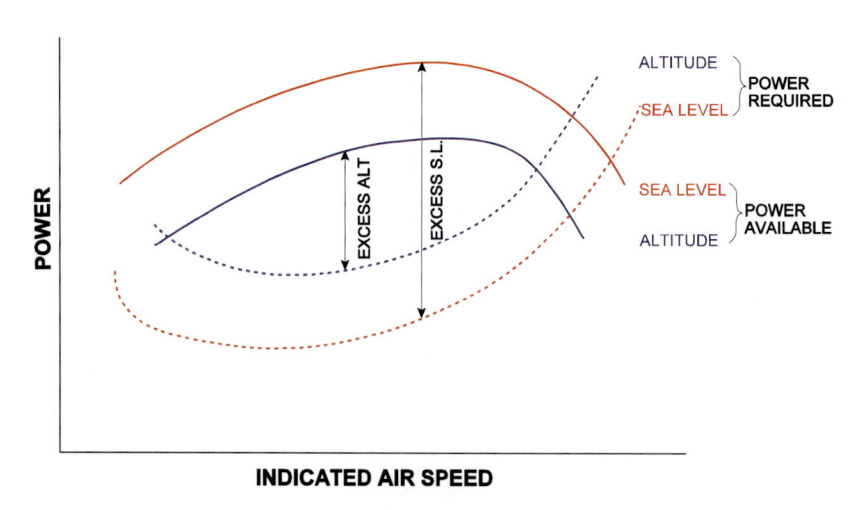

Figure 3.7

PERFORMANCE   **GENERAL PRINCIPLES: CLIMB AND DESCENT**

Wind.

Although the climb angle relative to the ground is affected by wind, the rate of climb is independent of the wind speed.

3.5 CEILING.

The gradient of climb and rate of climb decrease with altitude, and eventually at some altitude become zero.

This altitude is called the absolute ceiling.

The absolute ceiling decreases with increasing aeroplane mass, and increasing temperature. Although the absolute ceiling is theoretically the maximum height which an aircraft can reach, it would not be practicable for an aircraft to climb to its absolute ceiling, as the time taken would be very great and there is no range of speed available when the aircraft is at its absolute ceiling. The aeroplane operating data manual usually publishes a "service ceiling". This is the altitude at which a specified rate of climb occurs.

The service ceiling decreases with increasing mass and increasing temperature.

PERFORMANCE   GENERAL PRINCIPLES: CLIMB AND DESCENT

3.6   DESCENT

Balance of forces in the descent

In a steady descent the weight has a component along the flight path opposite to the drag and adding to any thrust (engines will usually be idling). In a glide the thrust will be zero. To maintain a steady speed the opposite forces along the flight path must be equal.

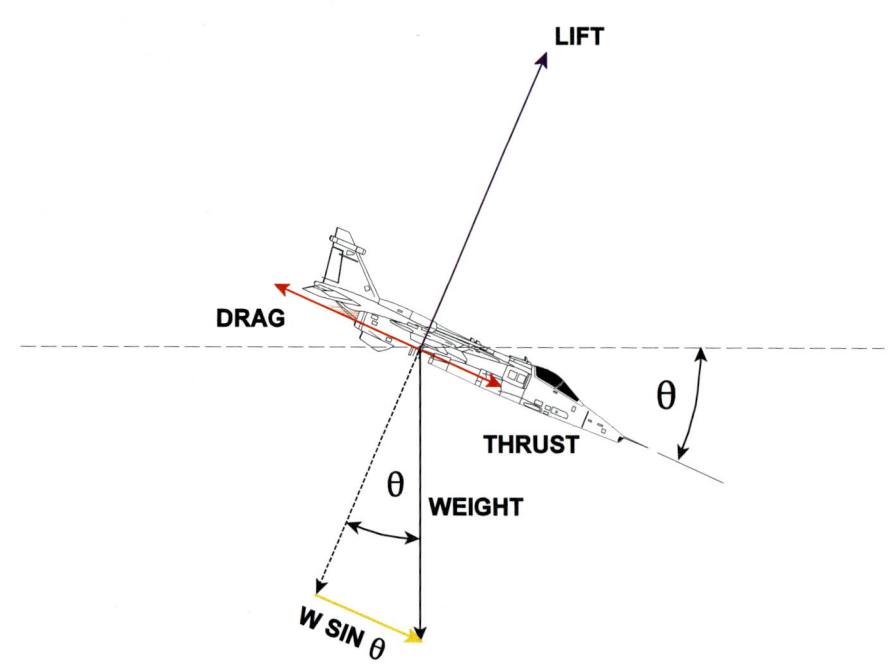

Figure 3.8   Forces in the descent

$$D = T + W \sin \theta \quad \text{...(Equation 3.7)}$$

$$\sin \theta = \frac{D - T}{W} \quad \text{...(Equation 3.8)}$$

It can be seen that equation (3.8) is the same as equation (3.2) except that the descent angle depends on the excess drag (D - T). The angle of descent needs to be considered when en route obstacle clearance is assessed, but more usually it is the rate of descent which is important, this is $V \sin \theta$

From (Equation 3.7) this is $V \quad \times \quad \dfrac{D - T}{W}$ ....(Equation 3.9)

Rate of descent = $\dfrac{VD - VT}{W}$ ....(Equation 3.10)

$$= \frac{\text{Power required - Power available}}{\text{Weight}} \quad \text{.........................(Equation 3.11)}$$

PERFORMANCE GENERAL PRINCIPLES: CLIMB AND DESCENT

For an emergency descent the maximum rate of descent is required. Rate of descent increases with increasing speed and increasing drag (power required) and so the maximum rate of descent occurs at the maximum permissible speed with the highest achievable drag. The maximum permissible speed for an aircraft is V_{MO}, but drag producing devices such as landing gear and flaps are restricted to speeds below V_{MO}. The optimum procedure for any aircraft will be given in the aircraft manual.

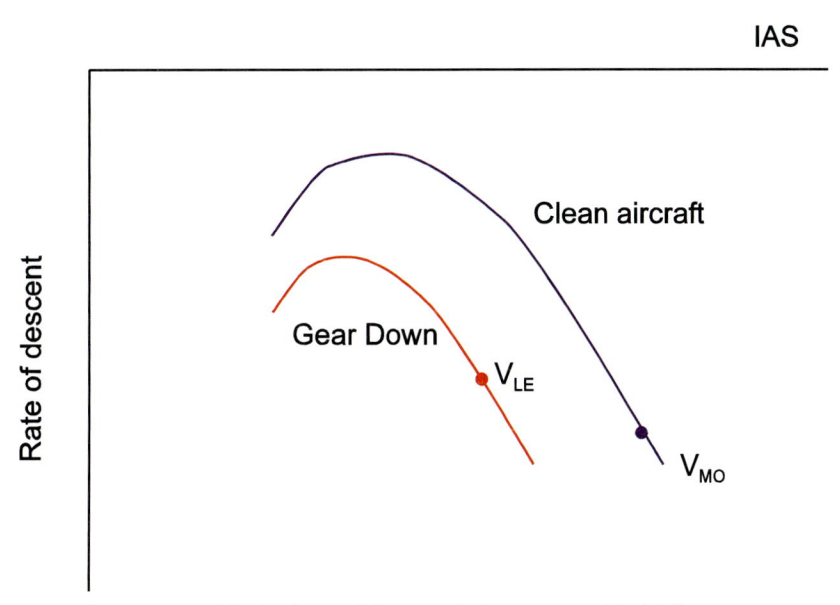

Figure 3.9 Variation of Rate of Descent with IAS

3.7 GLIDE

In a power-off glide, only the forces of lift, drag and weight are acting on the aircraft, and the equation for the glide angle becomes :

$$\tan \theta = \frac{D}{L} \quad \text{or} \quad \frac{C_D}{C_L} \quad \text{...................(Equation 3.12)}$$

The glide angle is least, and will give the maximum gliding distance when $\frac{C_L}{C_D}$ is a maximum

This occurs at the speed V_{md}. The value of $\frac{L}{D}$ max. is independent of weight.

The best (min) glide angle is therefore not affected by weight if the aircraft is flown at Vmd appropriate to the weight. (Vmd increases with weight).

The glide angle relative to the ground will be affected by wind, a head wind will give a steeper angle and a tail wind a flatter angle.

PERFORMANCE   **GENERAL PRINCIPLES: CLIMB AND DESCENT**

3.8 RATE OF DESCENT IN A GLIDE

The rate of descent is $V \sin \theta$

For small angles of descent, $\sin \theta \approx \tan \theta$

$$\tan \theta = \frac{D}{L}$$

$$\therefore \text{rate of descent} = V \times \frac{D}{L}$$

The rate of descent is a minimum, not when $\frac{L}{D}$ is a maximum (V_{md}) but when the product of $V \times \frac{D}{L}$ is a minimum, and this will occur at a speed below V_{md}.

The speed for the minimum rate of descent will be the minimum power speed V_{mp}. As V_{mp} will vary with weight the minimum rate of descent will vary with weight. Wind speed has no effect on the rate of descent in a glide.

PERFORMANCE **GENERAL PRINCIPLES: CLIMB AND DESCENT**

PAPER 1

1. Which combination of forces on the aircraft determine the climb gradient :

    a) Lift, weight, thrust
    b) Lift, drag, thrust
    c) Thrust, drag, weight
    d) Lift, weight, thrust, drag

2. The effect of increased aircraft mass on the climb gradient is :

    a) decrease due to increased drag
    b) increase due to increased lift required
    c) decrease due to increased drag and reduced ratio of excess thrust to weight
    d) increase due to increased speed required at optimum angle of attack

3. The speeds V_X and V_Y are, respectively

    a) Maximum achievable speed with max. continuous thrust and maximum take-off thrust
    b) Speed for best rate of climb and speed for best angle of climb
    c) Max. speed with flap extended and max. speed with gear extended
    d) Speed for best angle of climb and speed for best rate of climb

4. For a given aircraft mass, the climb gradient :

    a) increases if flap angle increases, and if temperature decreases.
    b) decreases if flap angle increases, and if temperature decreases.
    c) increases if flap angle increases, and if temperature increases.
    d) decreases if flap angle increases, and if temperature increases.

5. With a headwind, compared to still air conditions, the rate of climb (i) and the climb angle relative to the ground (ii) will :

	(i)	(ii)
a)	remain the same	increase
b)	increase	increase
c)	increase	remain the same
d)	remain the same	remain the same.

6. The rate of climb depends on :

    a) the excess thrust available
    b) the excess power available
    c) the excess lift available
    d) the C_{Lmax} of the wing.

PERFORMANCE — GENERAL PRINCIPLES: CLIMB AND DESCENT

7. The speed to give the maximum rate of climb will be :

   a) always the same as the speed for best angle of climb.
   b) as close to the stalling speed as possible
   c) higher than the speed for best angle of climb
   d) lower than the speed for best angle of climb.

8. With increasing altitude, the rate of climb :

   a) decreases because power available decreases and power required is constant.
   b) increases because density and drag decrease.
   c) decreases because power available decreases and power required increases.
   d) decreases because power available is constant and power required increases.

9. The maximum rate of descent will occur :

   a) at a speed close to the stalling speed with all permissible drag producing devices deployed.
   b) at V_{MO} with all permissible drag producing devices deployed.
   c) at V_{MO} with the aircraft in the clean configuration.
   d) at a speed corresponding to maximum L :D with the aircraft in the clean configuration.

10. In a power-off glide in still air, to obtain the maximum glide range, the aircraft should be flown:

    a) at a speed corresponding to maximum L : D
    b) at a speed close to the stall.
    c) at a speed corresponding to minimum C_D
    d) at a speed close to V_{NE}

PERFORMANCE   GENERAL PRINCIPLES: CLIMB AND DESCENT

PAPER 2

1. For a given aircraft mass the climb gradient is determined by :

    a) lift - weight
    b) thrust - drag
    c) lift - drag
    d) thrust - weight.

2. An aircraft has a mass of 5700 kg., the drag at best climb speed is 6675 N. and the thrust available is 15.6 kN. The gradient of climb would be :

    a) 1.56%
    b) 1.17%
    c) 8.9%
    d) 16%

3. For a given aircraft mass, the climb gradient :

    a) increases if the aircraft is accelerating and if the temperature increases.
    b) decreases if the aircraft is accelerating and if the temperature increases
    c) increases if the aircraft is accelerating and if the temperature decreases
    d) decreases if the aircraft is accelerating and if the temperature decreases.

4. With the flaps in the take-off position, compared to the aircraft clean, the climb gradient (i) and the speed for best climb angle (ii) will :

	(i)	(ii)
a)	decrease	increase
b)	increase	decrease
c)	decrease	decrease
d)	increase	increase

5. Compared to still air conditions, if an aircraft takes off with a headwind :

    a) clearance of obstacles will be increased, and the M-A-T limit mass will be increased.
    b) clearance of obstacles will be increased, and the M-A-T limit mass will not be affected.
    c) clearance of obstacles will not be affected, and the M-A-T limit mass will increase.
    d) clearance of obstacles will not be affected, and the M-A-T limit mass will not be affected.

PERFORMANCE — GENERAL PRINCIPLES: CLIMB AND DESCENT

6. The climb gradient will be reduced by :

    a) high mass, low temperature, high flap angle
    b) high pressure altitude, turning flight, low temperature
    c) high temperature, high pressure altitude, contaminated airframe.
    d) low pressure altitude, high mass, high temperature.

7. When an aircraft reaches its service ceiling :

    a) the excess power will be zero
    b) it will have a small positive rate of climb
    c) the rate of climb will be zero
    d) the lift will be insufficient to support the weight.

8. If the speed brakes are extended during the descent while maintaining a constant speed. The rate of descent (i) and the angle of descent (ii) will :

	(i)	(ii)
a)	increase	remain the same
b)	remain the same	increase
c)	increase	decrease
d)	increase	increase

9. In a power-off glide, an increase in aircraft mass will :

    a) increase the glide angle and increase the speed for minimum glide angle.
    b) not affect the glide angle, but increase the speed for minimum glide angle.
    c) increase the glide angle, but not affect the speed for minimum glide angle.
    d) not affect the glide angle, and not affect the speed for minimum glide angle.

10. For a jet aircraft, the speed to give the maximum rate of climb will be :

    a) the speed corresponding to maximum L:D
    b) the speed corresponding to minimum C_D
    c) a speed greater than that for maximum L:D
    d) a speed less than that for maximum L:D

PERFORMANCE **GENERAL PRINCIPLES: CLIMB AND DESCENT**

ANSWERS

PAPER 1

Questions	Answers
1	C
2	C
3	D
4	D
5	A
6	B
7	C
8	C
9	B
10	A

PAPER 2

Questions	Answers
1	B
2	D
3	B
4	C
5	B
6	C
7	B
8	D
9	B
10	C

CHAPTER FOUR - GENERAL PRINCIPLES: CRUISE

Contents

		Page
4.1	BALANCE OF FORCES IN LEVEL FLIGHT.	4 - 1
4.2	VARIATION OF DRAG.	4 - 1
4.3	VARIATION OF ANGLE OF ATTACK WITH IAS AND MASS	4 - 2
4.4	ACHIEVABLE SPEED.	4 - 2
4.5	RANGE	4 - 3
4.6	BEST RANGE SPEED	4 - 4
4.7	EFFECT OF WIND ON BEST RANGE SPEED	4 - 5
4.8	ENGINE SFC.	4 - 5
4.9	FACTORS AFFECTING RANGE	4 - 5
4.10	PAYLOAD - RANGE DIAGRAM	4 - 7
4.11	ENDURANCE	4 - 8
4.12	FACTORS AFFECTING ENDURANCE	4 - 8
	CRUISE PAPER 1	4 - 9
	CRUISE PAPER 2	4 - 11

© Oxford Aviation Services Limited

© Oxford Aviation Services Limited

PERFORMANCE   GENERAL PRINCIPLES: CRUISE

4.1 BALANCE OF FORCES IN LEVEL FLIGHT

To maintain a constant speed and height, the opposite forces must be equal;

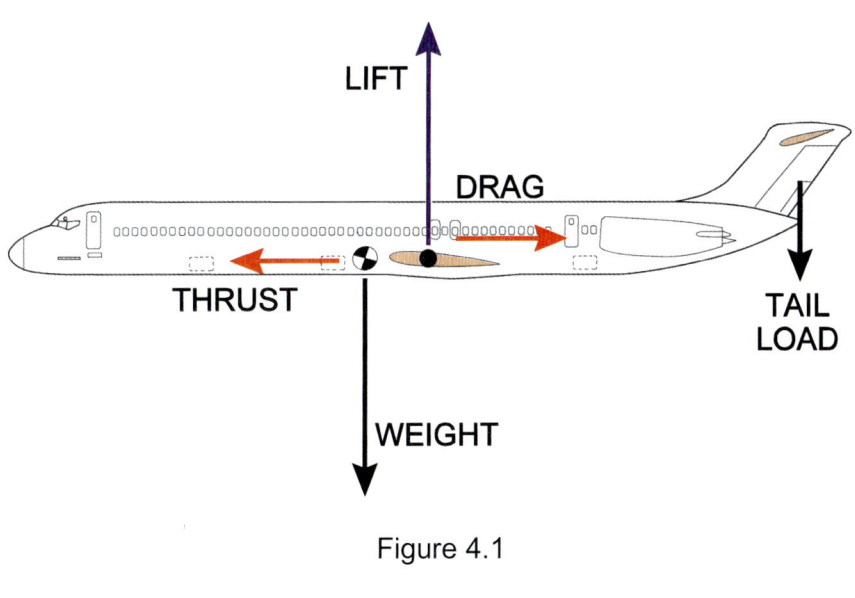

Figure 4.1

Thrust = drag ....(Equation 4.1)
Lift = Weight + tailplane downforce ....(Equation 4.2)

The aircraft's centre of gravity position will affect both of these equations, since the tailplane balancing load depends on the CG position. A more forward CG position will increase the download required for balance, and therefore increase the lift required. Increased lift will give an increase of induced drag, and hence increase the thrust required.

4.2 VARIATION OF DRAG.

For a given configuration, aircraft mass, and CG position, the drag will vary with speed as shown in Figure. 4.2

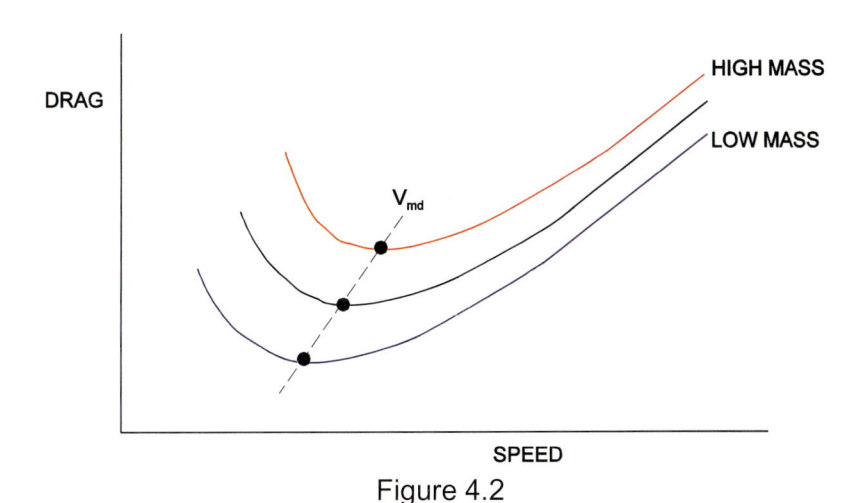

Figure 4.2

PERFORMANCE
GENERAL PRINCIPLES: CRUISE

The total drag is a minimum at V_{imd}, the minimum drag speed, which corresponds to $C_L : C_D$ max. For given Indicated Air Speeds, the total drag does not change with density (pressure or temperature) at a given mass. Increased aircraft mass will increase the drag, and will increase V_{imd}.

4.3 VARIATION OF ANGLE OF ATTACK WITH IAS AND MASS

For a given mass and CG position, the IAS at which the aircraft flies will determine the angle of attack required. Figure 4.3 shows the variation of angle of attack with IAS for a given aircraft mass.

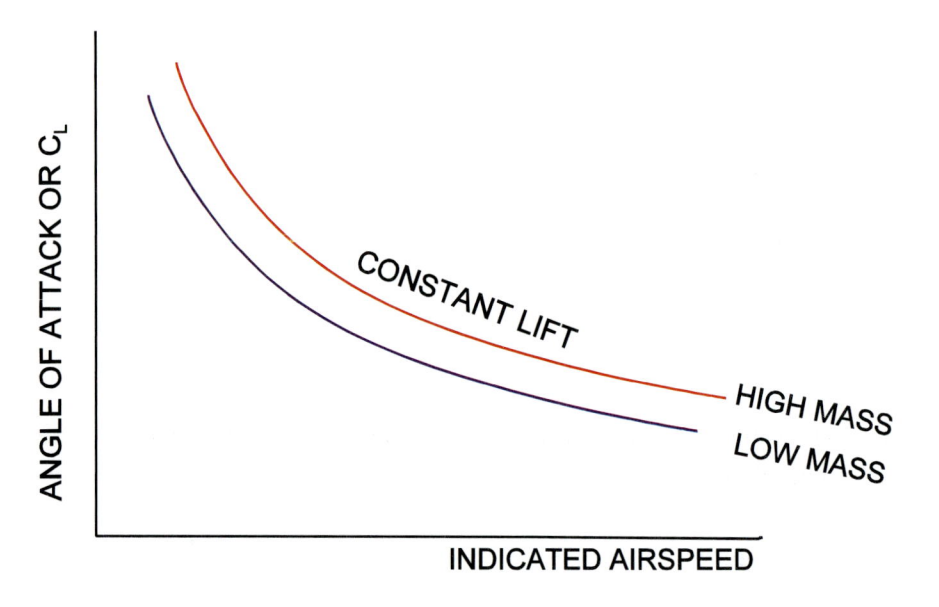

Figure 4.3

During the cruise the weight will decrease progressively as a result of fuel consumption. The lift required will therefore decrease. If the aircraft is being flown so as to achieve minimum drag, (maximum L:D ratio) it will have to be flown at a constant angle of attack, and this will require the IAS to be progressively decreased to give decreasing lift.

4.4 ACHIEVABLE SPEED.

The maximum speed which can be achieved with the thrust available is that at which the maximum thrust available is equal to the drag, or maximum power available is equal to power required. The maximum possible speed would be that achieved with take-off thrust, but use of this is limited, and for the cruise, maximum continuous thrust is normally used. For a jet aircraft, the thrust decreases with altitude, giving a decreasing achievable IAS. However the TAS will increase initially and then decrease.

PERFORMANCE   **GENERAL PRINCIPLES: CRUISE**

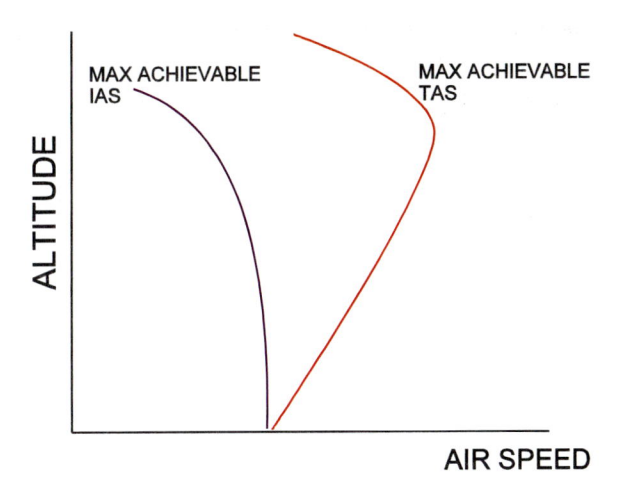

Figure 4.4 Maximum achievable TAS for a Jet Aircraft

The variation of achievable speed with altitude for a propeller engined aeroplane will depend on the type of engine, un-supercharged, turbo-charged or internally supercharged.
For an aircraft with an un-supercharged engine the achievable TAS will decrease with altitude. For an aircraft with a supercharged engine the achievable TAS will increase to full throttle height and then decrease.

4.5 RANGE

Range is the distance travelled with the fuel available. The quantity miles/kg is known as the specific range (SR).

$$SR = \frac{nm}{kg} = \frac{nm/hr}{kg/hr} = \frac{TAS}{fuel\ flow} \quad \text{.................(Equation 4.3)}$$

For a jet engine the fuel flow is a function of engine thrust, and for a propeller engine a function of engine power. In both cases they are related by the specific fuel consumption (sfc)

    sfc = fuel flow per unit of thrust (jet)
    or fuel flow per unit of power (propeller)

It is usually required to fly so that the maximum range is achieved, that is, to cover the greatest distance on the fuel carried, or to use the least fuel for the distance which it is required to travel. The range will depend on the amount of fuel carried, and the number of miles flown per kilogramme of fuel, the specific range.

For the jet, $SR = \dfrac{TAS}{sfc \times T}$ Where T = Thrust

For the propeller $SR = \dfrac{TAS}{sfc \times P}$ Where P = Engine Power

PERFORMANCE   **GENERAL PRINCIPLES: CRUISE**

Since for steady level flight, T = D, and P available = P required, then

$$SR = \frac{TAS}{sfc \times D} \text{ (jet)} \quad \text{and} \quad \frac{TAS}{sfc \times P \text{ req}} \text{ (prop)}$$

The specific range thus depends on the engine efficiency (the sfc) and the airframe efficiency (the ratio of V/D or V/Pr)

The maximum range will occur when the product of 1/sfc and V/D or V/P is a maximum, ie the sfc is as low as possible and the V/D or V/P as high as possible. These optima do not necessarily occur at the same time, and the maximum range will be obtained when they are coincident.

The ratio V/D or V/P will vary with speed, and there will be an optimum speed at which these ratios are a maximum.

4.6 BEST RANGE SPEED

The speed which gives the maximum range (ignoring variations of engine efficiency) is that at which the ratio of TAS: Drag (jet) and TAS: Power required (propeller) is a maximum.

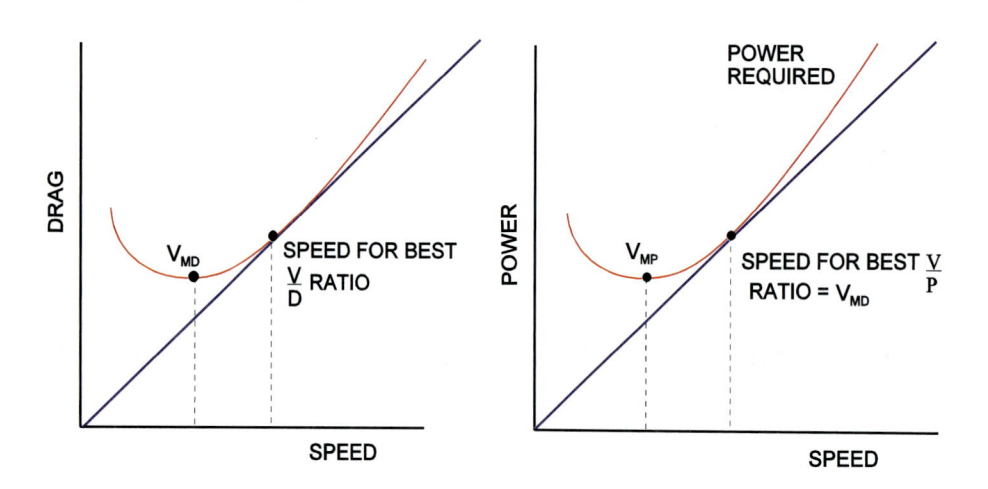

Figure 4.5

This occurs where the tangent from the origin touches the drag or power curve, and is a speed of 1.32 V_{imd} for the jet and V_{imd} for the propeller aircraft. These speeds are the ones that will give the maximum possible range, but the recommended range speeds may be faster. This is because:

a)   Speed stability is poor at V_{imd}

b)   Speeds higher than the optimum values do not initially give serious loss of range, and give better sector times.

PERFORMANCE   **GENERAL PRINCIPLES: CRUISE**

4.7 EFFECT OF WIND ON BEST RANGE SPEED

The speed at the point where the tangent touches the drag and power curves gives the speed for best range in still air. If there is a wind the speed to give the best range will be higher in a headwind and lower in a tailwind. It may be found by drawing the tangent from a point displaced to the right by the magnitude of the headwind and to the left by the magnitude of the tailwind. For a jet aircraft increasing the speed may take it into the compressibility drag rise region, and this may offset the benefits. The changes in speed are usually small unless the wind is very strong, and in this case it would be preferable to find an altitude where the wind strength is less.

4.8 ENGINE SFC.

For a given $\frac{V}{D}$ or $\frac{V}{P}$ the range will depend on the sfc.

The sfc varies with altitude and rpm for the jet engine, being lowest at high altitude and at the design rpm.

For the turbo-prop aircraft the sfc varies in a similar way to the jet but for the piston engine the sfc depends on the combination of rpm and manifold pressure. Sfc is lowest at the combination of lowest rpm and highest manifold pressure that will give the power required. Piston engine sfc decreases with altitude, mainly due to increasing thermal efficiency, up to full throttle height, but above that may increase again if power is maintained by increasing rpm.

4.9 FACTORS AFFECTING RANGE

Aeroplane Mass

Increased mass increases drag and power required. This requires greater thrust which increases the fuel flow and reduces the specific range.

The drag will increase due to increased induced drag (greater lift required) and increased profile drag (higher speed at the same angle of attack). The proportion of total drag due to induced drag will vary with speed. At V_{md} the induced drag is 50% of the total, whereas at 1.32 V_{md} the induced drag is approximately 25% of the total. For the piston engined aircraft flying at V_{md} an increase in mass of 10% will give a 5% increase in IAS (and TAS) giving an increase in power required of 15% and an increase in fuel flow of 15%, giving a decrease in specific range of 9%.

$$(S.R. = \frac{V}{F} \times \frac{1.05}{1.15} = \frac{V}{F} \frac{1}{1.09})$$

For the jet aircraft flying at 1.32 V_{md}, a 10% increase in mass will give a 5% increase in speed, a 10% increase in thrust and fuel flow, and a decrease in specific range of 5%.

PERFORMANCE GENERAL PRINCIPLES: CRUISE

Altitude

For the jet aircraft, increasing altitude increases the range up to an optimum altitude, and then decreases again. The improvement with altitude is due to:

i)   Increasing TAS at the optimum IAS
ii) Increasing engine efficiency (decreasing sfc)

Above the optimum altitude, the effects of compressibility cause the drag to increase and reduce the specific range. The optimum altitude for range increases as weight decreases. The procedure to give maximum range would therefore be to allow the aircraft to climb as the weight decreases. This procedure is not normally acceptable for civil aircraft operations, but it is possible to take advantage of the effect of decreasing weight by performing a step cruise.

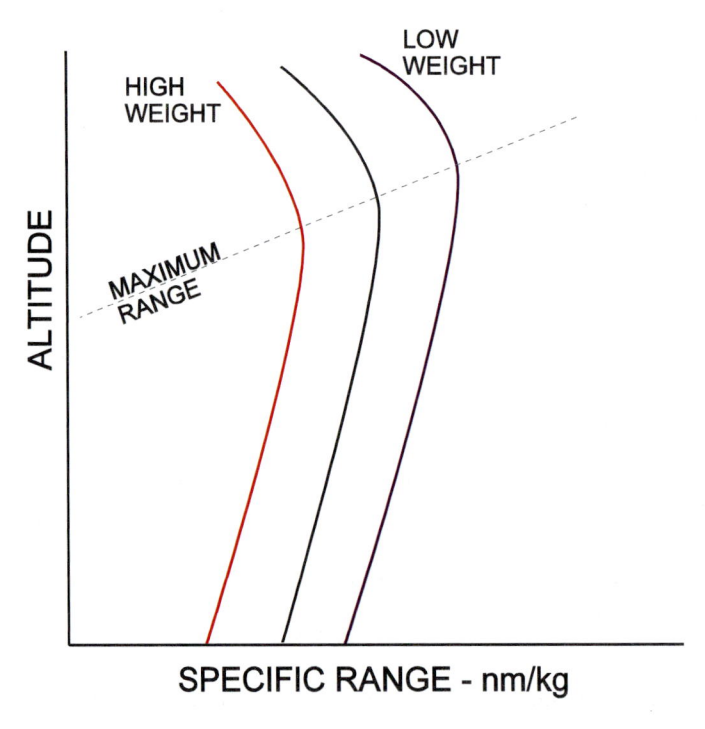

Figure 4.6 Variation of specific range for jet aircraft

For the propeller engined aircraft, altitude has less effect on the range, since both the TAS at the optimum IAS, and the power required increase with altitude, maintaining the ratio of TAS : power required. Range is only affected by changes of engine sfc with altitude, and this depends on engine type and supercharger type.

Temperature

Temperature has a relatively small effect on range. Increasing temperature increases the TAS for the optimum IAS, but reduces engine efficiency.

PERFORMANCE   GENERAL PRINCIPLES: CRUISE

Wind

The specific range and usable fuel will determine the still air range, but the ground distance travelled will depend on the wind speed. A head wind will decrease the ground range and a tailwind will increase it. Flying at the optimum altitude may not give the most favourable wind. "Wind trade" data is often given to show the wind change required to compensate for a change of altitude from the optimum.

4.10 PAYLOAD - RANGE DIAGRAM

The amount of payload that can be carried for a given range is usually shown in a payload - range diagram, illustrated in Figure 4.7

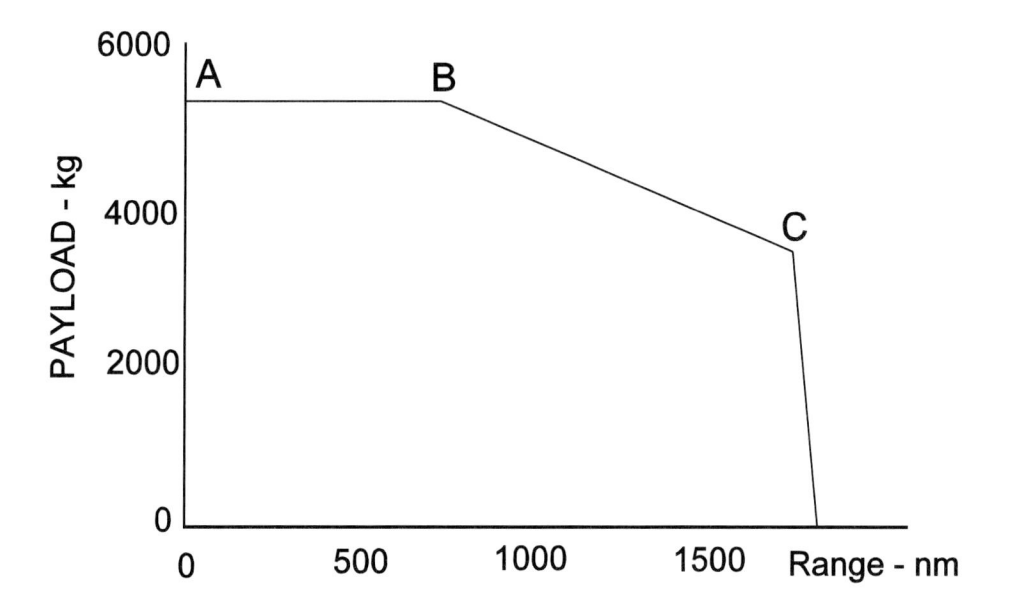

Figure 4.7. Payload - Range Diagram

Point 'A' is the maximum permissible payload, which may be limited by maximum zero fuel weight, or by capacity of baggage holds and seating. This payload can be carried for increasing range by increasing fuel carried until Maximum Permissible Take off Weight is reached (Point B). The fuel tanks are not full at this point, and so range can be increased further by increasing the fuel carried, but only at the expense of decreasing the payload, such that the Maximum Permissible Take off Weight is not exceeded. At point 'C' the tanks are full, and range can only be increased by reducing aircraft weight to give a better specific range.

PERFORMANCE   GENERAL PRINCIPLES: CRUISE

4.11 ENDURANCE

Endurance is the time that the aircraft can remain airborne on the fuel available. It will be greatest when the fuel is used at the lowest possible rate, that is, the fuel flow is a minimum. Fuel flow depends on thrust and sfc for a jet, and power and sfc for a propeller aircraft. Thrust and power will be least when drag and power required are least, and the speeds at which this occurs are V_{imd} (jet) and V_{imp} (propeller). Again, these speeds are the theoretical optimum, but in practice, higher speeds may be recommended to improve speed stability.

4.12 FACTORS AFFECTING ENDURANCE

Mass

Drag and power required increase with increased mass, and so endurance will decrease with increased mass, due to the higher thrust and power required causing increased fuel flow.

Altitude

For the jet at V_{imd}, increasing altitude does not affect the drag, but the engine sfc improves, giving greater endurance. At very high altitude, compressibility will increase the drag, causing increased fuel flow, and reduced endurance.

For the propeller aircraft, at V_{imp}, increasing altitude increases the power required, tending to reduce the endurance, but this may be counteracted by improvement in sfc. Generally endurance does not change significantly with altitude.

Temperature

For the jet, changes of temperature do not affect the drag at V_{imd}, and so changes in endurance result only from changes in sfc, which increases with increasing temperature, giving reduced endurance.

For the propeller aircraft, higher temperatures will increase the power required at V_{imp}, increasing the fuel flow, and reducing endurance.

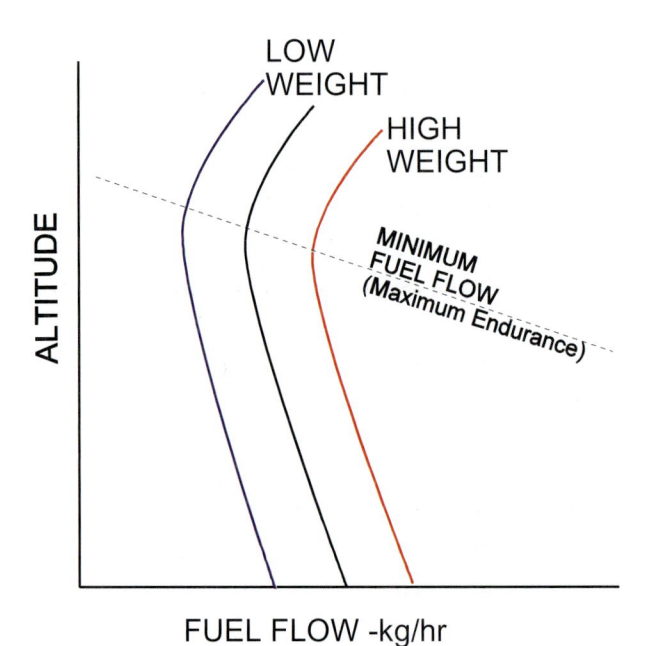

Figure 4.8 Jet Aircraft fuel flow - altitude

PERFORMANCE   GENERAL PRINCIPLES: CRUISE

PAPER 1

1. For an aircraft in steady straight and level flight, if the tail-load is acting downwards for balance,

    a) The wing lift will equal the weight and the induced drag will be the same as with zero tail-load.
    b) The wing lift will be greater than the weight and the induced drag will be greater than with zero tail-load
    c) The wing lift will be less than the weight and the induced drag will be less than with zero tail-load
    d) The wing lift will be greater than the weight and the induced drag will be the same as with zero tail-load.

2. The minimum total drag of an aircraft in level flight occurs :

    a) when the induced drag is a minimum
    b) when the parasite drag is a minimum
    c) when $C_L : C_D$ is a maximum
    d) when C_{Dtotal} is a minimum

3. For a jet aircraft the maximum achievable True Air Speed will occur :

    a) at sea level
    b) at the absolute ceiling
    c) at the same altitude that the maximum Indicated Air Speed occurs
    d) at high altitude but below the absolute ceiling.

4. The specific fuel consumption (sfc) for a jet engine is :

    a) the miles flown per kilogram of fuel used
    b) the fuel flow per unit of thrust
    c) the fuel flow at maximum take off thrust
    d) the thrust produced per kilogram of fuel used.

5. For a jet aircraft to travel a given distance, the speed to be flown that would result in the least amount of fuel being used would be :

    a) the speed for maximum range
    b) the speed for minimum fuel flow
    c) the speed for minimum drag
    d) the speed for minimum power.

© Oxford Aviation Services Limited

PERFORMANCE   GENERAL PRINCIPLES: CRUISE

6. The tangent from the origin to the Power ~ Speed curve will give for a jet aircraft the speed for:

    a) maximum range
    b) minimum drag
    c) minimum power
    d) maximum speed

7. To obtain the maximum possible range when flying into a headwind, the speed should be :

    a) the same as the optimum speed for still air
    b) increased by the magnitude of the wind speed
    c) higher than the optimum speed for still air
    d) lower than the optimum speed for still air

8. If the mass of an aircraft is increased :

    a) the range is reduced and the altitude for maximum range is higher
    b) the range is increased and the altitude for maximum range is higher
    c) the range is reduced and the altitude for maximum range is lower
    d) the range is increased and the altitude for maximum range is lower.

9. To obtain the maximum range for a jet aircraft it should be flown :

    a) at the altitude which is the optimum for the mass at the top of climb
    b) at an increasing altitude as the mass decreases
    c) at a decreasing altitude as the mass decreases
    d) at the altitude which is the optimum for the mass at the top of descent.

10. For the Payload ~ Range diagram in Figure 1. For the sector labelled B - C :

    a) the aircraft mass is decreasing
    b) the aircraft mass is increasing
    c) the aircraft mass is constant and equal to the APS weight
    d) the aircraft mass is constant and equal to the Maximum Authorised Take-off Weight

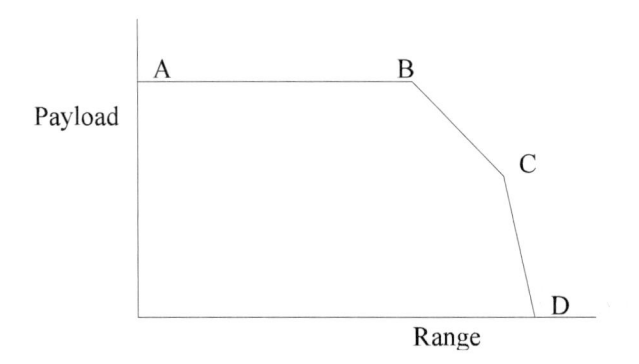

PERFORMANCE   GENERAL PRINCIPLES: CRUISE

PAPER 2

1. If the aircraft's C.G. is moved into a more forward position :

   a) the drag and lift will be increased and the range will be increased
   b) the drag and lift will be increased and the range will be decreased
   c) the drag will be increased, lift will be decreased and the range will be decreased
   d) the drag will be decreased, the lift will be increased and the range will be increased.

2. For a given aircraft mass, as altitude increases, the minimum total drag will :

   a) decrease, and the V_{imd} will decrease
   b) increase, and the V_{imd} will increase
   c) remain the same and the V_{imd} will remain the same
   d) remain the same and the V_{imd} will increase.

3. The maximum speed that can be achieved in level flight will occur :

   a) when the drag is a minimum
   b) when $C_L : C_D$ is a maximum
   c) when C_D is a minimum
   d) when the drag is equal to the maximum thrust available

4. The specific range (S R) is :

   a) the distance that the aircraft would fly with full fuel
   b) the distance that the aircraft would fly without using the reserve fuel
   c) the distance that the aircraft would fly per kilogram of fuel
   d) the distance that the aircraft could fly with the capacity payload

5. The tangent from the origin to the Drag ~ Speed curve will give, for a jet aircraft, the speed for:

   a) maximum range
   b) maximum endurance
   c) minimum drag
   d) minimum power required

6. The recommended range speed may exceed the theoretical speed for maximum range because of:

   a) compressibility effects
   b) reduced speed stability at low speed
   c) reduced directional control at low speed
   d) reduced engine life.

7. The specific fuel consumption (sfc) for a jet engine is best (lowest) at :

   a) high altitude and high rpm
   b) high altitude and low rpm
   c) low altitude and low rpm
   d) low altitude and high rpm

8. Maximum range for a jet aircraft will occur :

   a) at sea level
   b) at the absolute ceiling
   c) at the altitude where the maximum True Air Speed is achieved
   d) at a high altitude, but below the absolute ceiling

9. It may be advantageous to fly below the optimum range altitude because :

   a) a more favourable wind may give a greater ground distance
   b) cruising at the optimum altitude may give high speed buffet
   c) speed stability is better
   d) higher temperatures give better engine efficiency.

10. Endurance for a jet aircraft is a maximum :

    a) at low altitude, and increases with increasing aircraft mass
    b) at low altitude, and decreases with increasing aircraft mass
    c) at high altitude, and increases with increasing aircraft mass
    d) at high altitude, and decreases with increasing aircraft mass

PERFORMANCE GENERAL PRINCIPLES: CRUISE

ANSWERS

PAPER 1

Questions	Answers
1	B
2	C
3	D
4	B
5	A
6	B
7	C
8	C
9	B
10	D

PAPER 2

Questions	Answers
1	B
2	C
3	D
4	C
5	A
6	B
7	A
8	D
9	A
10	D

CHAPTER FIVE - GENERAL PRINCIPLES: LANDING

Contents

		Page
5.1	LANDING DISTANCE	5 - 1
5.2	DISTANCE REQUIRED.	5 - 1
5.3	AERODYNAMIC DRAG.	5 - 2
5.4	BRAKE DRAG.	5 - 2
5.5	REVERSE THRUST	5 - 2
5.6	LANDING SPEED.	5 - 2
5.7	EFFECT OF VARIABLE FACTORS ON LANDING DISTANCE	5 - 3
5.8	HYDROPLANING	5 - 4
	LANDING PAPER 1	5 - 5

© Oxford Aviation Services Limited

© Oxford Aviation Services Limited

PERFORMANCE   GENERAL PRINCIPLES: LANDING

5.1 LANDING DISTANCE

The landing distance is the distance from a screen of defined height to touch down and come to a complete stop. It comprises an airborne segment from the screen to touchdown and a ground roll.
The airborne segment will consist of a steady descent gradient followed by the landing flare.

The ground run will consist of a short period of free roll until braking is applied, and then a braked run to the point where the aircraft stops.

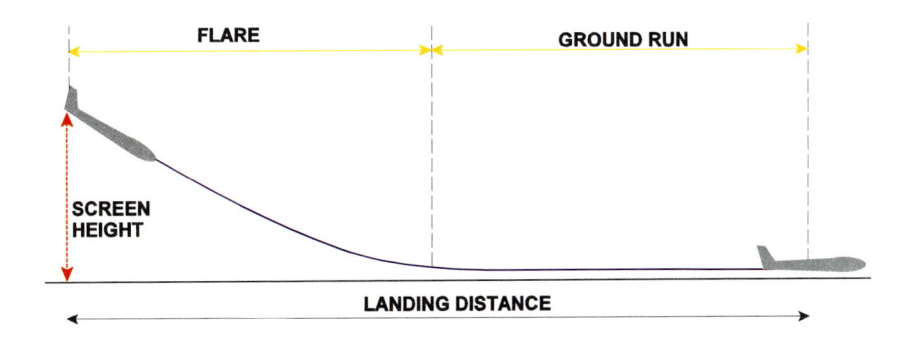

Figure 5.1 Landing Distance

5.2 DISTANCE REQUIRED.

The airborne distance will be determined mainly by the descent gradient, and this will depend on the drag, thrust and weight of the aeroplane. The ground run distance depends on the speed at the start of the run, the decelerating force and the mass of the aircraft.
The distance to decelerate from a speed 'V' to a stop, with a constant deceleration d

$$\text{Distance} = \frac{V^2}{2d} \quad \text{Where 'V' is True Ground Speed and 'd' is deceleration.} \quad \text{..........(Equation 5.1)}$$

The deceleration depends on the mass (M) and the net decelerating force (F)

$$d = \frac{F}{M} \quad \text{..........(Equation 5.2)}$$

The decelerating force is the difference between the drag and the thrust. The thrust is usually small on landing, with the engines idling, but any thrust produced will increase the landing distance. The drag will come from the aerodynamic drag of the aeroplane, the brake drag and any reverse thrust available from thrust reversers or braking propellers.

PERFORMANCE **GENERAL PRINCIPLES: LANDING**

5.3 AERODYNAMIC DRAG.

The aerodynamic drag depends on the IAS, the configuration and aircraft attitude. Configuration will normally be with landing flap and wing spoilers deployed. Attitude will be that with the nose wheel in contact with the runway. As the speed decreases the drag will decrease with speed squared.

5.4 BRAKE DRAG.

The brake drag depends on the brake torque available, the coefficient of braking friction, and the load on the wheels. The load on the wheels is the weight - lift.

$$\text{Braking drag} = \mu (W-L). \qquad \text{(Equation 5.3)}$$
Where μ is the coefficient of braking friction.

The wing lift may be reduced if lift dumpers are operated, and this will increase the brake drag. Brake drag may be increased by increasing brake pressure until the wheel locks (limiting friction) or until the maximum brake torque is reached. Optimum braking occurs just before wheel lock is reached, and an anti-skid system will control the brake pressure so that this condition is maintained. The braking coefficient of friction will depend on the runway surface, being greatest on a dry hard runway, and less on grass or on a wet or icy runway. If hydroplaning occurs, the coefficient of friction will be very low.

5.5 REVERSE THRUST

If the engines have the capability of giving reverse thrust, by reversing the propeller pitch or the jet efflux, this will add to the braking force available. Unlike the wheel brake drag, reverse thrust is not affected by the runway coefficient of friction and so is particularly valuable if the runway is wet or icy.

Reverse thrust is most effective at high speed, and so it should be selected as early as possible during the landing. At low speed it is less effective, and begins to give problems due to re-ingestion, leading to over-heating, and to possible engine damage due to debris being blown up from the runway. There is usually a minimum speed given for the use of reverse thrust, and so it should be cancelled before this speed, unless in an emergency. Because of the increased noise associated with the use of reverse thrust, its use may be restricted at some aerodromes, particularly at night.

5.6 LANDING SPEED.

The speed at the screen (V_{REF}) is determined by the Regulations, and must be safely above the stalling speed and the minimum control speed. The scheduled speed will be an IAS appropriate to the mass, but the distance depends on ground speed, and so the IAS must be corrected for density and wind when calculating the distance required.

PERFORMANCE   **GENERAL PRINCIPLES: LANDING**

5.7 EFFECT OF VARIABLE FACTORS ON LANDING DISTANCE

Aeroplane mass

The mass of the aeroplane affects:

a) the stalling speed and hence the screen speed

b) the deceleration for a given decelerating force

c) the wheel drag

Increased mass increases stalling speed, and reduces the deceleration for a given decelerating force, both effects increasing the landing distance. Increased mass increases the brake drag available (if not torque limited) and this decreases the landing distance. The net effect is that the landing distance will increase with increasing mass, but to a lesser degree than the increase of take off distance with increasing mass.

Air density

The air density affects:

a) the TAS for a given IAS

b) the thrust or power of the engine

As thrust is small the main effect will be on the TAS.

Low density (high temperature or low pressure) will give an increase in the landing distance due to the higher TAS, but again to a lesser degree than for the take-off distance.

Wind

The landing distance depends on the True Ground Speed, and so for a given scheduled IAS, a headwind will reduce the ground speed and reduce the landing distance, and a tailwind will increase the ground speed and increase the landing distance. The Regulations require that 50% of a headwind component is used, and 150% of a tailwind component.

PERFORMANCE   **GENERAL PRINCIPLES: LANDING**

Runway slope

If the runway is sloping, the weight component along the runway will add to or subtract from the decelerating force. A downhill slope will increase the landing distance required and an uphill slope will reduce the landing distance.

Runway surface

The brake drag depends on the runway coefficient of friction, and this depends on the runway surface, a hard dry surface giving the highest coefficient of friction, with a wet surface or grass giving a lesser coefficient. Icy runways or runways on which hydroplaning occurs will have coefficients of friction as low as 0.05.

5.8 HYDROPLANING

As in the case of an abandoned take-off, hydroplaning can occur on landing at speeds above the hydroplaning speed. Three forms of hydroplaning have been identified:

a) Dynamic hydroplaning. Due to standing water or slush supporting the tyres clear of the runway surface.

b) Viscous hydroplaning. Caused by a thin film of fluid covering a very smooth surface, which is not penetrated by the tyre.

c) Reverted rubber. Caused by friction between the tyre and the wet runway surface causing the water to boil and revert the tyre rubber to its uncured state. This forms a seal, trapping the steam between the tyre and the runway.

PERFORMANCE   **GENERAL PRINCIPLES: LANDING**

PAPER 1

1. The scheduled landing distance required is the distance :

    a) from touch down to the point at which the aircraft has decelerated to a speed of 10 knots
    b) from touch down to the point at which the aircraft has come to a complete stop
    c) from a screen of a designated height to the point at which the aircraft has come to a complete stop.
    d) from the point at which the aircraft is 50 metres above the runway to the point at which the aircraft has come to a complete stop.

2. The landing distance required will be increased as a result of all of the following :

    a) increased temperature, increased pressure altitude, uphill runway slope.
    b) increased temperature, increased pressure altitude, downhill runway slope.
    c) decreased temperature, decreased pressure altitude, uphill runway slope
    d) increased temperature, decreased pressure altitude, downhill runway slope.

3. When calculating the landing distance, what percentage of the reported wind component must be allowed for ?

    a) 50% of a headwind, and 150% of a tailwind.
    b) 100% of a headwind, and 100% of a tailwind.
    c) 50% of a headwind, and 100% of a tailwind.
    d) 150% of a headwind, and 50% of a tailwind.

4. Which of the following statements is correct ?

    a) A reduced flap setting for landing will give a shorter landing distance, as a result of reduced lift and greater load on the wheels.
    b) Wheel braking is most effective when the wheels are locked.
    c) Landing distance required on a grass runway will be shorter than on tarmac because of the rougher surface.
    d) Deployment of lift dumpers will increase the effectiveness of the wheel brakes.

5. During the landing run with wheel brakes applied and engines in reverse thrust, as the speed decreases :

    a) the wheel brake drag remains constant, reverse thrust becomes less effective.
    b) the wheel brakes become more effective, reverse thrust becomes less effective.
    c) the wheel brakes become less effective, reverse thrust becomes more effective.
    d) the wheel brakes become less effective, reverse thrust becomes less effective.

PERFORMANCE   GENERAL PRINCIPLES: LANDING

6. Which of the following statements relating to hydroplaning is true :

   a) Hydroplaning can only occur if the depth of the contaminant exceeds 3 mm.
   b) When the wheel begins to hydroplane the wheel drag decreases.
   c) Hydroplaning can only occur if the brakes are applied and releasing the brakes will stop the hydroplaning.
   d) Hydroplaning can be delayed by reducing the tyre pressure.

7. Decreasing air density will give :

   a) increased landing distance due to increased TAS, and increased idling thrust.
   b) reduced landing distance due to reduced TAS, and reduced idling thrust.
   c) increased landing distance due to increased TAS, and reduced idling thrust.
   d) reduced landing distance due to reduced TAS, and increased idling thrust.

8. Dynamic hydroplaning is likely to occur as a result of a combination of :

   a) high speed, low tyre pressure, high contaminant density.
   b) high speed, high tyre pressure, high contaminant density.
   c) low speed, high tyre pressure, low contaminant density.
   d) high speed, high tyre pressure, low contaminant density.

9. The effect of increasing aircraft mass on the landing distance is :

   a) screen speed increases, brake drag decreases, landing distance increases.
   b) screen speed increases, brake drag increases, landing distance increases.
   c) screen speed decreases, brake drag increases, landing distance decreases.
   d) screen speed decreases, brake drag decreases, landing distance decreases.

10. Which of the following statements regarding reverse thrust is true ?

    a) The braking effect of reverse thrust is greatest at low speeds.
    b) Reverse thrust may be used for landing but not for an aborted take off.
    c) Reverse thrust may not be used on a slippery runway.
    d) At low speeds re-ingestion of the jet efflux may occur, causing over-heating.

ANSWERS

Questions	Answers
1	C
2	B
3	A
4	D
5	B
6	B
7	C
8	A
9	B
10	D

CHAPTER SIX - SINGLE ENGINE CLASS B AIRCRAFT: TAKE OFF

Contents

		Page
6.1	PERFORMANCE CLASS B	6 - 1
6.2	GENERAL (JAR - OPS 1.525)	6 - 1
6.3	TAKE-OFF	6 - 1
6.4	TAKE-OFF DISTANCE (JAR 23.53)	6 - 1
6.5	FACTORS TO BE ACCOUNTED FOR	6 - 3
6.6	PRESENTATION OF DATA	6 - 4
6.7	DETERMINATION OF MAXIMUM PERMISSIBLE TAKE-OFF MASS	6 - 5
6.8	TAKE-OFF SPEEDS	6 - 6
	SINGLE ENGINE CLASS B AIRCRAFT: TAKE OFF	6 - 7

© Oxford Aviation Services Limited

© Oxford Aviation Services Limited

PERFORMANCE   SINGLE ENGINE CLASS B: TAKE OFF

SINGLE ENGINE CLASS B AIRCRAFT

6.1 PERFORMANCE CLASS B

Propeller driven aeroplanes with a maximum approved passenger seating configuration of 9 or less, and a maximum take-off mass of 5700 kg. or less.   (JAR - OPS 1.470).

JAR 23 contains requirements for normal, utility, and aerobatic category aeroplanes, and also for commuter category aeroplanes

Chapters 6 to 9 relate to single engined aeroplanes in Performance class B.

6.2 GENERAL (JAR - OPS 1.525)

An operator shall not operate a single engined aeroplane:

a)   at night

b)   in IMC conditions except under Special VFR

c)   unless en route surfaces are available which permit a safe forced landing (JAR-OPS 1.240 (a) (6))

6.3 TAKE-OFF

In assessing the take-off performance, two requirements must be considered:

a)   the ability to take-off within the aerodrome distances available

b)   the ability to climb out with an adequate gradient after take-off.
     (obstacle clearance need not be demonstrated for single-engined Class B aeroplanes)

6.4 TAKE-OFF DISTANCE (JAR 23.53)

The gross take-off distance for class B aeroplanes (other than those in the commuter category) is the distance from the start of take-off to a point 50ft. above the take-off surface, with take-off power on each engine, rotating at V_R and achieving the specified speed at the screen.

V_R The rotation speed, must not be less than V_{S1}

PERFORMANCE SINGLE ENGINE CLASS B: TAKE OFF

The speed at 50 ft. must be not less than :

a) a speed that is safe under all reasonably expected conditions

b) $1.2 V_{S1}$

whichever is higher

The unfactored (gross) take-off distance is specified in the aeroplane Flight Manual

Take-off distance Required (JAR-OPS 1.530 (b))

a) if there is no stopway or clearway

Gross TOD x 1.25 must not exceed the TORA

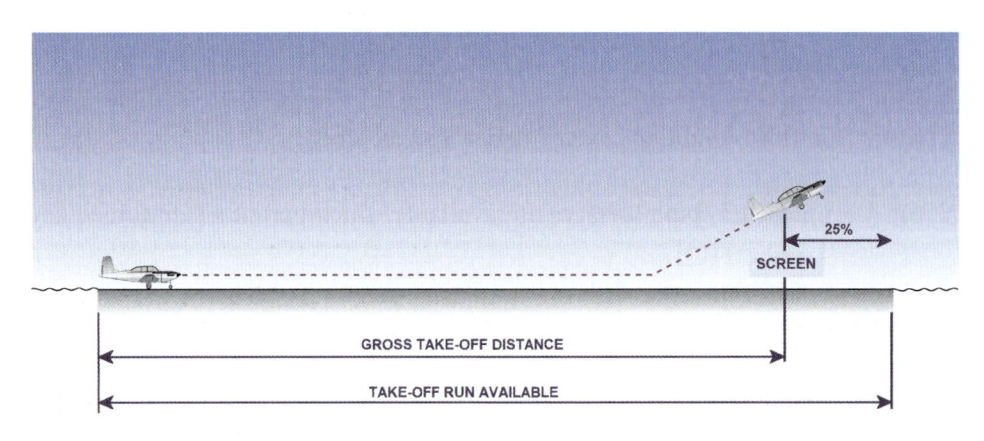

Figure 6.1

b) If stopway and/or clearway exists:

i) Gross TOD must not exceed the TORA

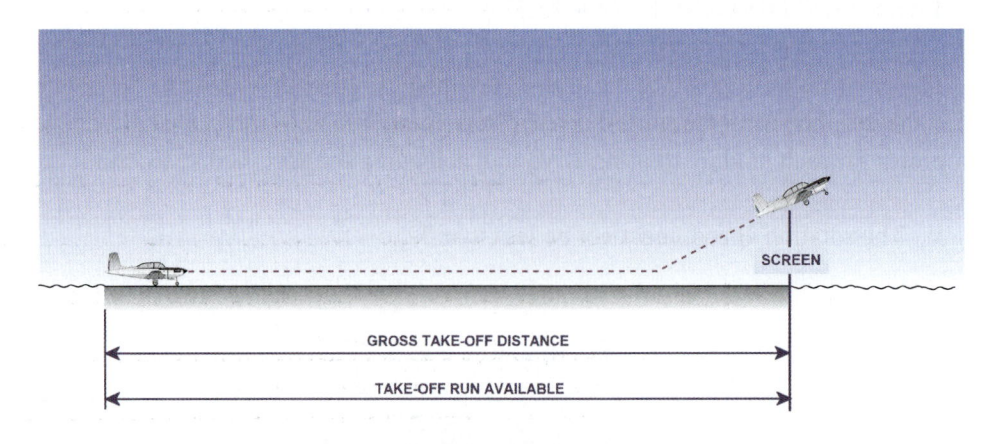

Figure 6.2

ii) Gross TOD x 1.15 must not exceed the TODA

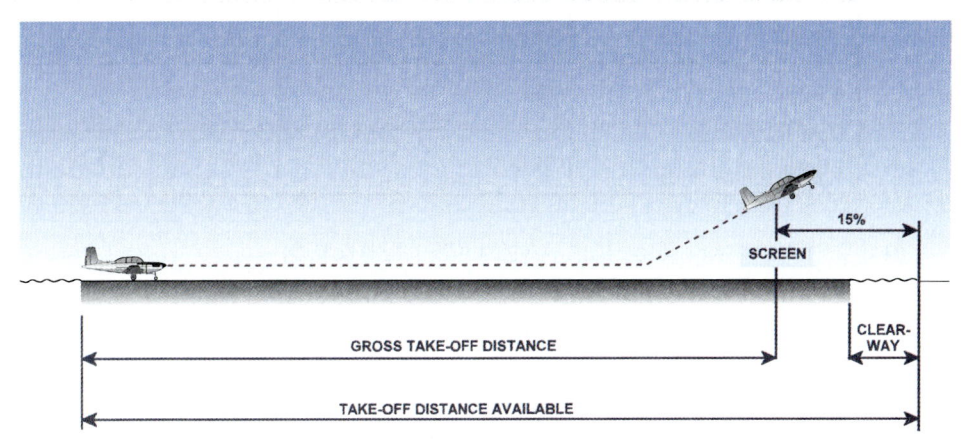

Figure 6.3

iii) Gross TOD x 1.3 must not exceed the ASDA

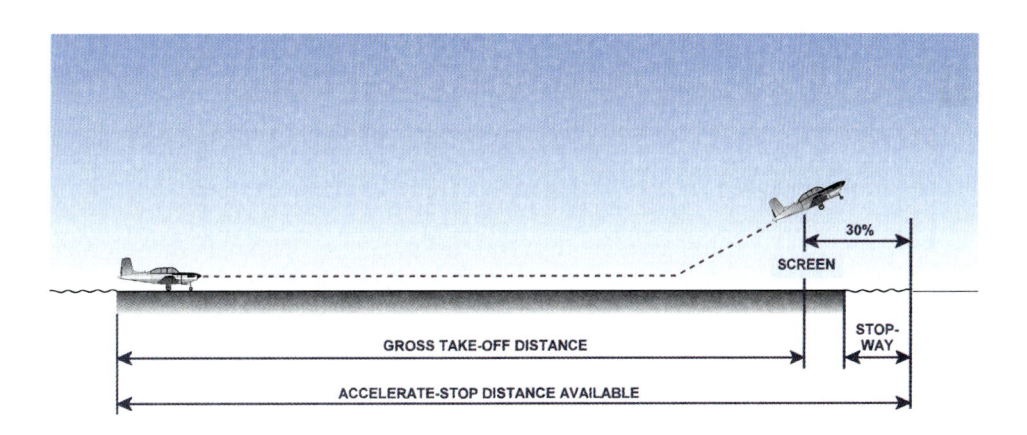

Figure 6.4

6.5 FACTORS TO BE ACCOUNTED FOR

The take-off distance required shall take account of:

a) the mass of the aeroplane at the start of the take-off run

b) the pressure altitude at the aerodrome

c) the ambient temperature at the aerodrome

d) the runway surface conditions and the type of runway surface Ref. AMC. OPS 1.530 (c) (4)

PERFORMANCE   **SINGLE ENGINE CLASS B: TAKE OFF**

Surface	Condition	Factor
Paved	Wet	1.0
Grass (On firm soil) Up to 20cm long	Dry	1.2
	Wet	1.3

e)   the runway slope Ref. AMC OPS 1.530 (c) (5)
    The TOD should be increased by 5% for each 1% of uphill slope. (Max. slope 2% without acceptance by the Authority). Do not correct for downhill slope

f)   not more than 50% of the reported headwind component or not less than 150% of the reported tailwind component (Aircraft data may already be factored)

6.6 PRESENTATION OF DATA

The take-off distance required is usually presented graphically in the form illustrated in Figure 6.5. The use of the graph is illustrated by the arrowed broken line.

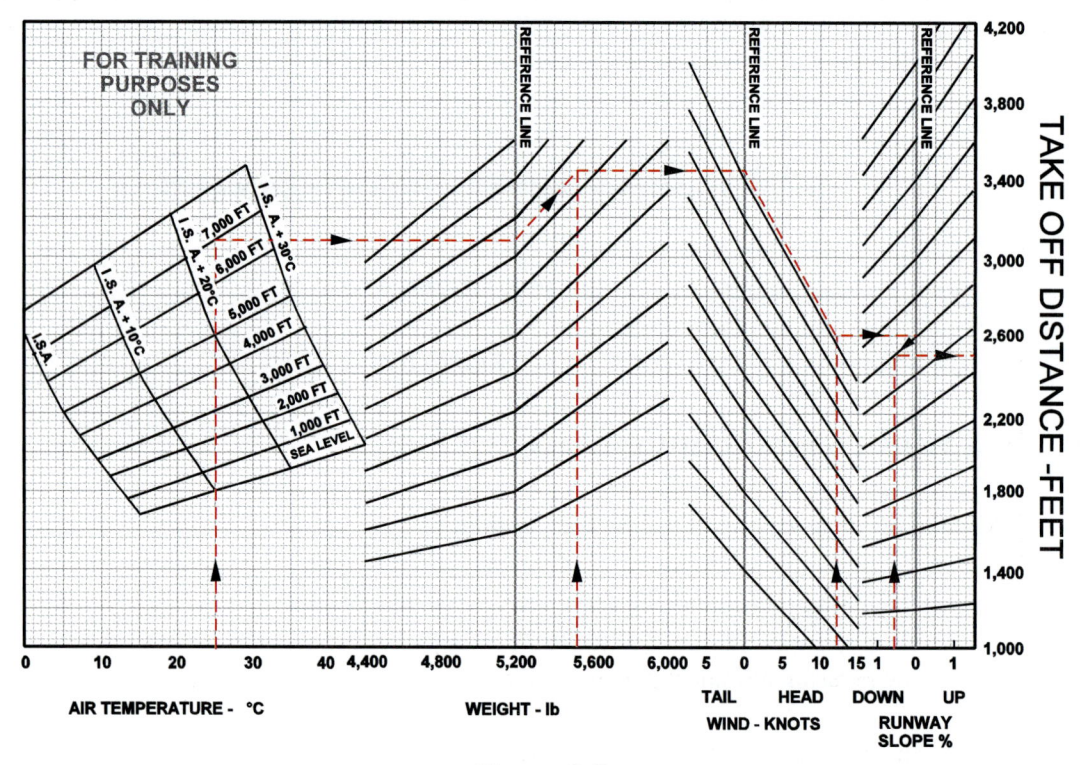

Figure 6.5

PERFORMANCE   SINGLE ENGINE CLASS B: TAKE OFF

The example shows that for an air temperature of 25°C at a pressure altitude of 7000 ft., a weight of 5525 lb., with a factored headwind component of 12 kt. and a downhill runway slope of 0.6% the take-off distance required is 2500 ft.

This distance, with the appropriate factors applied must be compared to the distances available at the aerodrome for take off.

If there is no stopway or clearway,   2500 x 1.25 = 3125 ft.
So the runway length must be at least 3125 ft.

If there is stopway and clearway :

a)   Runway length must be at least 2500 ft.

b)   TODA must be at least 2500 x 1.15 = 2875 ft.

c)   ASDA must be at least 2500 x 1.3 = 3250 ft.

6.7   DETERMINATION OF MAXIMUM PERMISSIBLE TAKE-OFF MASS

If it is required to find the maximum permissible take-off mass for the aerodrome distances available, the graph must be entered with the most limiting distance available. Figure. 6.6 illustrates the use of the graph in this situation.

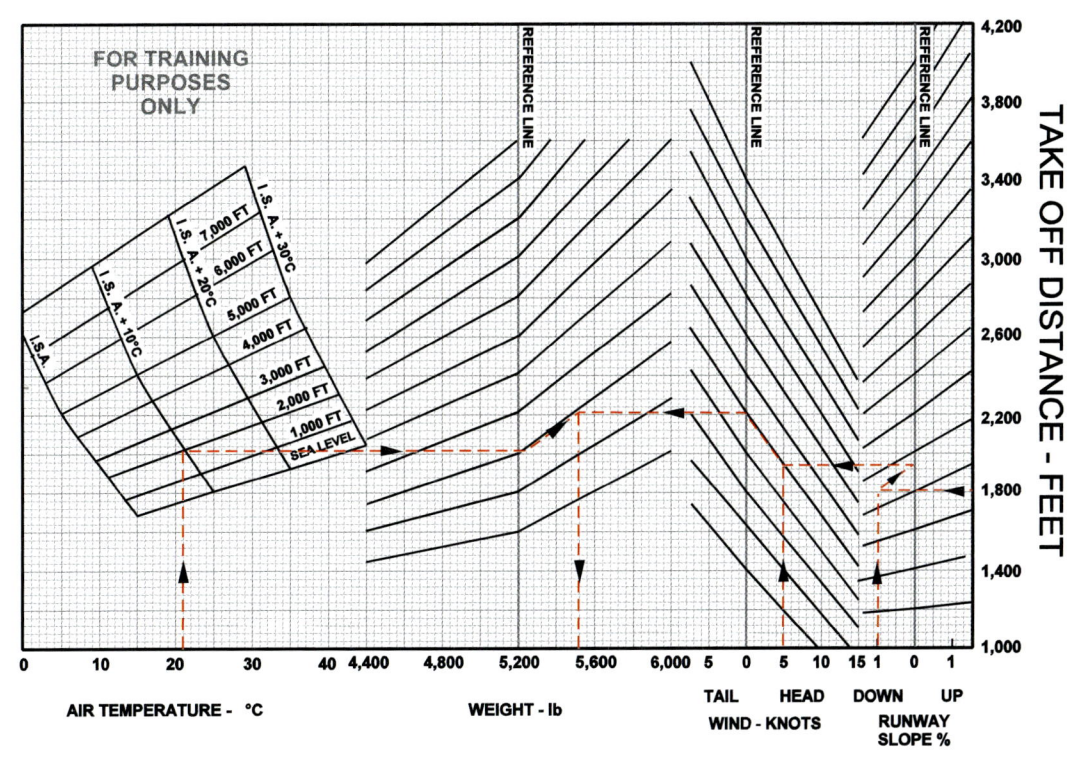

Figure 6.6

PERFORMANCE   **SINGLE ENGINE CLASS B: TAKE OFF**

The distance with which the graph should be entered will depend on whether or not there is clearway or stopway.

a)   If there is no stopway or clearway, enter the graph with the $\dfrac{\text{TORA}}{1.25}$

b)   If stopway or clearway exists, enter with the lesser of:

       i)   TORA

       ii) $\dfrac{\text{TODA}}{1.15}$

       iii) $\dfrac{\text{ASDA}}{1.3}$

The example shows that for an aerodrome at a pressure altitude of 2000 ft. and a temperature of 21° C with a wind of 5 kt. factored headwind if the runway length available is 2250ft. with a downhill slope of 1% and there is no stopway or clearway, the graph would be entered with a distance of $\dfrac{2250}{1.25} = 1800$ ft. and this gives a weight of 5500 lb

6.8   **TAKE-OFF SPEEDS**

For take-off it will be necessary to know the rotation speed V_R and the speed which should be achieved at the 50ft screen. This will usually be given as a function of weight.

Weight (lb)	Rotation speed (kt)	50 ft. Speed (kt)
5200 and below	48	58
5600	49	60
6000	51	62

PERFORMANCE   SINGLE ENGINE CLASS B: TAKE OFF

PAPER 1

1. The gross take off distance required for a single engine Class B aircraft is the distance

   a) from the start of the run to the point at which the wheels are just clear of the ground.
   b) from the start of the run to the point at which the wheels are just clear of the ground multiplied by a factor of 1.25
   c) from the start of the run to a screen height of 35 feet.
   d) from the start of the run to a screen height of 50 feet.

2. Given that control requirements are adequate, the speed at the screen must not be less than :

   a) 10% above the stall speed
   b) 20% above the stall speed
   c) 25% above the stall speed
   d) 30% above the stall speed

For the following numerical examples refer to CAP 698 Figure 2.1

3. Find the gross take off distance required for the following conditions :
   Pressure altitude   2000 ft.
   Temperature   25° C
   Weight   3300 lb.
   Reported wind   10 kt. Head
   Runway slope   level, paved dry surface

   a) 2000 ft.
   b) 2550 ft.
   c) 1600 ft.
   d) 950 ft.

4. Find the gross take off distance required for the following conditions :
   Pressure altitude   4000 ft.
   Temperature   0° C
   Weight   3150 lb.
   Reported wind   5 kt. Tail
   Runway slope   level, paved dry surface

   a) 1200 ft.
   b) 2200 ft.
   c) 2625 ft.
   d) 2730 ft.

PERFORMANCE SINGLE ENGINE CLASS B: TAKE OFF

5. Find the gross take off distance required for the following conditions :
   Pressure altitude   5000 ft.
   Temperature   I.S.A.
   Weight   3020 lb.
   Reported wind   still air
   Runway slope   1.0% uphill

   a)   1155 ft.
   b)   2150 ft.
   c)   2625 ft.
   d)   3000 ft.

6. For take off from an aerodrome where there is no stopway or clearway, what minimum runway length is required to comply with the regulations, given :
   Pressure altitude   6000 ft.
   Temperature   10° C
   Weight   3450 lb.
   Reported wind   5 kt. Head
   Runway slope   level

   a)   3200 ft.
   b)   1750 ft.
   c)   2190 ft.
   d)   3875 ft.

7. If the take off distance required on a paved dry runway is 3000 ft. the distance on a dry grass runway for the same conditions would be :

   a)   2500 ft.
   b)   3600 ft.
   c)   3750 ft.
   d)   3900 ft.

8. What is the maximum permissible take off weight to comply with the regulations for the following conditions :
   Pressure altitude   4000 ft.
   Temperature   0° C
   Reported wind   15 Head
   Runway slope   level
   Runway length   2500 ft. with no stopway or clearway

   a)   2800 lb.
   b)   3040 lb.
   c)   3380 lb.
   d)   3650 lb.

PERFORMANCE   **SINGLE ENGINE CLASS B: TAKE OFF**

9. The following conditions apply to a take off aerodrome :
   Pressure altitude   6000 ft.
   Temperature   8° C
   Reported wind   10 kt. Head
   Runway slope   level
   Runway length   2800 ft. with 500 ft. of stopway and 500 ft. of clearway

   The maximum permissible weight for take off would be :

   a)   3400 lb.
   b)   3300 lb.
   c)   2900 lb.
   d)   3600 lb.

10. An aerodrome has a dry grass level runway of 3600 ft. with no stopway or clearway. The pressure altitude is 2000 ft., the OAT is 20° C and the wind component is zero. The maximum permissible weight for take off would be :

    a)   1980 lb.
    b)   2980 lb.
    c)   3450 lb.
    d)   3650 lb.

PERFORMANCE SINGLE ENGINE CLASS B: TAKE OFF

ANSWERS

Questions	Answers
1	D
2	B
3	A
4	B
5	B
6	D
7	B
8	C
9	B
10	C

CHAPTER SEVEN - SINGLE ENGINE CLASS B: CLIMB

Contents

		Page
7.1	CLIMB PERFORMANCE	7 - 1
7.2	TAKE-OFF AND LANDING CLIMB LIMITS	7 - 1
7.3	OBSTACLE CLEARANCE	7 - 3
7.4	CLIMB TO CRUISING ALTITUDE	7 - 3
7.5	EN-ROUTE CLIMB PERFORMANCE (JAR - OPS 1.542)	7 - 4
7.6	EN-ROUTE CEILING	7 - 5
	SINGLE ENGINE CLASS B AIRCRAFT: CLIMB	7 - 7

© Oxford Aviation Services Limited

© Oxford Aviation Services Limited

PERFORMANCE   SINGLE ENGINE CLASS B: CLIMB

7.1 CLIMB PERFORMANCE

The climb performance of the aircraft will determine:

a) the take-off and landing MAT limitations
b) the clearance of obstacles after take-off
c) the time, distance and fuel during the climb to the cruising altitude
d) the en-route limitations

7.2 TAKE-OFF AND LANDING CLIMB LIMITS   Appendix 1 to (JAR-OPS 1.525 (b)

a) Take off climb

The steady gradient of climb after take-off must be at least 4% with:

i) take-off power
ii) landing gear extended unless it can be retracted in not more than 7 seconds
iii) flaps in the take-off position
iv) a climb speed of not less than 1.2 V_{S1}

The data may be presented as a mass for altitude and temperature, which achieves the required gradient, or a gradient from which the limiting weight can be determined. Figures 7.1 and 7.2 show typical examples of these types of graph.

The example in figure 7.1 shows that for a pressure altitude of 7000 ft. and a temperature of 25°C, the maximum permissible weight is 5520 lb.

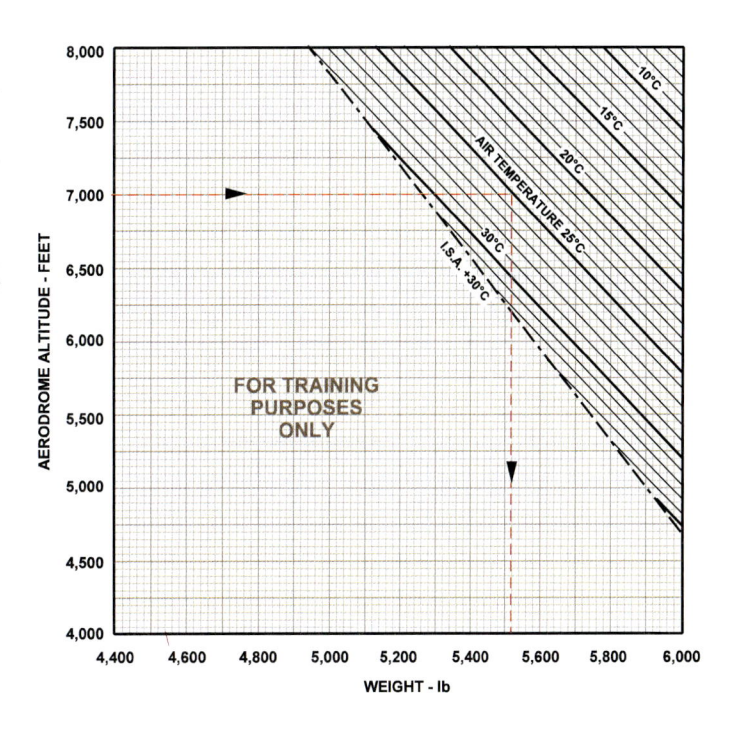

Figure 7.1 Maximum Take-off and Landing Weight for Altitude and Temperature

PERFORMANCE

SINGLE ENGINE CLASS B: CLIMB

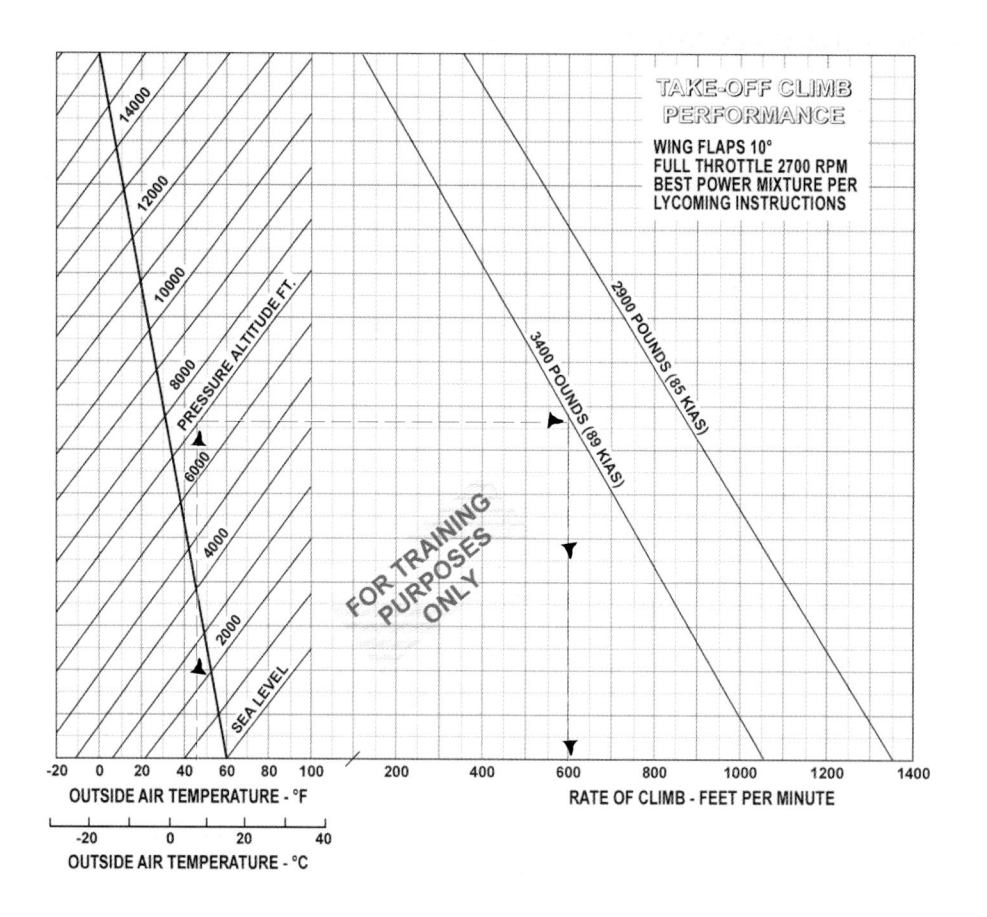

Figure 7.2. Take-off Climb Performance

The example shows that at a pressure altitude of 7000 ft. and a temperature of 45°F, (7°C) at a weight of 3400 lb. The rate of climb is 600 ft./min. at a speed of 89kt IAS.

At 7000 ft 7°C, 89 kt IAS   =   100 kt TAS

The climb gradient would be: $\dfrac{600}{100}$ x $\dfrac{60}{6080}$ x 100   =   5.9%

b)   Landing Climb.

The gradient of climb must be at least 2.5% with :

i)   Not more than the power that is available 8 seconds after initiation of movement of the power control from the minimum flight idle position.
ii) The landing gear extended.
iii) The wing flaps in the landing position.
iv) A climb speed equal to V_{REF}

PERFORMANCE   SINGLE ENGINE CLASS B: CLIMB

7.3 OBSTACLE CLEARANCE

There are no requirements for obstacle clearance for single engined aircraft in JAR Class B, but the climb data could be used to assess the height which the aircraft will reach at the distance of an obstacle from the take-off distance. For the gradient available, the height gain in a distance D from the end of the take-off distance will be:

$$D \times \frac{\% \text{ gradient}}{100}$$

The total height at this point will be: Screen height + height gain

This height can be compared to the height of the obstacle, to verify that adequate clearance is obtained. (Clearance for multi-engined class B aircraft is required to be 50ft).

If there is a wind the gradient should be factored by $\frac{\text{True Air Speed}}{\text{True Ground Speed}}$, using the factored wind speed. (50% headwind, 150% tailwind)

7.4 CLIMB TO CRUISING ALTITUDE

For flight planning it will be necessary to know the time, distance and fuel used from the take-off altitude to the cruise altitude.
Fig 7.3 shows a typical presentation for this data.

ASSOCIATED CONDITIONS

POWER .... FUEL THROTTLE - 2,500 RPM
FUEL DENSITY .... 6.0 LBS/GAL
MIXTURE .... FULL RICH
COWL FLAPS .... AS REQUIRED

EXAMPLE

AERODROME PRESSURE ALTITUDE   5650 FT
AERODROME OAT   15°C
CRUISE PRESSURE ALTITUDE   11500 FT
CRUISE   -5°C
INITIAL CLIMB RATE   3650 LB
TIME TO CLIMB = 18 - 6.5 = 11.5 MIN
FUEL TO CLIMB = 6 - 2.5 = 3.5 GALL
DISTANCE TO CLIMB = 36 - 12.5 = 23.5 NM

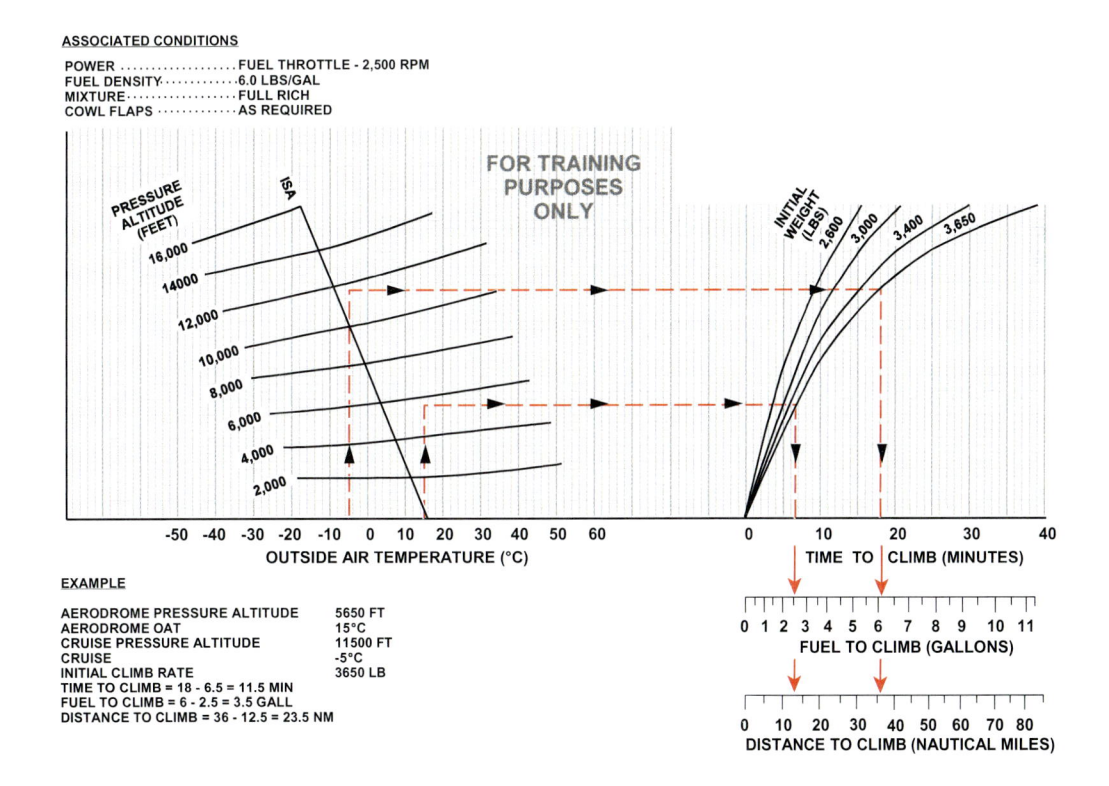

Figure 7.3 Time, Fuel and Distance to Climb

PERFORMANCE — SINGLE ENGINE CLASS B: CLIMB

Example: Take-off aerodrome pressure altitude : 5600 ft.   OAT : 15°C
Cruise pressure altitude : 11500 ft.   OAT : -5°C
Time to climb = 18 - 6.5 = 11.5 min.
Fuel to climb = 6 - 2.5 = 3.5 gallons
Distance to climb = 36 - 12.5 = 23.5 nm.

7.5 EN-ROUTE CLIMB PERFORMANCE (JAR - OPS 1.542)

An operator shall ensure that the aeroplane, in the meteorological conditions expected for the flight, and in the event of engine failure, is capable of reaching a place at which a safe forced landing can be made.

a) the aeroplane must not be assumed to be flying, with the engine operating within the maximum continuous power conditions specified, at an altitude exceeding that at which the rate of climb equals 300 ft/min.

b) the assumed en-route gradient shall be the gross gradient of descent increased by a gradient of 0.5%.

To calculate the distance in the glide descent, divide the height loss by the descent gradient increased by 0.5%.

Example: An aircraft has a glide ratio of 1 in 20 at the best glide speed. What distance is travelled in descending from 5000ft to 1000ft pressure altitude.

$$\text{Glide ratio} = \frac{1}{20} = \frac{1}{20} \times 100\% = 5\% \text{ gross}$$

Net descent gradient = 5 + 0.5% = 5.5%

$$\text{Descent distance} = \frac{4000}{5.5} \times \frac{100}{6080} \text{ nm} = 11.96 \text{ nm}$$

This is a still air distance and must be adjusted if there is a wind.

For example if the TAS is 90 kt and there is a 20 kt headwind, the distance would be :
$$11.96 \times \frac{70}{90} = 9.3 \text{ nm}$$

PERFORMANCE   SINGLE ENGINE CLASS B: CLIMB

7.6   EN-ROUTE CEILING

The ceiling is the pressure altitude at which the gradient or rate of climb is zero (absolute ceiling) or a specified value (service ceiling).

To find the ceiling, enter the climb chart with the gradient of climb required, correct for weight, and read the pressure altitude at the ambient temperature.

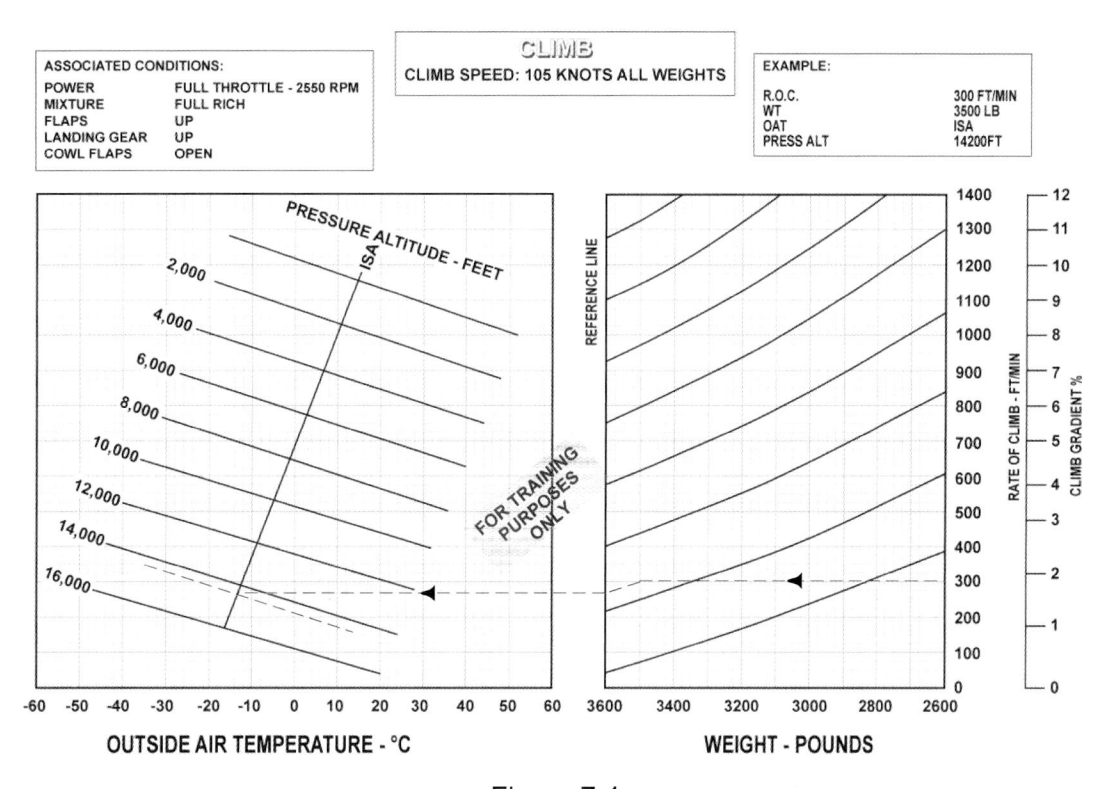

Figure 7.4

Example :   Using Figure 7.4 find the 300 ft./min ceiling at a weight of 3500 lb. in ISA conditions.

Figure 7.4   14200 ft.

PERFORMANCE SINGLE ENGINE CLASS B: CLIMB

PAPER 1

1. The take off climb gradient requirement for a single engine Class B aircraft is :

   a) 4.0% with maximum continuous power and flaps up
   b) 2.5% with maximum take off power and flaps in the take off position
   c) 4.0% with maximum take off power and flaps in the take off position
   d) a gradient of not less than zero with maximum continuous power and flaps in the take off position.

2. At sea level I.S.A. an aircraft has a rate of climb of 650 ft./min. at an indicated air speed of 85 knots. The climb gradient in still air is :

   a) 0.75%
   b) 1.3%
   c) 7.5%
   d) 13%

3. With reference to Appendix C the maximum permissible weight for take off at an aerodrome with a pressure altitude of 6400 ft. and an OAT of 27°C would be :

   a) 5920 lb.
   b) 5840 lb.
   c) 5740 lb.
   d) 5660 lb.

4. With reference to Appendix C the maximum permissible weight for take off at an aerodrome with a pressure altitude of 5500 ft .and an OAT of 24° C would be :

   a) 6120 lb.
   b) 6000 lb.
   c) 5740 lb.
   d) take off would not be permissible.

5. An aircraft has a gradient of climb of 7% after take off. There is an obstacle having a height of 320 ft. at a distance of 5400 ft. from the screen. The clearance of the obstacle would be :

   a) 58 ft.
   b) 93 ft.
   c) 108 ft.
   d) 378 ft.

PERFORMANCE   SINGLE ENGINE CLASS B: CLIMB

6. When assessing the en-route performance, it may not be assumed that the aircraft is flying at an altitude exceeding :

    a)   the altitude at which the rate of climb is 300 ft./min. with maximum continuous power.
    b)   the absolute ceiling with maximum continuous power.
    c)   the altitude at which the rate of climb is 300 ft./min. with maximum take-off power.
    d)   the altitude at which a gradient of 0.5% is achieved with maximum continuous power.

7. A single engined aircraft has a Lift : Drag ratio of 12 : 1 at the scheduled glide speed. The net gradient of descent to be assumed for the purpose of assessing the en-route performance after engine failure is :

    a)   7.8%
    b)   8.3%
    c)   8.8%
    d)   12%

8. An aircraft has a descent gradient of 4.8% with the engine inoperative. For the purpose of compliance with the en-route requirements, the distance travelled during a height loss of 2000 ft. would be :

    a)   6.2 n.m.
    b)   6.8 n.m.
    c)   7.65 n.m.
    d)   96 n.m.

9. An aircraft glides for a distance of 25 n.m. in still air at a speed of 75 knots I.A.S. which is a True air speed of 85 knots in the prevailing conditions. In a headwind of 15 knots it would glide for :

    a)   20 n.m.
    b)   20.6 n.m.
    c)   23 n.m.
    d)   31 n.m.

PERFORMANCE SINGLE ENGINE CLASS B: CLIMB

Appendix C

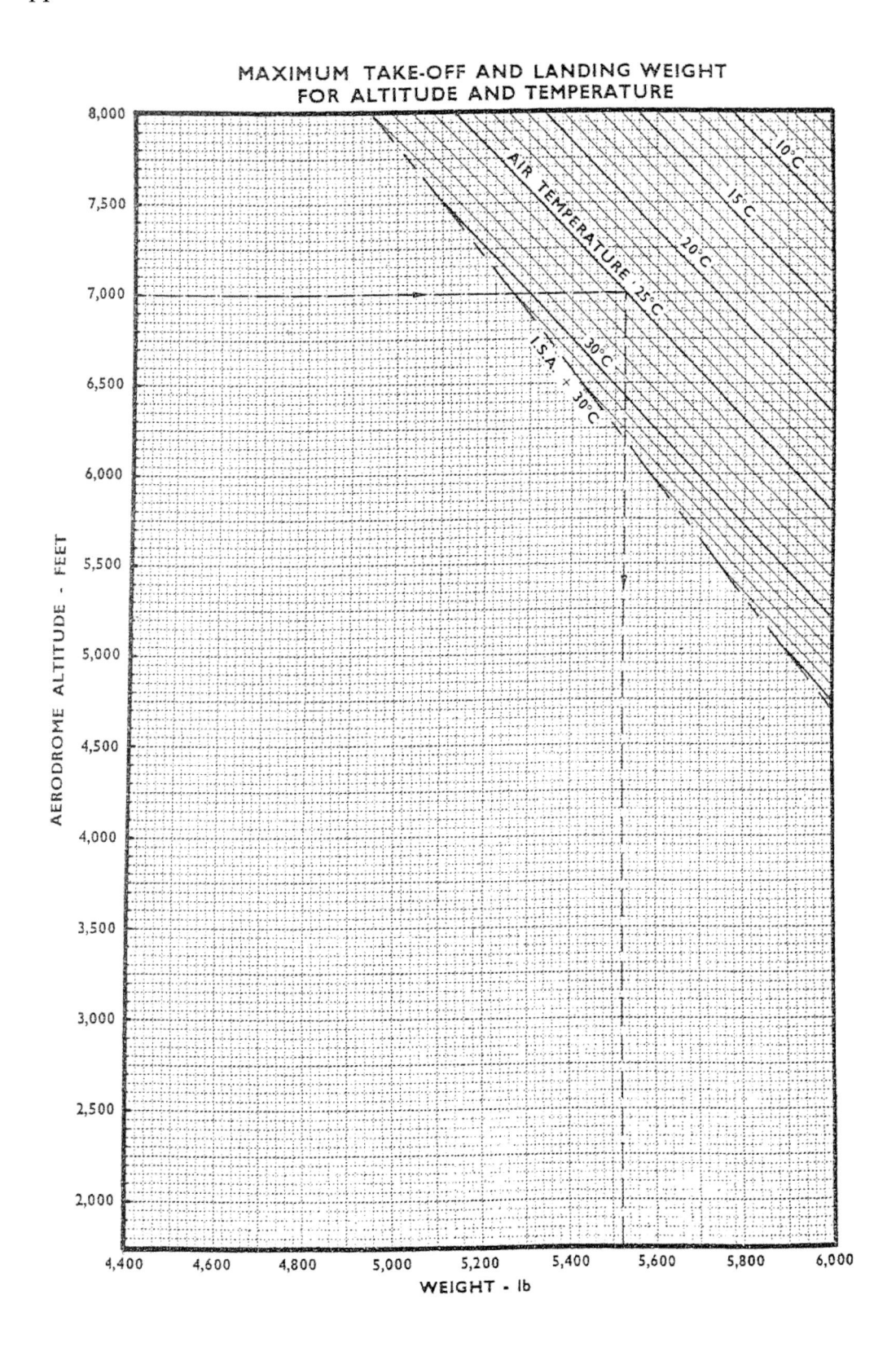

MAXIMUM TAKE-OFF AND LANDING WEIGHT FOR ALTITUDE AND TEMPERATURE

PERFORMANCE **SINGLE ENGINE CLASS B: CLIMB**

ANSWERS

Questions	Answers
1	C
2	C
3	D
4	B
5	C
6	A
7	C
8	A
9	B

CHAPTER EIGHT - SINGLE ENGINE CLASS B: CRUISE

Contents

		Page
8.1	CRUISE PERFORMANCE	8 - 1
8.2	MAXIMUM SPEED ACHIEVABLE	8 - 1
8.3	RANGE	8 - 2
8.4	ENDURANCE	8 - 3
8.5	DESCENT	8 - 3
	SINGLE ENGINE CLASS B AIRCRAFT: CRUISE	8 - 5

© Oxford Aviation Services Limited

© Oxford Aviation Services Limited

PERFORMANCE   SINGLE ENGINE CLASS B: CRUISE

8.1   CRUISE PERFORMANCE

Data is usually given in the Aircraft Manual to permit the determination of:

a)   the maximum speed achievable

b)   the range

c)   the endurance

8.2   MAXIMUM SPEED ACHIEVABLE

The speed achieved will depend on the power setting and rpm selected, and the altitude flown. Figure 8.1 shows a typical presentation of achievable speeds.

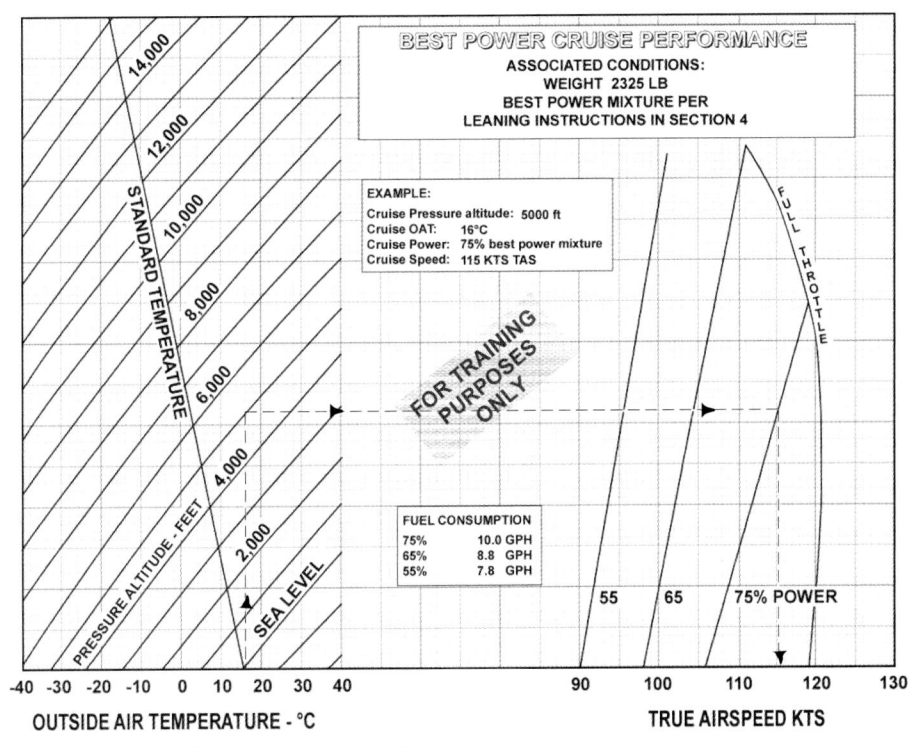

Figure 8.1 Best Power Cruise Performance

The example shows that at 5000 ft. pressure altitude, OAT 16°C, with 75% power, the speed achieved would be 115 kt. TAS.

At 11000 ft. ISA, the maximum achievable speed would be 116 kt TAS with full throttle. The altitude at which the maximum possible speed could be achieved with full throttle in ISA conditions is 6000 ft.

PERFORMANCE
SINGLE ENGINE CLASS B: CRUISE

All these figures relate to a weight of 2325 lb. At lower weights the speeds would be higher.

It can be seen there is an optimum altitude at which the maximum True Speed is achieved, and for a supercharged engine this will usually occur at the full throttle height.

8.3 RANGE

The maximum range for a propeller engined aeroplane is obtained by flying at V_{imd}, but the recommended range speed is usually faster, to give better speed stability and better sector times. Figure 8.2 shows a typical presentation of range data.

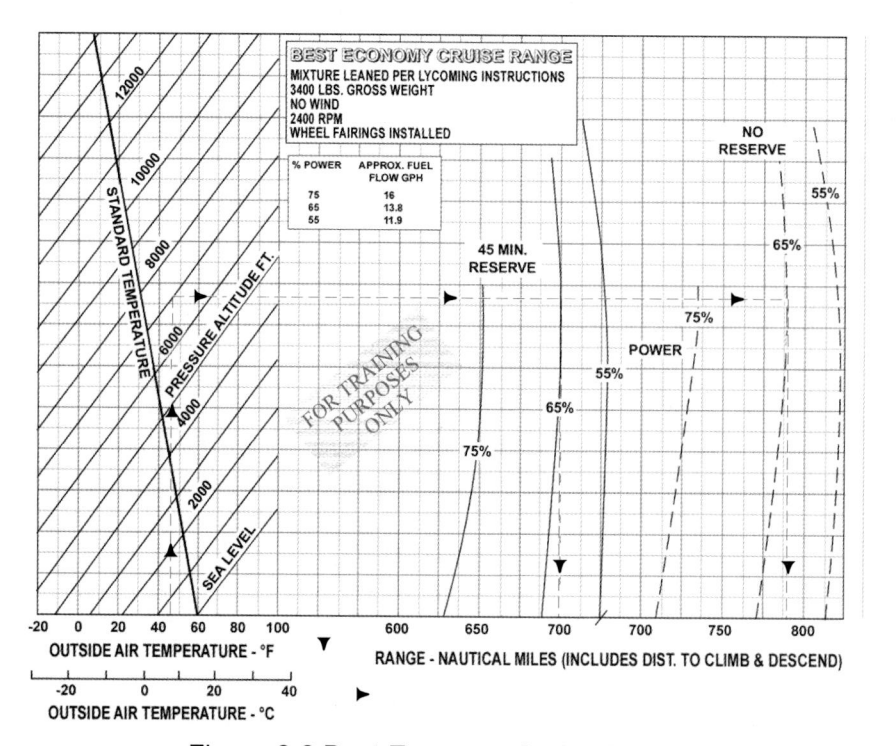

Figure 8.2 Best Economy Cruise Range

The example shows that at 7000 ft. OAT 45°F (7°C) weight 3400 lb. the range with 45 minutes reserves at 65% power would be 700 nm.

The pressure altitude at which the maximum range would be achieved in ISA conditions with 55% power would be 6200 ft. and this would give a range of 728nm with 45 minutes reserve fuel.

The maximum range will be obtained by flying at a particular altitude, depending on the power setting used.

PERFORMANCE   SINGLE ENGINE CLASS B: CRUISE

8.4 ENDURANCE

The maximum endurance for a propeller engined aircraft is obtained by flying at V_{imp}, but again, this speed is not suitable practically because of poor speed stability. Figure 8.3 shows typical endurance data.

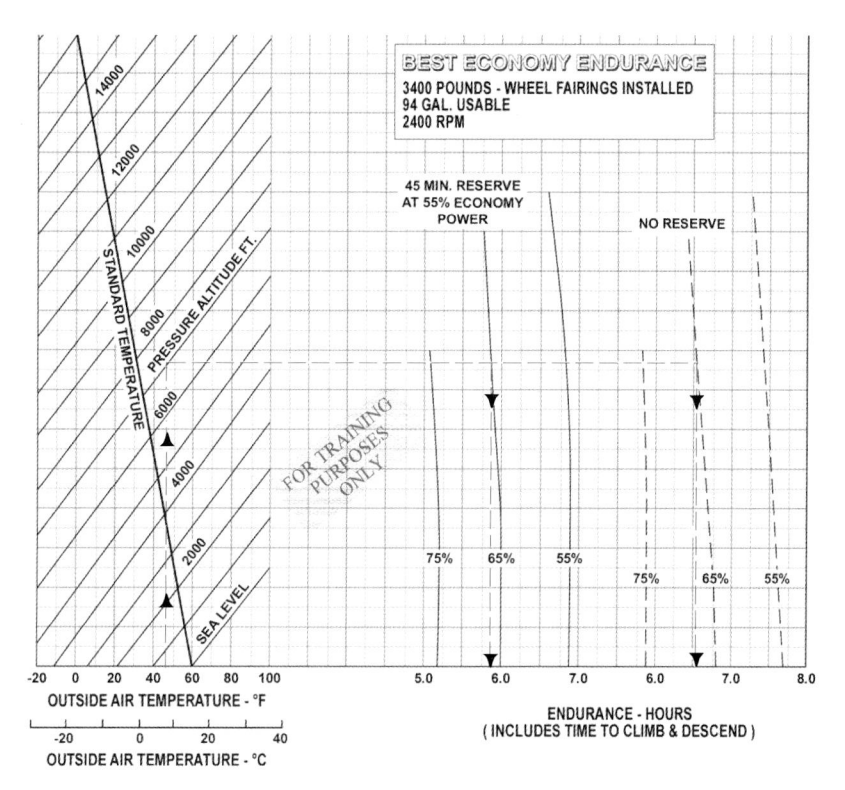

Figure 8.3

The example shows that the endurance with 94 usable gallons of fuel, leaving no reserves, at a pressure altitude of 7000 ft. 45°F at a weight of 3400 lb. would be 6.6 hours, using 65% power. Lower weights and power settings would increase the endurance.

8.5 DESCENT

The performance manual for single engined Class B aircraft does not normally present descent data. The distance covered in a power-off descent (glide) depends on the lift:drag ratio

$$\text{Distance in glide} = \text{height decrement} \times \frac{\text{lift}}{\text{drag}}$$

This distance will be greater if the engine is producing thrust, and the distance travelled over the ground will depend on the wind speed.

PERFORMANCE SINGLE ENGINE CLASS B: CRUISE

PAPER 1

Refer to Appendix E for questions 1 and 2

1. For an aircraft cruising at a pressure altitude of 6000 ft. I.S.A. at a weight of 3400 lb. with the r.p.m. set at 2100 rpm. the True Air Speed would be :

   a) 144 kt. with 21" hg. MAP or 159.5 kt. with 25" hg MAP
   b) 148 kt. with 21" hg. MAP or 176 kt. with 25" hg. MAP
   c) 153.5 kt. with 21" hg. MAP or 161 kt. with 25" hg. MAP
   d) 148 kt .with 21" hg. MAP or 159.5 kt. with 25" hg. MAP

2. To obtain the maximum True Air Speed with 2300 rpm and 23" hg. MAP the aircraft should be flown at a pressure altitude of :

   a) 7300 ft. and the speed would be 174.5 kt.
   b) 5000 ft. and the speed would be 161 kt.
   c) 4900 ft. and the speed would be 177 kt.
   d) 7300 ft. and the speed would be 167.5 kt.

3. **Refer to Appendix F**
   The still air range with 45 minutes reserve fuel allowance, with 2300 rpm and 23" hg. MAP at a pressure altitude of 5000 ft. I.S.A. would be :

   a) 830 n.m.
   b) 840 n.m.
   c) 845 n.m.
   d) 907 n.m.

4. **Refer to Appendix G**
   The endurance at full throttle height with 2300 rpm allowing 45 mins. reserve fuel would be :

   a) 5.16 hours
   b) 5.91 hours
   c) 6.0 hours
   d) 6.25 hours

5. At what distance from the destination should the descent begin for an aircraft cruising at 5000 ft. if the Lift : Drag ratio is 12.5 at the descent speed of 100 kt. TAS and there is a headwind component of 20 kt.

   a) 6.25 n.m.
   b) 8.2 n.m.
   c) 10.3 n.m.
   d) 12.8 n.m.

PERFORMANCE   **SINGLE ENGINE CLASS B: CRUISE**

Appendix E

CRUISE SPEEDS

20°C RICH OF PEAK EGT

ASSOCIATED CONDITIONS:
- AVERAGE CRUISE WT .... 3400 LBS
- TEMPERATURE .... STD DAY (ISA)

EXAMPLE:
- CRUISE ALTITUDE .... 11,500 FT
- POWER SETTING .... FULL THROTTLE, 2500 RPM
- TRUE AIRSPEED .... 168 KTS

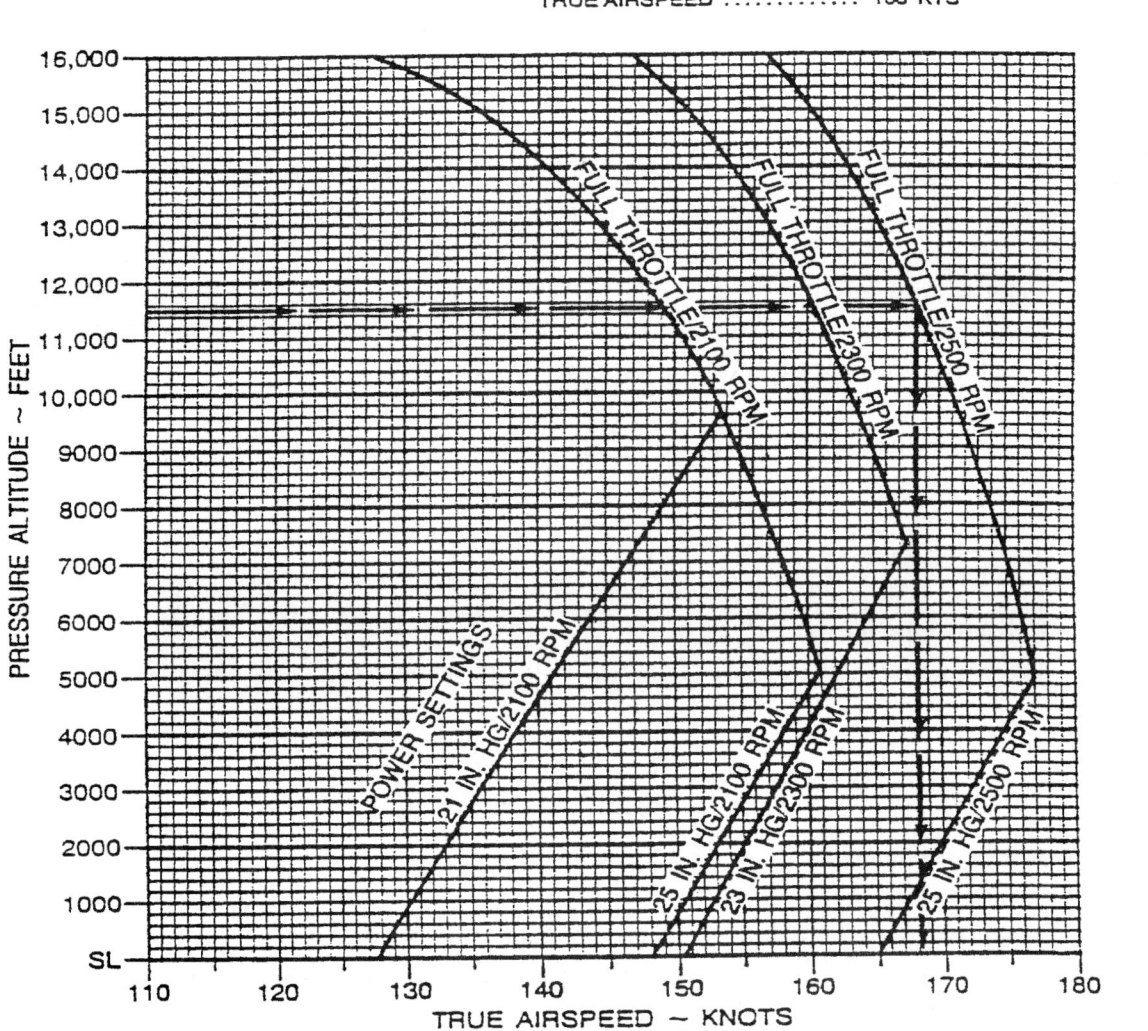

Power settings shown: 21 IN. HG/2100 RPM, 23 IN. HG/2300 RPM, 25 IN. HG/2500 RPM, FULL THROTTLE 2100 RPM, FULL THROTTLE 2300 RPM, FULL THROTTLE 2500 RPM.

X-axis: TRUE AIRSPEED ~ KNOTS (110 to 180)
Y-axis: PRESSURE ALTITUDE ~ FEET (SL to 16,000)

PERFORMANCE SINGLE ENGINE CLASS B: CRUISE

Appendix F

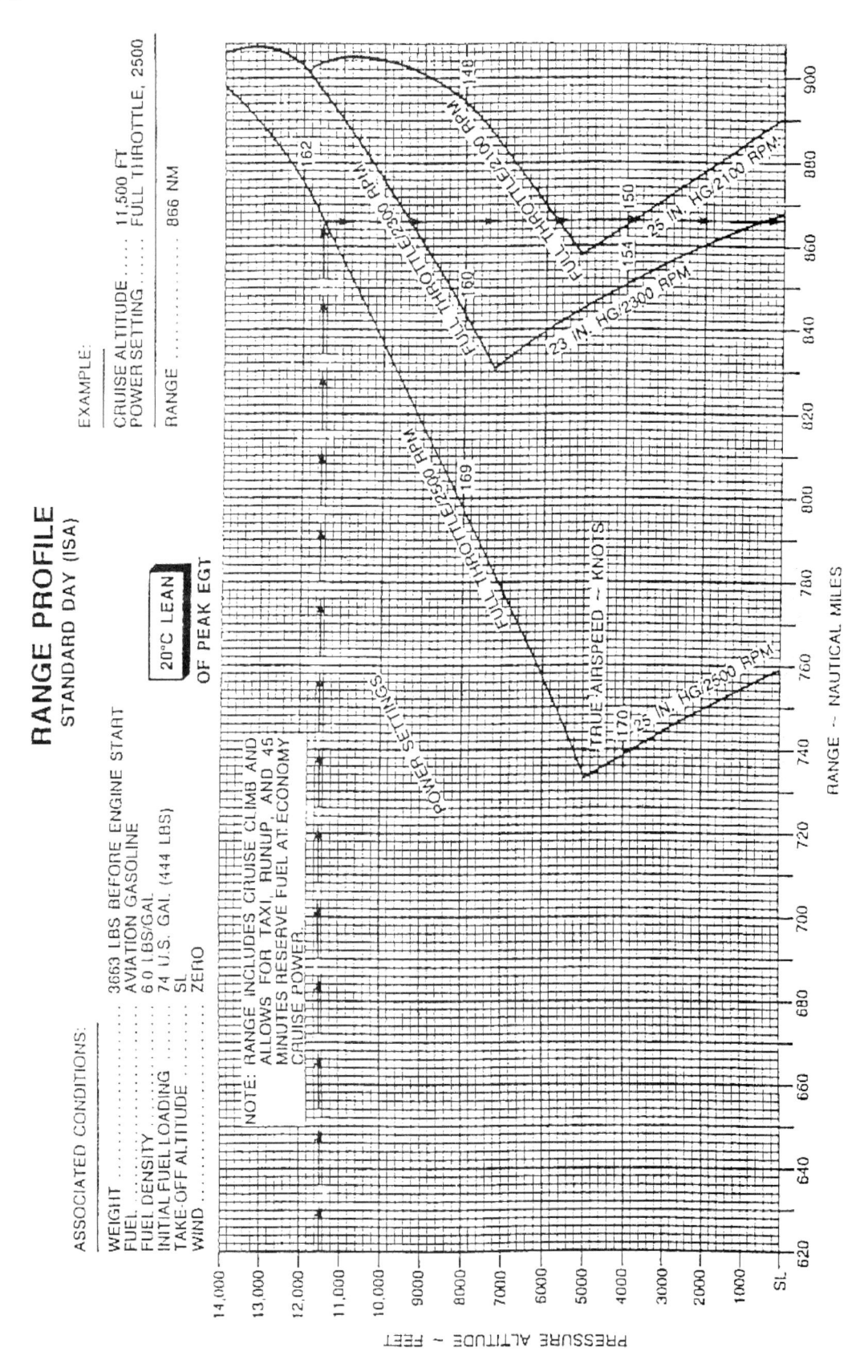

RANGE PROFILE
STANDARD DAY (ISA)

20°C LEAN OF PEAK EGT

ASSOCIATED CONDITIONS:

WEIGHT	3663 LBS BEFORE ENGINE START
FUEL	AVIATION GASOLINE
FUEL DENSITY	6.0 LBS/GAL
INITIAL FUEL LOADING	74 U.S. GAL. (444 LBS)
TAKE-OFF ALTITUDE	SL
WIND	ZERO

NOTE: RANGE INCLUDES CRUISE CLIMB AND ALLOWS FOR TAXI, RUNUP, AND 45 MINUTES RESERVE FUEL AT ECONOMY CRUISE POWER.

EXAMPLE:

CRUISE ALTITUDE	11,500 FT
POWER SETTING	FULL THROTTLE, 2500
RANGE	866 NM

8 - 7 © Oxford Aviation Services Limited

PERFORMANCE SINGLE ENGINE CLASS B: CRUISE

Appendix G

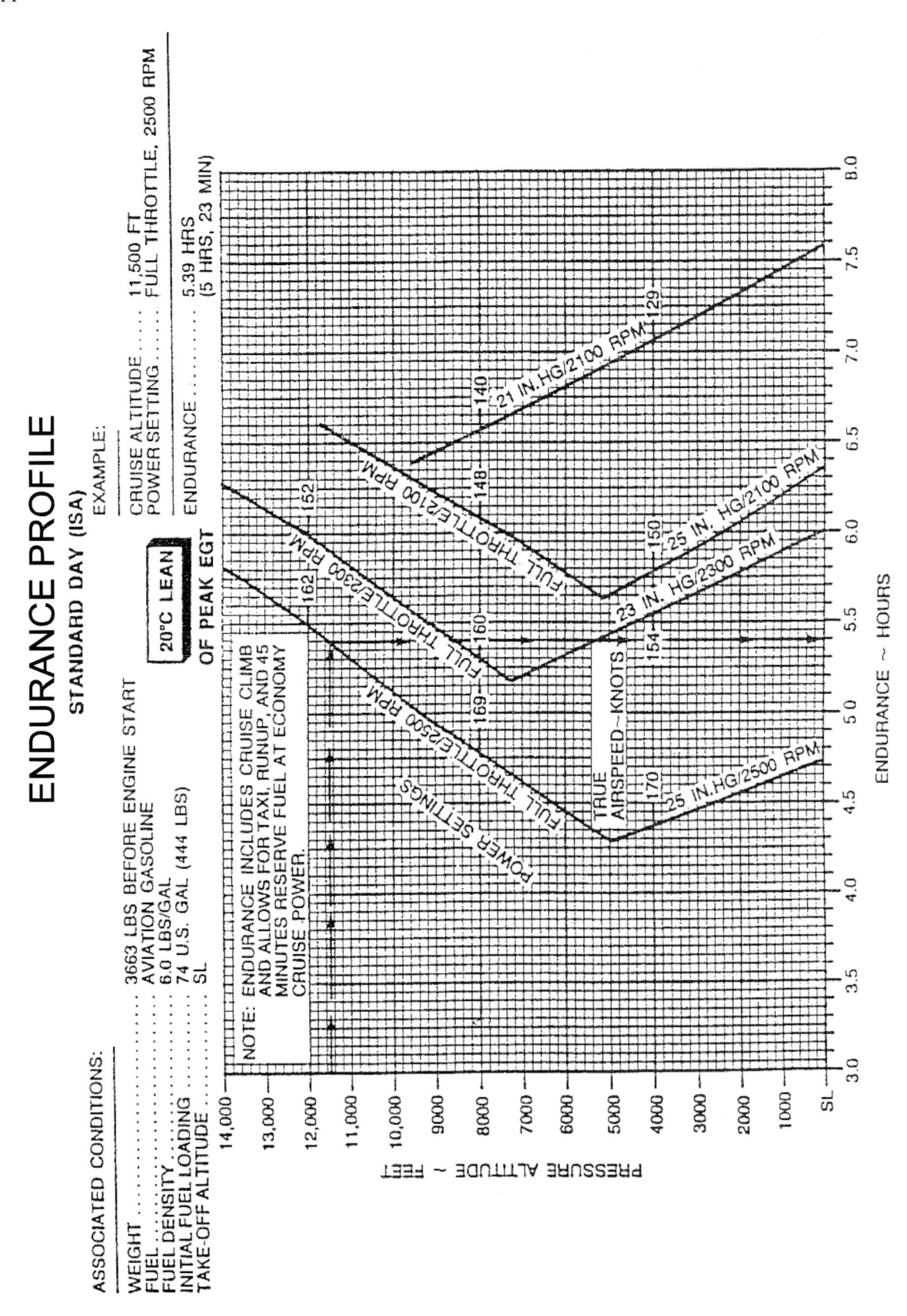

ENDURANCE PROFILE
STANDARD DAY (ISA)

EXAMPLE:
CRUISE ALTITUDE ...... 11,500 FT
POWER SETTING .... FULL THROTTLE, 2500 RPM
ENDURANCE .... 5.39 HRS
(5 HRS, 23 MIN)

20°C LEAN OF PEAK EGT

ASSOCIATED CONDITIONS:
WEIGHT .... 3663 LBS BEFORE ENGINE START
FUEL .... AVIATION GASOLINE
FUEL DENSITY .... 6.0 LBS/GAL
INITIAL FUEL LOADING . 74 U.S. GAL (444 LBS)
TAKE-OFF ALTITUDE .... SL

NOTE: ENDURANCE INCLUDES CRUISE CLIMB AND ALLOWS FOR TAXI, RUNUP, AND 45 MINUTES RESERVE FUEL AT ECONOMY CRUISE POWER.

8 - 8 © Oxford Aviation Services Limited

PERFORMANCE — SINGLE ENGINE CLASS B: CRUISE

ANSWERS

Questions	Answers
1	A
2	D
3	C
4	A
5	B

CHAPTER NINE - SINGLE ENGINE CLASS B: LANDING

Contents

		Page
9.1	LANDING	9 - 1
9.2	LANDING DISTANCE REQUIRED - DRY RUNWAY (JAR-OPS 1.550)	9 - 1
9.3	LANDING WET AND CONTAMINATED RUNWAYS	9 - 2
9.4	PRESENTATION OF DATA	9 - 3
9.5	LANDING CLIMB REQUIREMENT	9 - 3
	SINGLE ENGINE CLASS B AIRCRAFT: LANDING	9 - 5

© Oxford Aviation Services Limited

PERFORMANCE   **SINGLE ENGINE CLASS B: LANDING**

9.1 LANDING

The requirements for landing ensure that:

a) the landing distance available is not exceeded

b) the climb performance is adequate in the event of a go-around

9.2 LANDING DISTANCE REQUIRED - DRY RUNWAY (JAR-OPS 1.550)

The aeroplane must land from 50 feet above the threshold to a full stop within 70% of the landing distance available at the destination and at any alternate aerodrome.

The speed at 50 ft. (V_{REF}) must not be less than 1.3 V_{S0}.

Steep Approach Procedures JAR-OPS 1.550 Appendix

The Authority may approve the application of Steep Approach procedures using glide slope angles of 4.5° or more, and with screen heights of less than 50 feet but not less than 35 feet.

To determine the landing distance required, account must be taken of:

a) the altitude of the aerodrome

b) the standard day temperature

c) not more than 50% of the headwind component and not less than 150% of the tailwind component.

d) the runway surface condition and the type of runway surface. For grass up to 20cm long on firm soil, the distance should be increased by a factor of 1.15 (AMC-OPS 1.550 b3)

e) the runway slope in the direction of landing. The Aircraft Performance manual may not give a correction for slope. In this case, increase the landing distance required by 5% for each 1% of downhill slope (up to a max. of 2%)
Do not correct for uphill slope.

f) wind ... for despatching an aeroplane it must be assumed that:

    i) The aeroplane will land on the most favourable runway in still air.

    ii) The aeroplane will land on the runway most likely to be used considering the probable wind speed and direction, or other requirements.

The lesser mass given by these two considerations should be selected. If it not possible to meet the forecast wind requirement, the aeroplane may be despatched if an alternate aerodrome is designated which permits full compliance with the above requirements.

9.3   LANDING. WET AND CONTAMINATED RUNWAYS.

If the runway is expected to be wet at the time of arrival the landing distance must be increased by a factor of 1.15.

If the Aeroplane Manual gives additional information on landing on wet runways, this may be used even if it gives a lesser distance than that from the above paragraph.

If the runway is expected to be contaminated at the time of arrival, the landing distance required, determined by data acceptable to the Authority, must not exceed the landing distance available.

PERFORMANCE   SINGLE ENGINE CLASS B: LANDING

9.4 PRESENTATION OF DATA

The landing distance required is usually presented graphically as illustrated in Fig. 9.1.

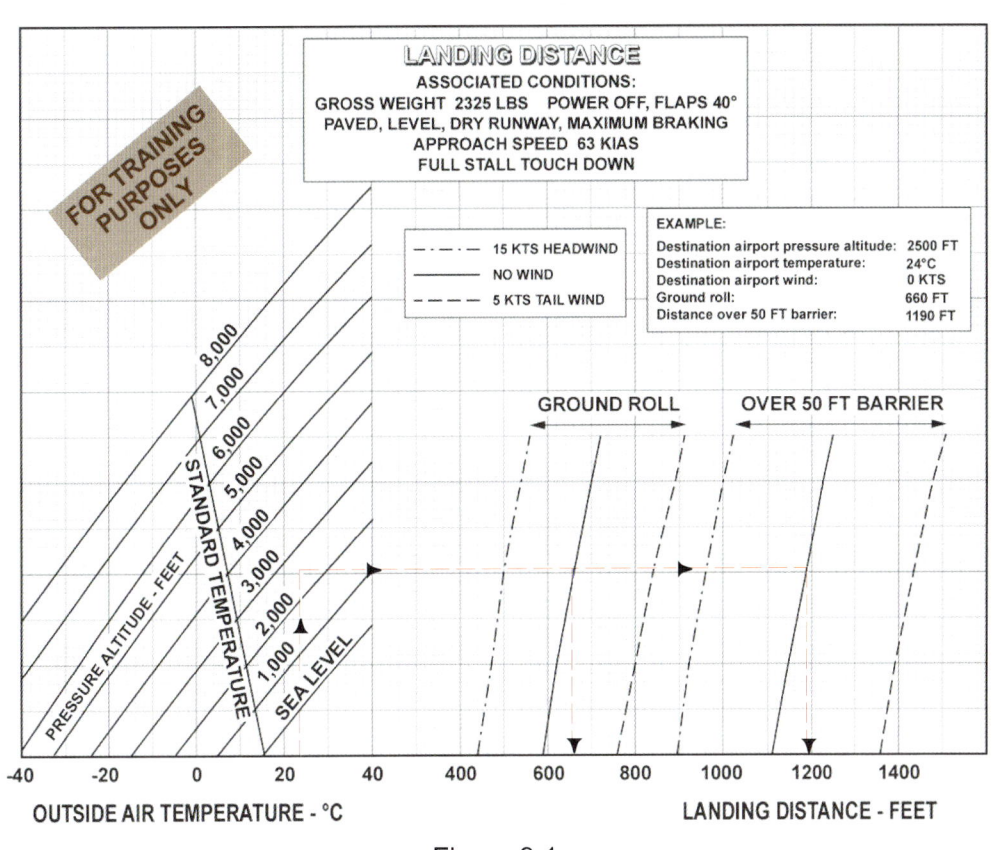

Figure 9.1

The example shows that at an aerodrome pressure altitude of 2500 ft. OAT 24°C the landing distance from 50 ft. in still air at a weight of 2325 lb. would be 1190 ft.

NOTE:
1. For planning purposes the standard day temperature would be used.
2. Both still air and forecast wind would need to be considered.
3. The wind factors have been applied to the data shown.

9.5 LANDING CLIMB REQUIREMENT

The climb gradient required in the landing configuration was considered in Chapter 7, and will determine a maximum landing mass for altitude and temperature.

The required gradient is 2.5% in the landing configuration.

PERFORMANCE SINGLE ENGINE CLASS B: LANDING

PAPER 1

1. Given that the control requirement is met, the speed at the 50 ft. screen must not be less than :

    a) 20% above the stalling speed with landing flap
    b) 30% above the stalling speed with flaps up
    c) 30% above the stalling speed with landing flap
    d) 50% above the stalling speed with flaps up.

2. The landing distance required must not exceed :

    a) 70% of the landing distance available at a destination aerodrome, and 60% of the landing distance available at an alternate aerodrome.
    b) 60% of the landing distance available at a destination aerodrome, and 70% of the landing distance available at an alternate aerodrome.
    c) 60% of the landing distance available at destination and alternate aerodromes.
    d) 70% of the landing distance available at destination and alternate aerodromes.

3. To determine the landing distance required

    a) the reported wind components do not have to be factored.
    b) the reported wind components must be factored by 1.5
    c) the reported headwind component must be factored by 0.5 and the reported tailwind component by 1.5
    d) the reported headwind component must be factored by 1.5 and the reported tailwind component by 0.5

4. If the runway is wet, the landing distance required for a dry runway :

    a) may be used unfactored
    b) must be increased by 50%
    c) must be increased by 15%
    d) must be increased by 5%

5. The following landing weights have been determined for an aerodrome with two runways 'A' and 'B'

	Still Air Weight	Weight with forecast wind
Runway A	4000 lb.	4100 lb.
Runway B	3900 lb.	4200 lb.

    The maximum permissible planned landing weight would be :

    a) 3900 lb.
    b) 4000 lb.
    c) 4100 lb.
    d) 4200 lb.

PERFORMANCE — SINGLE ENGINE CLASS B: LANDING

Refer to CAP 698 Figure 2.4 for questions 6, 7, and 8.

6. Given the following data for a destination aerodrome :

   Pressure altitude   2000 ft.
   Temperature   30° C
   Landing Weight   3400 lb.
   Reported wind comp.   0

   What landing distance must be available to comply with the landing regulations :

   a)   1050 ft.
   b)   1500 ft.
   c)   2143 ft.
   d)   2286 ft.

7. Determine the gross landing distance required for the following conditions :

   Pressure altitude   4000 ft.
   Temperature   I.S.A.
   Landing Weight   3600 lb.
   Reported wind comp.   0
   Runway slope   1.0% downhill

   a)   1680 ft.
   b)   1600 ft.
   c)   1050 ft.
   d)   1100 ft.

8. If the landing distance required from 50 ft. for the prevailing conditions is 900 ft., the ground roll would be :

   a)   500 ft.
   b)   600 ft.
   c)   750 ft.
   d)   800 ft.

9. Which of the following statements is true :

   a)   The landing distance required is not affected by temperature.
   b)   If the runway has an uphill slope of 1.0%, the landing distance required should be decreased by 5%
   c)   For planning purposes the landing distance required should be based on the I.S.A. temperature appropriate to the pressure altitude.
   d)   The landing distance required is not affected by weight as the effect of the increased speed is cancelled by the increased braking effect.

PERFORMANCE SINGLE ENGINE CLASS B: LANDING

10. For a grass runway, the landing distance required for a paved runway should be :

 a) decreased by 5%
 b) increased by 5%
 c) increased by 15%
 d) increased by 20%

PERFORMANCE SINGLE ENGINE CLASS B: LANDING

ANSWERS

Questions	Answers
1	C
2	D
3	C
4	C
5	B
6	C
7	A
8	A
9	C
10	C

CHAPTER TEN - MULTI-ENGINED CLASS B: TAKE OFF

Contents

		Page
10.1	TAKE-OFF	10 - 1
10.2	ACCELERATE - STOP DISTANCE	10 - 1
10.3	MAXIMUM MASS FOR TAKE OFF	10 - 1
10.4	GRADIENT REQUIREMENT	10 - 1
10.5	TAKE-OFF DISTANCE REQUIREMENTS	10 - 2
10.6	ACCELERATE STOP DISTANCE REQUIREMENTS	10 - 4
10.7	OBSTACLE CLEARANCE REQUIREMENTS (JAR-OPS 1.535)	10 - 5
10.8	TAKE OFF FLIGHT PATH	10 - 6
10.9	CONSTRUCTION OF THE FLIGHT PATH	10 - 6
	MULTI-ENGINED CLASS B AIRCRAFT: TAKE OFF	10 - 9

© Oxford Aviation Services Limited

10.1 TAKE-OFF

The take-off requirements for multi-engined Class B aircraft (other than those in the commuter category) are the same as for single engined aircraft with the addition of requirements for clearance of obstacles after take-off.

Commuter aircraft are propeller driven twin engined aircraft that have a seating configuration, excluding the pilot seat(s), of 19 or fewer and a maximum certificated take off weight of 8618 kg (19000 lb) or less (JAR 23.1)

The requirements for commuter category aircraft are more similar to those for Class A aircraft, requiring consideration of engine failure on take-off and accelerate-stop performance.

10.2 ACCELERATE - STOP DISTANCE

Although there are no requirements for the accelerate-stop distance required, (for non-commuter aircraft) the Aeroplane Flight Manual may present data for this distance.

10.3 MAXIMUM MASS FOR TAKE OFF

To determine the maximum mass for take-off, it is necessary to consider:

a) the mass for altitude and temperature (gradient requirement)

b) the take-off distance requirement

c) the obstacle clearance requirement

d) the structural mass limit

10.4 GRADIENT REQUIREMENT   JAR-OPS 1.530

Three climb gradient requirements must be considered, and the most limiting will determine the maximum permissible mass.

a) **All engines operating**

   A climb gradient of 4% is required with:

   i) take-off power on each engine

   ii) landing gear extended, except that if the landing gear can be retracted in not more than 7 seconds, it may be assumed to be retracted

   iii) the wing flaps in the take-off position

   iv) a climb speed of not less than the greater of $1.1V_{MC}$ and $1.2V_{S1}$

b) **One engine inoperative**

   i) The climb gradient at an altitude of 400ft above the take-off surface must be measurably positive with:

   1. the critical engine inoperative and its propeller in the minimum drag position.

   2. the remaining engine at take-off power

   3. the landing gear retracted

   4. the wing flaps in the take-off position

   5. a climb speed equal to that achieved at 50 ft

   and

   ii) The climb gradient must not be less than 0.75% at an altitude of 1500 ft above the take-off surface, with:

   1. the critical engine inoperative and its propeller in the minimum drag position

   2. the remaining engine at not more than maximum continuous power

   3. the landing gear retracted

   4. the wing flaps retracted

   5. a climb speed not less than $1.2 V_{SI}$

10.5 TAKE-OFF DISTANCE REQUIREMENTS

Other than for commuter category aircraft, these are the same as for the single engined Class B aircraft.

The gross take-off distance required is the distance from the start of take-off to a point 50 ft. above the take-off surface, with take-off power on each engine, rotating at V_R and achieving the specified speed at the screen.

PERFORMANCE — MULTI-ENGINED CLASS B AIRCRAFT: TAKE-OFF

V_R The rotation speed, must not be less than :

a)   1.05 V_{MC}

b)   1.1 V_{S1}

The speed at 50 ft. must not be less than :

a)   a speed that is safe under all reasonably expected conditions

b)   1.1 V_{MC}

c)   1.2 V_{S1}

V_{MC} The minimum control speed, is the calibrated airspeed at which, when the critical engine is suddenly made inoperative, it is possible to maintain control of the aeroplane, and thereafter maintain straight flight at the same speed with an angle of bank of not more than 5°

V_{MC} for take-off must not exceed 1.2 V_{S1}

For all aeroplanes except piston engined aeroplanes of 2730 kg. or less V_{MC} must also be determined for the landing configuration.

Take-off distance factors

These are the same as for the single engined Class B aircraft.

To summarise:

If no stopway or clearway
Gross TOD x 1.25 ≯ TORA

If stopway or clearway is available

Gross TOD ≯ TORA
Gross TOD x 1.15 ≯ TODA
Gross TOD x 1.3 ≯ ASDA

The factors for surface condition and runway slope are the same as for the single engined Class B aircraft.

Figure 10.1 shows a typical presentation of Take-off distance. The use of the graph is illustrated by the broken arrowed line.

PERFORMANCE   MULTI-ENGINED CLASS B AIRCRAFT: TAKE-OFF

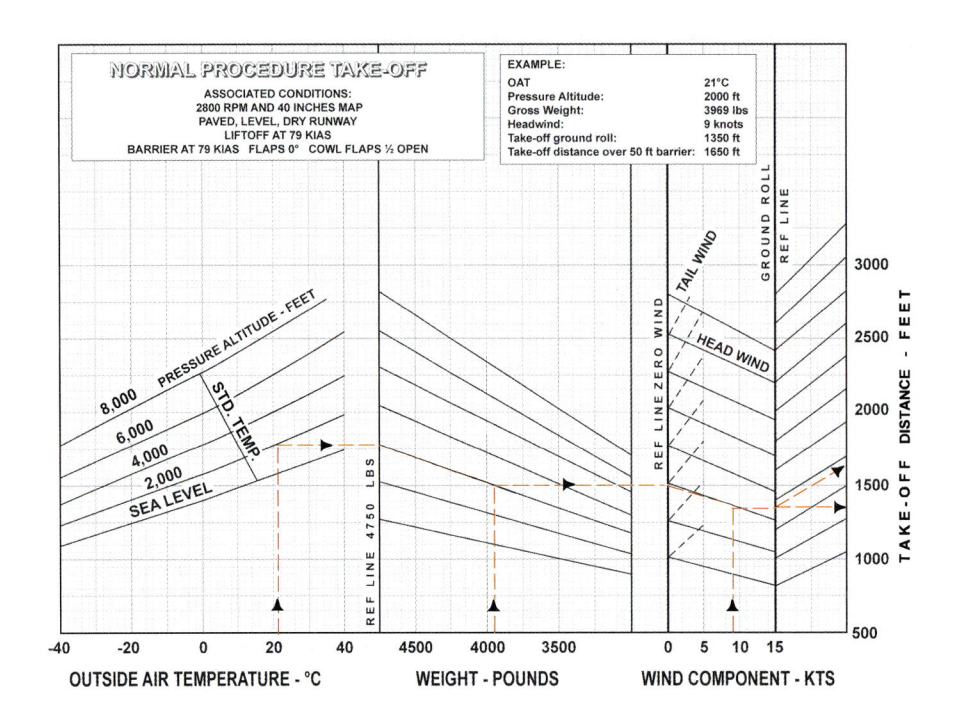

Figure 10.1 Normal Procedure Take-off

10.6   ACCELERATE-STOP DISTANCE REQUIREMENTS.

Other than for commuter category aircraft, there is no requirement for accelerate-stop distance, but the data may be given. Figure 10.2 shows a typical presentation of accelerate-stop distance.

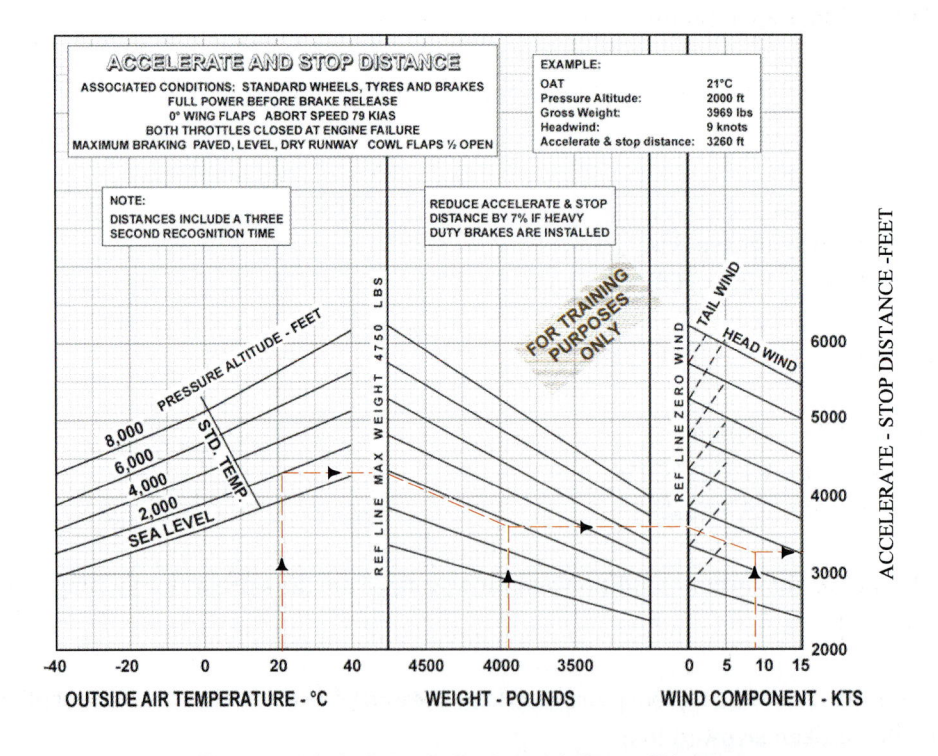

Figure 10.2 Accelerate and Stop Distance

PERFORMANCE   **MULTI-ENGINED CLASS B AIRCRAFT: TAKE-OFF**

For commuter category aircraft, the accelerate-stop distance is the sum of the distances necessary to :

a)   Accelerate the aircraft to V_{EF} with all engines operating.

b)   Accelerate from V_{EF} to V_1 assuming the critical engine fails at V_{EF}

c)   Come to a full stop from the point at which V_1 is reached.

(JAR 23.55)

10.7   OBSTACLE CLEARANCE REQUIREMENTS (JAR-OPS 1.535)

Multi-engined Class B aircraft must demonstrate clearance of obstacles after take-off up to a height of 1500ft.

Obstacles must be cleared by:

a)   a vertical margin of at least 50ft

or

b)   a horizontal distance of at least 90m + 0.125 D where D is the distance from the end of the TODA, or the end of the TOD if a turn is scheduled before the end of the TODA. For aeroplanes with a wingspan of less that 60m the horizontal distance may be taken as 60m + half the wingspan + 0.125D.

The following conditions must be assumed:

a)   the flight path begins at a height of 50 ft above the surface at the end of the TOD required and ends at a height of 1500 ft above the surface.

b)   the aeroplane is not banked before it has reached the height of 50 ft, and thereafter that the angle of bank does not exceed 15°.

c)   failure of the critical engine occurs at the point on the all engine take-off flight path where visual reference for the purpose of avoiding obstacles is expected to be lost.

d)   the gradient to be assumed from 50 ft to the point of engine failure is equal to the average all engine gradient during climb and transition to the en-route configuration, multiplied by a factor of 0.77

e)   the gradient from the point of engine failure to 1500 ft. is equal to the one engine inoperative en-route gradient.

If the flight path does not require track changes of more then 15°, obstacles do not need to be considered if the lateral distance is greater than 300m if in VMC or 600m for all other conditions. If the flight path requires track changes of more than 15°, obstacles need not be considered if the lateral distance is greater than 600m in VMC or 900m for all other conditions.

10.8 TAKE OFF FLIGHT PATH

The flight path profile performance should take account of:

a) the mass of the aeroplane at the commencement of the take-off run

b) the pressure altitude at the aerodrome

c) the ambient temperature

d) not more than 50% of the reported headwind component and not less than 150% of the reported tailwind component.

10.9 CONSTRUCTION OF THE FLIGHT PATH

The flight path profile will depend on whether or not visual reference is lost before reaching 1500 ft.

a) **Visibility clear to 1500 ft**

i) Determine the TOD required for the take-off mass

ii) Determine the all engines net gradient (gross gradient x 0.77)

iii) Divide the height gain (1450 ft) by the gradient to determine the distance travelled (feet) from 50 ft to 1500 ft.

iv) This profile may be plotted as shown in figure 10.3 and clearance of obstacles assessed.

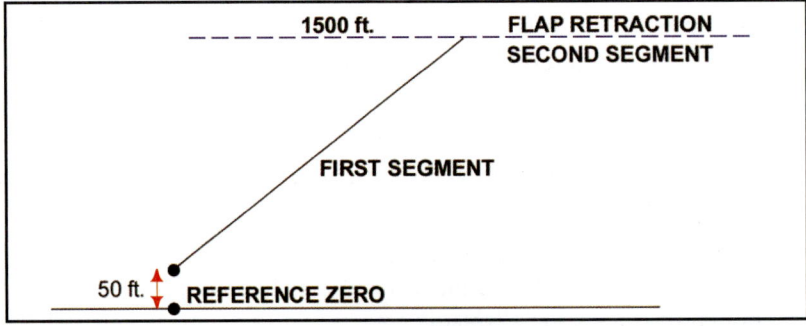

Figure 10.3

PERFORMANCE — MULTI-ENGINED CLASS B AIRCRAFT: TAKE-OFF

Alternatively for a single obstacle, find the TOD req. and gradient as above, then multiply the distance from the TODR to the obstacle by the gradient to find the height gain, and add 50 ft to find the aeroplane height at the obstacle distance. This must exceed the obstacle height by 50 ft.

If the obstacle is not cleared by 50 ft, a lower take-off mass must be assumed and a revised height calculated.

The maximum mass which will just clear the obstacle by 50 ft can then determined by interpolation.

b) **Cloud base below 1500 ft**
If visual reference is lost before 1500 ft, the flight path will consist of two segments.

Segment 1   From 50 ft to cloud base

Distance from 50 ft to cloud base   = $\dfrac{\text{height gain}}{\text{all eng. Net gradient}} \times 100$

where height gain = cloud base - 50 ft

Segment 2.   From cloud base to 1500 ft

Distance = $\dfrac{\text{height gain}}{\text{Gross gradient with one engine inoperative}} \times 100$

The profile may be plotted as shown in Figure 10.4 and clearance of obstacles assessed. If the required clearance is not achieved a reduced take-off mass must be assumed and a second flight path calculated. As before the maximum permissible weight may be determined by interpolation.

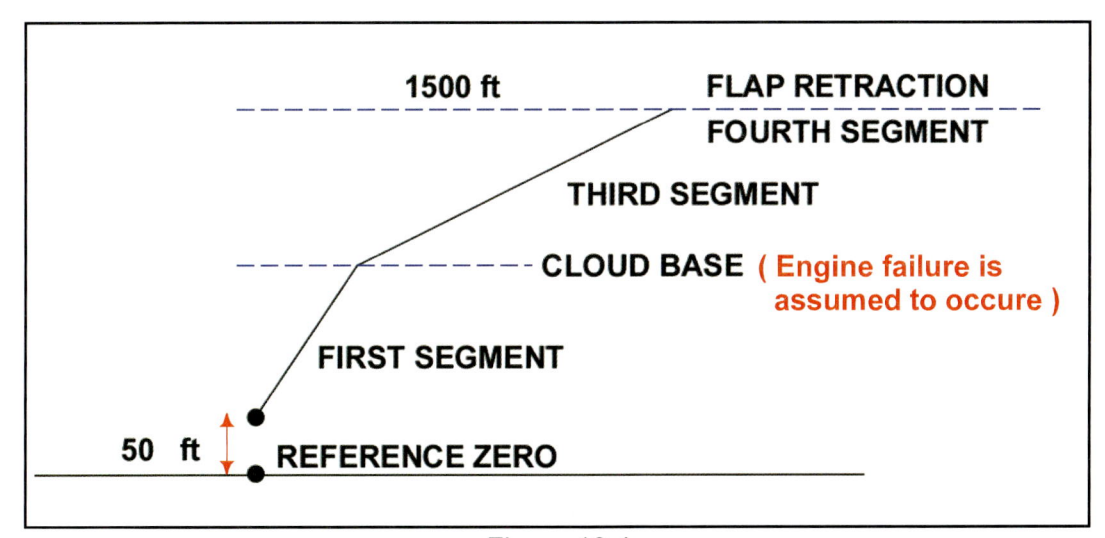

Figure 10.4

PERFORMANCE — MULTI-ENGINED CLASS B AIRCRAFT: TAKE-OFF

If the climb data is given in terms of rate of climb, this can be converted to gradient:

$$\% \text{ Gradient} = \frac{\text{Rate of climb (ft/min)}}{\text{Aircraft true groundspeed}} \times 100$$

and proceed as above.

Alternatively the time on each segment can be calculated:

$$\text{Time (mins)} = \frac{\text{height gain (ft)}}{\text{Rate of climb (ft/min)}}$$

and the distance on each segment obtained from:

$$\text{Distance (ft)} = \text{Aircraft True Ground Speed (ft/min)} \times \text{Time (min)}$$

PERFORMANCE — MULTI-ENGINED CLASS B AIRCRAFT: TAKE-OFF

PAPER 1

1. For a twin engined Class B aircraft the rotation speed V_R must not be less than :

   a) the stalling speed with the flaps in the take off position V_{S1}
   b) $1.05\ V_{MC}$ or $1.1\ V_{S1}$
   c) $1.05\ V_{MC}$ or $1.2\ V_{S1}$
   d) $1.1\ V_{MC}$ or $1.2\ V_{S1}$

2. Select the correct sequence of speeds :

   a) $V_{S1}\ \ V_R\ \ V_{MC}\ \ V_2$
   b) $V_{S1}\ \ V_R\ \ V_2\ \ V_{MC}$
   c) $V_{S1}\ \ V_{MC}\ \ V_R\ \ V_2$
   d) $V_{S1}\ \ V_{MC}\ \ V_2\ \ V_R$

Refer to CAP 698 Figure 3.3 for questions 3, 4 and 5

3. What length of Take off Run must be available to meet the requirements, if there is no stopway or clearway at the aerodrome, given the following conditions :

   Pressure altitude   3000 ft.
   Temperature   5°C
   Weight   4150 lb.
   Reported wind comp.   10 kt. Head

   a) 780 ft.
   b) 975 ft.
   c) 1000 ft.
   d) 1250 ft.

4. Find the maximum permissible weight for take off from an aerodrome with a runway length of 1000 ft. having no stopway or clearway, given :

   Pressure altitude   Sea level
   Temperature   10°C
   Reported wind comp.   5 kt. Head

   a) 4100 lb.
   b) 4600 lb.
   c) 3700 lb.
   d) 4750 lb.

PERFORMANCE — MULTI-ENGINED CLASS B AIRCRAFT: TAKE-OFF

5. Find the maximum permissible weight for take off from an aerodrome with a runway length of 1200 ft., stopway of 200 ft. and clearway of 300 ft. Given :

    Pressure altitude   4000 ft.
    Temperature   15 °C
    Reported wind comp.   10 kt. Headwind

    a) 3400 lb.
    b) 4000 lb.
    c) 4300 lb.
    d) 4750 lb.

Refer to CAP 698 Figure 3.4 for question 6

6. In the event of an engine failure during the take off run, find the gross accelerate-stop distance required if the take off is aborted.   Given :

    Pressure altitude   2000 ft.
    Temperature   18°C
    Weight   4600 lb.
    Reported wind comp.   4 kt. Tail

    a) 2400 ft.
    b) 2850 ft.
    c) 3000 ft.
    d) 3900 ft.

7. For an aircraft having a wing span of 20 m., an obstacle at a distance of 1800 m. from the end of the TODA, would not have to be considered if its distance from the track exceeded :

    a) 225 m.
    b) 233 m.
    c) 285 m.
    d) 295 m.

8. To determine the flight path after take off, if the cloud base at the aerodrome is 1000 ft. it must be assumed that :

    a) an engine fails at 50 ft.
    b) an engine fails at 950 ft.
    c) an engine fails at 1000 ft.
    d) an engine fails at 1500 ft.

PERFORMANCE   **MULTI-ENGINED CLASS B AIRCRAFT: TAKE-OFF**

9. Which of the following statements is true with respect to the take off flight path :

    a) The angle of bank may not be assumed to exceed 15° after the 50 ft. screen
    b) If the change of heading exceeds 15°, obstacles at a lateral distance from the track in IMC of greater than 600 m. need not be considered.
    c) Obstacles must be cleared by at least 35 ft.
    d) Reported winds are not required to be factored.

10. If the net gradient of climb after take off with all engines operating is 14.0% in the prevailing conditions, the horizontal distance required to reach a height of 1500 ft. from the screen, if the cloud base is 2000 ft., would be :

    a) 13914 ft.
    b) 13450 ft
    c) 10714 ft.
    d) 10357 ft.

PERFORMANCE　　　　　　　　**MULTI-ENGINED CLASS B AIRCRAFT: TAKE-OFF**

ANSWERS

Questions	Answers
1	B
2	C
3	D
4	A
5	B
6	C
7	D
8	C
9	A
10	D

CHAPTER ELEVEN - MULTI ENGINED CLASS B: CLIMB AND CRUISE

Contents

		Page
11.1	EN-ROUTE CLIMB REQUIREMENTS	11 - 1
11.2	DRIFT DOWN	11 - 1
11.3	CONSTRUCTION OF THE DRIFT DOWN PROFILE	11 - 1
11.4	CLIMB TO CRUISE ALTITUDE	11 - 2
11.5	ACHIEVABLE SPEED ON THE CRUISE	11 - 2
11.6	RANGE	11 - 4
11.7	ENDURANCE	11 - 4
11.8	DESCENT	11 - 5
	MULTI - ENGINE CLASS B AIRCRAFT: EN-ROUTE	11 - 7

© Oxford Aviation Services Limited

PERFORMANCE — MULTI-ENGINED CLASS B: CLIMB AND CRUISE

11.1 EN-ROUTE CLIMB REQUIREMENTS
JAR-OPS 1.540 STATES:

a) An operator shall ensure that the aeroplane, in the meteorological conditions expected for the flight, and in the event of the failure of one engine, with the remaining engines operating within the maximum continuous power conditions specified, is capable of continuing flight at or above the relevant minimum altitudes for safe flight stated in the Operations Manual to a point 1000 ft above an aerodrome at which the performance requirements for landing can be met.

b) When showing compliance with paragraph (a) above.

   i) The aeroplane must not be assumed to be flying at an altitude exceeding that at which the rate of climb equals 300 ft/min with all engines operating within the maximum continuous power conditions specified; and

   ii) The assumed en-route gradient with one engine inoperative shall be the gross gradient of descent or climb, as appropriate, respectively increased by a gradient of 0.5%, or decreased by a gradient of 0.5%.

Note that the 300 ft/min ceiling is not a restriction on the maximum cruising altitude at which the aeroplane can fly in practice, it is merely the maximum altitude from which the drift down procedure can be planned to start.

11.2 DRIFT DOWN

To demonstrate compliance with the requirement to remain above the minimum altitude for safe flight whilst proceeding to a suitable aerodrome with one engine inoperative, it may be necessary to determine the descent profile of the aeroplane. This is called a drift down.

11.3 CONSTRUCTION OF THE DRIFT DOWN PROFILE

a) Select small decrements in height from the cruising altitude down to the obstacle height.

b) From the aircraft data with one engine inoperative determine the gradient at the mid height of each height interval.

c) Calculate the distance travelled for each height interval from:

$$\text{Distance (n.m.)} = \frac{\text{Height interval (ft)}}{\text{\% gradient}} \times \frac{100}{6080}$$

The intervals chosen should be sufficiently small to give a smooth profile.

© Oxford Aviation Services Limited

PERFORMANCE   **MULTI-ENGINED CLASS B: CLIMB AND CRUISE**

11.4 CLIMB TO CRUISE ALTITUDE

Figure 11.1 shows a typical presentation of the fuel used, time taken and distance travelled during the climb to cruising altitude. The broken arrowed line illustrates the use of the graph.

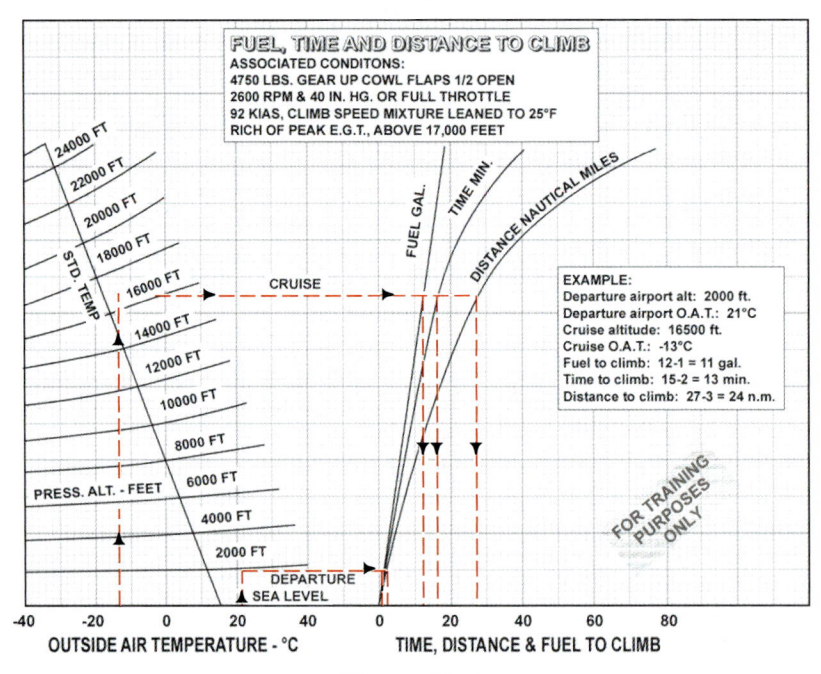

Figure 11.1

11.5 ACHIEVABLE SPEED ON THE CRUISE

Figure 11.2 shows a typical presentation of the TAS achieved with various power settings. The combinations of manifold pressure and rpm which give the required power are shown in Figure 11.3.

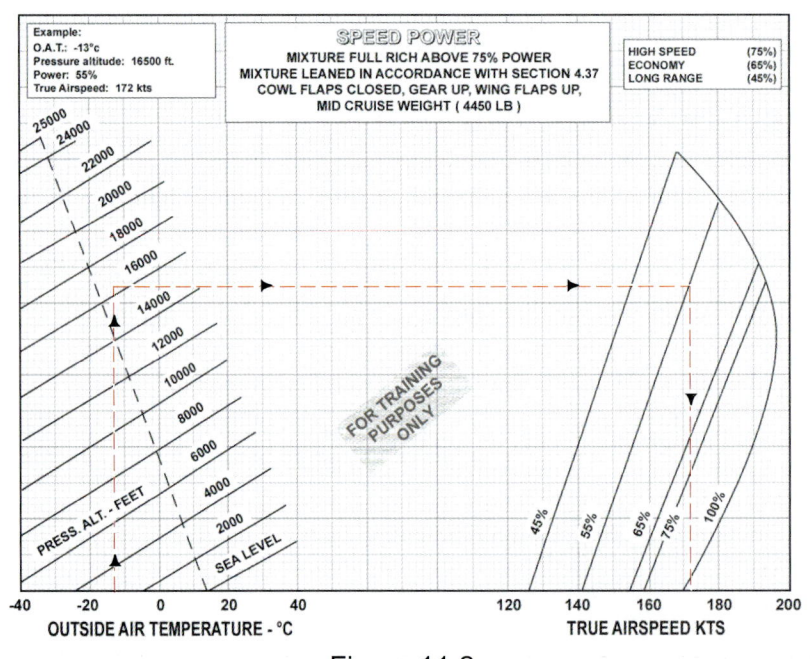

Figure 11.2

FIGURE 11.3 POWER SETTING TABLE - T.C.M. - 360K SERIES PA- 34-22-OT

Press Alt Feet	Std Alt Temp C°	45% Power Approx. Fuel 16 G.P.H. RPM AND MAN. PRESS						55% Power Approx. Fuel 18.7 G.P.H. RPM AND MAN. PRESS.						65% Power Approx. Fuel 23.2 G.P.H. RPM AND MAN PRESS			75% Power Approx. Fuel 29.0 G.P.H. RPM AND MAN PRESS	
		2100	2200	2300	2400	2500	2600	2100	2200	2300	2400	2500	2600	2400	2500	2600	2500	2600
S.L.	15	27.1	26.4	25.5	24.3	23.3	22.5	31.2	30.3	29.4	28.2	27.2	26.3	33.8	32.0	31.0	34.0	33.0
2000	11	26.4	25.8	24.6	23.7	22.8	22.1	30.5	29.7	28.8	27.8	26.8	26.0	33.2	31.7	30.7	33.8	32.7
4000	7	25.8	25.0	24.0	23.2	22.3	21.8	30.0	29.2	28.3	27.4	26.4	25.6	32.8	31.5	30.5	33.6	32.4
6000	3	25.3	24.5	23.5	22.8	21.9	21.5	29.7	28.8	28.0	27.0	26.2	25.3	32.5	31.2	30.3	33.4	32.2
8000	-1	24.8	24.0	23.0	22.4	21.6	21.2	29.4	28.4	27.7	26.8	25.7	25.0	32.3	31.0	30.1	33.1	32.0
10000	-5	24.4	23.7	22.0	22.0	21.4	21.0		28.3	27.5	26.5	25.5	24.7	32.0	30.9	30.0	33.0	31.9
12000	-9	24.0	23.3	22.5	21.7	21.2	20.9		28.3	27.2	26.3	25.3	24.6	31.8	30.7	29.8	32.5	31.8
14000	-13		23.0	22.3	21.4	21.1	20.8			27.1	26.1	25.2	24.4		30.5	29.7		31.7
16000	-17			22.0	21.3	21.0	20.6				25.9	25.0	24.3		30.4	29.5		31.6
18000	-21				21.2	20.9	20.5					25.0	24.2			29.4		
20000	-25				21.2	20.8	20.4						24.2			29.3		
22000	-28						20.4						24.1					
24000	-33						20.4											
25000	-34						20.4											

1650°F MAX E.G.T. (See P.O.H. Section 4)

1525°F MAX E.G.T. (See P.O.H. Section 4)

To maintain constant power add approximately 1% for each 6°C above standard. Subtract approximately 1% for each 6°C below standard. Do not exceed 34" MAP in cruise.

PERFORMANCE MULTI-ENGINED CLASS B: CLIMB AND CRUISE

11.6 RANGE

Figure 11.4 shows a typical presentation of still air range for various power settings. The distances include the climb and descent distances.

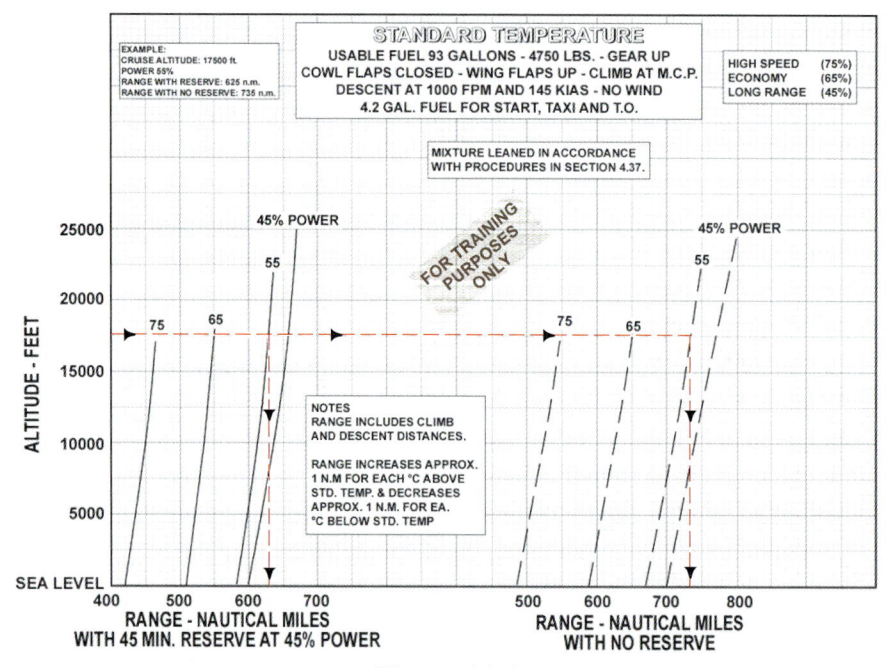

Figure 11.4

11.7 ENDURANCE

Figure 11.5 shows the endurance for various power settings. The times include the climb and descent times.

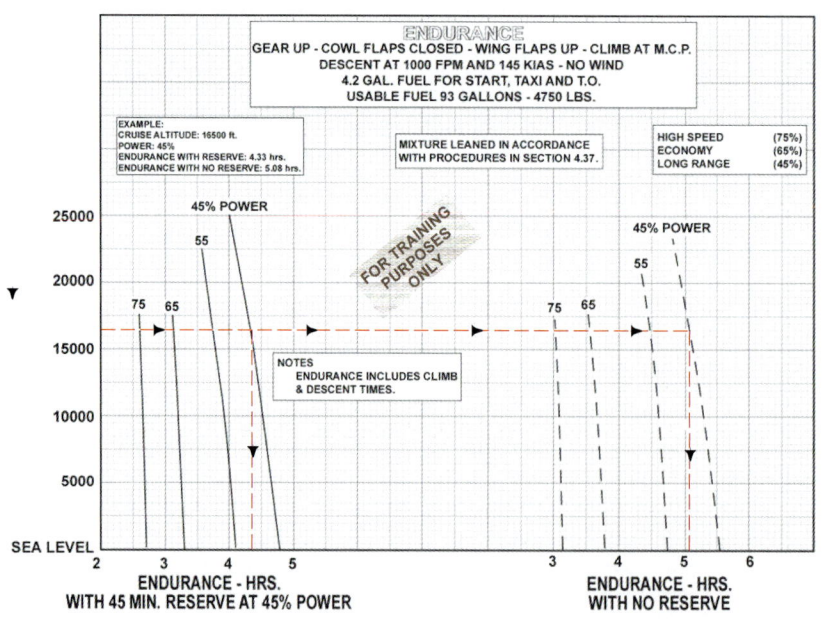

Figure 11.5

PERFORMANCE   MULTI-ENGINED CLASS B: CLIMB AND CRUISE

11.8 DESCENT

Figure 11.6 shows the fuel used, time taken and distance travelled during the descent from the cruising altitude to the destination aerodrome.

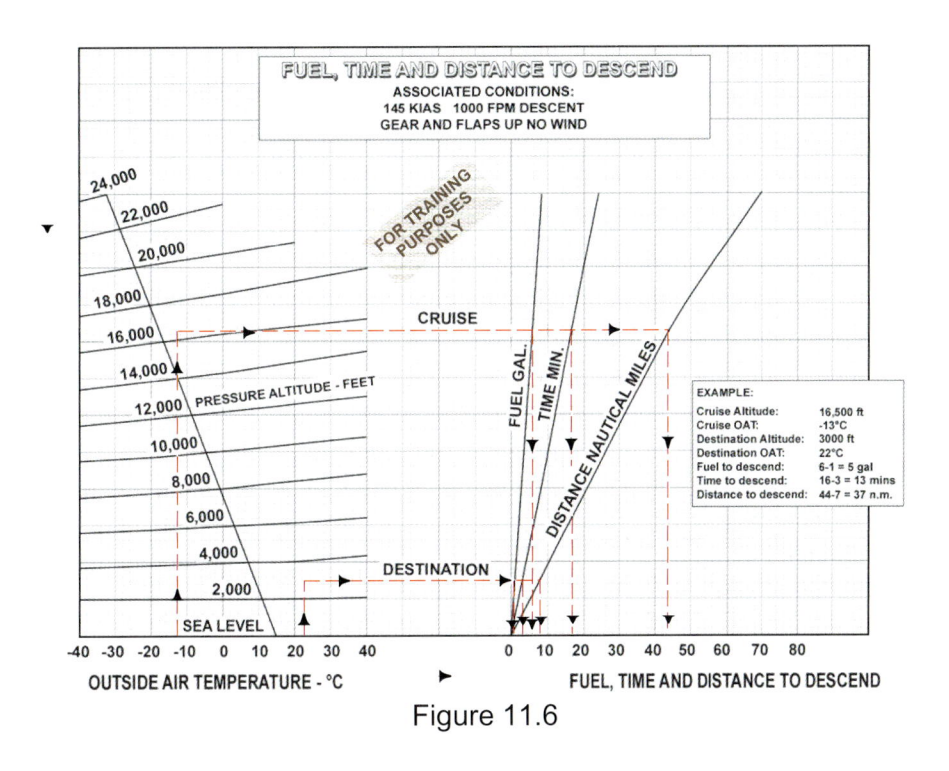

Figure 11.6

PERFORMANCE   MULTI-ENGINED CLASS B: CLIMB AND CRUISE

PAPER 1

1. For a twin-engined Class B aircraft, in the event of a single engine failure en route:

    a) It must be able to remain at least 1000 ft. above the minimum altitude for safe flight in continuing to a suitable landing aerodrome.
    b) The gross gradient of climb may be used in assessing obstacle clearance.
    c) It must be able to continue the flight to a point 1000 ft. above an aerodrome at which the performance requirements for landing can be met.
    d) It must not be assumed to be flying above an altitude at which the rate of climb equals 500 ft. / min.

2. The gross climb gradient with one engine inoperative for the prevailing conditions has been determined and is shown below:

    Pressure altitude   Gradient
    16000 ft.   - 2.8%
    15000 ft.   - 2.2%
    14000 ft.   - 1.8%

    Calculate the net horizontal distance travelled during the descent from 16000 ft. to 14000 ft.

    a) 10 n.m.
    b) 12.2 n.m.
    c) 14.9 n.m.
    d) 19.3 n.m.

3. Which of the following statements is true with regard to the en-route regulations:

    a) The aircraft may not cruise at an altitude above that at which a 300 ft./min. rate of climb can be achieved.
    b) In assessing the en-route performance it may not be assumed that the aircraft is initially above an altitude at which a 300 ft./min rate of climb can be achieved with one engine inoperative.
    c) In assessing the en-route performance it may not be assumed that the aircraft is initially above an altitude at which a rate of climb of 300 ft./min. can be achieved with all engines operating.
    d) The maximum altitude to be assumed may be determined with maximum take off power.

PERFORMANCE
MULTI-ENGINED CLASS B: CLIMB AND CRUISE

4. **Using Appendix K** determine the time (i) the distance (ii) and the fuel required (iii) from take off at a pressure altitude of 4000 ft., ISA + 10°C to a cruising altitude of 14000 ft., ISA + 10°C

	(i)	(ii)	(iii)
a)	17.5 min.	31 n.m.	9.5 gall.
b)	9.5 min.	17.5 n.m.	31 gall.
c)	24 min.	42 n.m.	13 gall.
d)	30.5 min.	53 n.m.	16.5 gall

   Refer to Appendix L for questions 4 and 5.

5. To achieve a True Air Speed of 180 kt. in ISA conditions, the aircraft must be flown at 65% power:

   a) above an altitude of 12800 ft.
   b) at an altitude of 12800 ft. only
   c) at an altitude of 14000 ft.
   d) at an altitude of 18400 ft.

6. To achieve the maximum possible TAS in ISA conditions, the aircraft should be flown :

   a) at 14500 ft. with 100% power, and would achieve 188 kts.
   b) at 14500 ft. with 100% power, and would achieve 196 kts.
   c) at 17200 ft. with 75% power, and would achieve 193 kts.
   d) at 12000 ft. with 100% power, and would achieve 196 kts.

7. **Refer to Appendix M**

   For an aircraft flying at a pressure altitude of 12000 ft. I.S.A. with 65% power, the still air range with 45 minutes reserve would be :

   a) 915 n.m.
   b) 860 n.m.
   c) 765 n.m.
   d) 760 n.m.

8. **Refer to Appendix N**

   What endurance would be obtained with 55% power, cruising at a pressure altitude of 5000 ft., allowing 45 minutes reserve at 45% power.

   a) 5½ hours
   b) 6¼ hours
   c) 6½ hours
   d) 5 hr. 36 mins.

PERFORMANCE — MULTI-ENGINED CLASS B: CLIMB AND CRUISE

9. **Refer to Appendix O**

   For the data given in the previous question, what rpm / manifold pressure should be set for the cruise.

   a) 2500 / 33.5" Hg
   b) 2400 / 27.6" Hg.
   c) 2300 / 23.7" Hg.
   d) 2200 / 29" Hg.

10. **Refer to Appendix P**

    An aircraft is cruising at 10000 ft. I.S.A. The destination aerodrome is at a pressure altitude of 4000 ft. and the temperature is 20°C. At what distance from the aerodrome should the descent begin in still air conditions.

    a) 5 n.m.
    b) 10 n.m.
    c) 14.5 n.m.
    d) 22.5 n.m.

PERFORMANCE MULTI-ENGINED CLASS B: CLIMB AND CRUISE

Appendix K

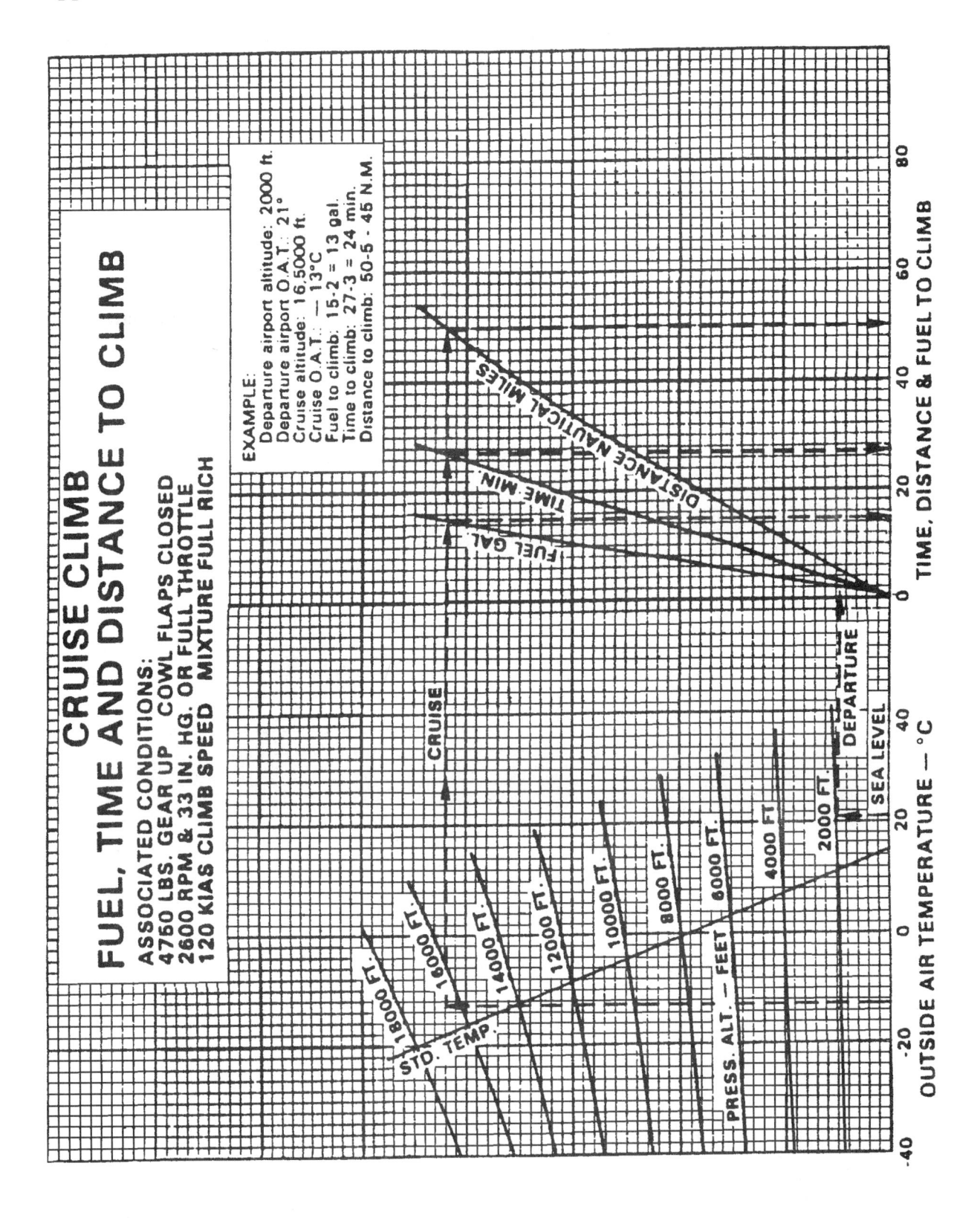

CRUISE CLIMB
FUEL, TIME AND DISTANCE TO CLIMB

ASSOCIATED CONDITIONS:
4750 LBS. GEAR UP   COWL FLAPS CLOSED
2600 RPM & 33 IN. HG. OR FULL THROTTLE
120 KIAS CLIMB SPEED   MIXTURE FULL RICH

EXAMPLE:
Departure airport altitude: 2000 ft.
Departure airport O.A.T.: 21°
Cruise altitude: 16,500 ft.
Cruise O.A.T.: −13°C
Fuel to climb: 15·2 = 13 gal.
Time to climb: 27·3 = 24 min.
Distance to climb: 50·5 − 45 N.M.

PERFORMANCE   **MULTI-ENGINED CLASS B: CLIMB AND CRUISE**

Appendix L

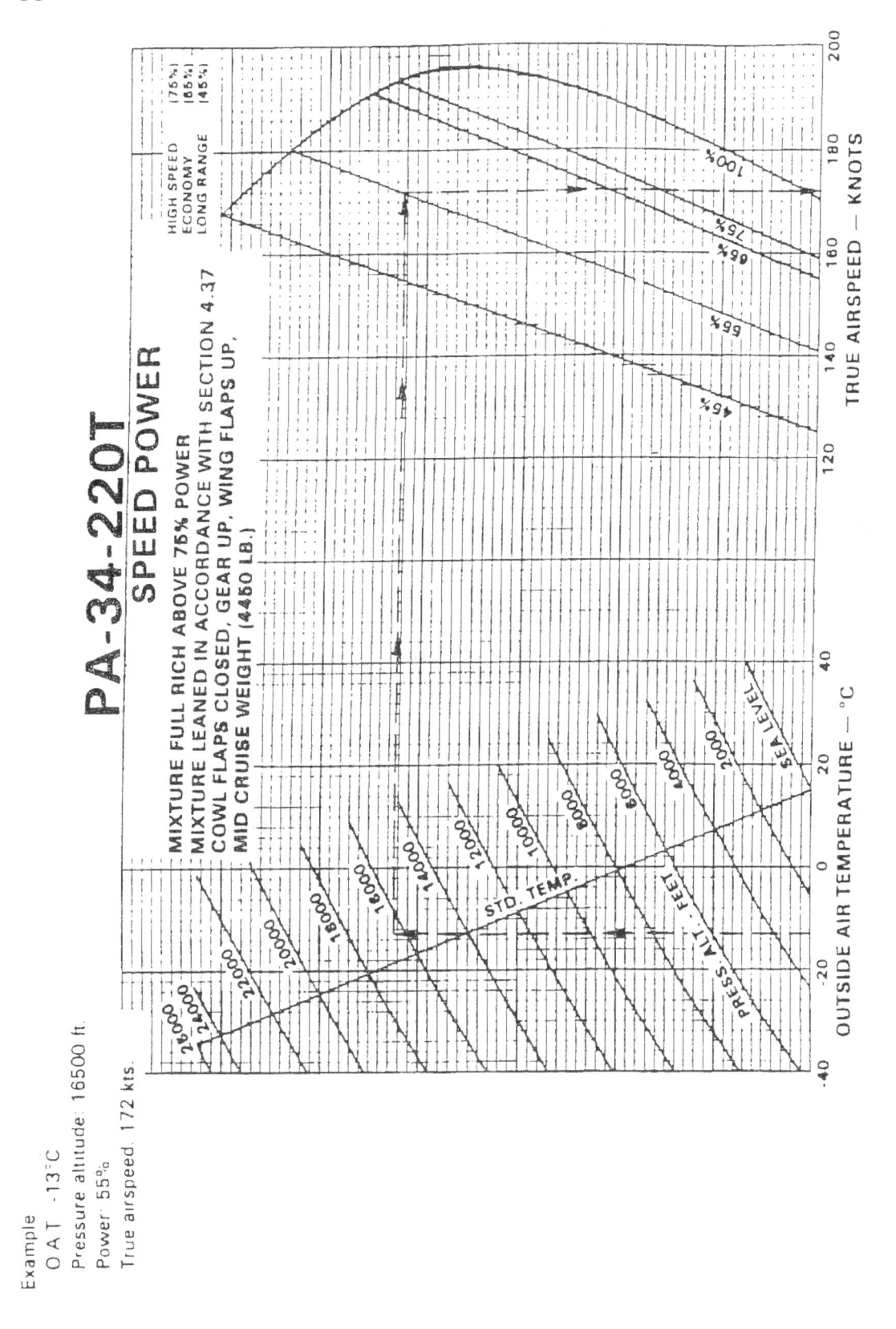

PA-34-220T SPEED POWER

MIXTURE FULL RICH ABOVE 75% POWER
MIXTURE LEANED IN ACCORDANCE WITH SECTION 4.37
COWL FLAPS CLOSED, GEAR UP, WING FLAPS UP,
MID CRUISE WEIGHT (4450 LB.)

Example
O.A.T. -13°C
Pressure altitude: 16500 ft.
Power: 55%
True airspeed: 172 kts.

PERFORMANCE   **MULTI-ENGINED CLASS B: CLIMB AND CRUISE**

Appendix M

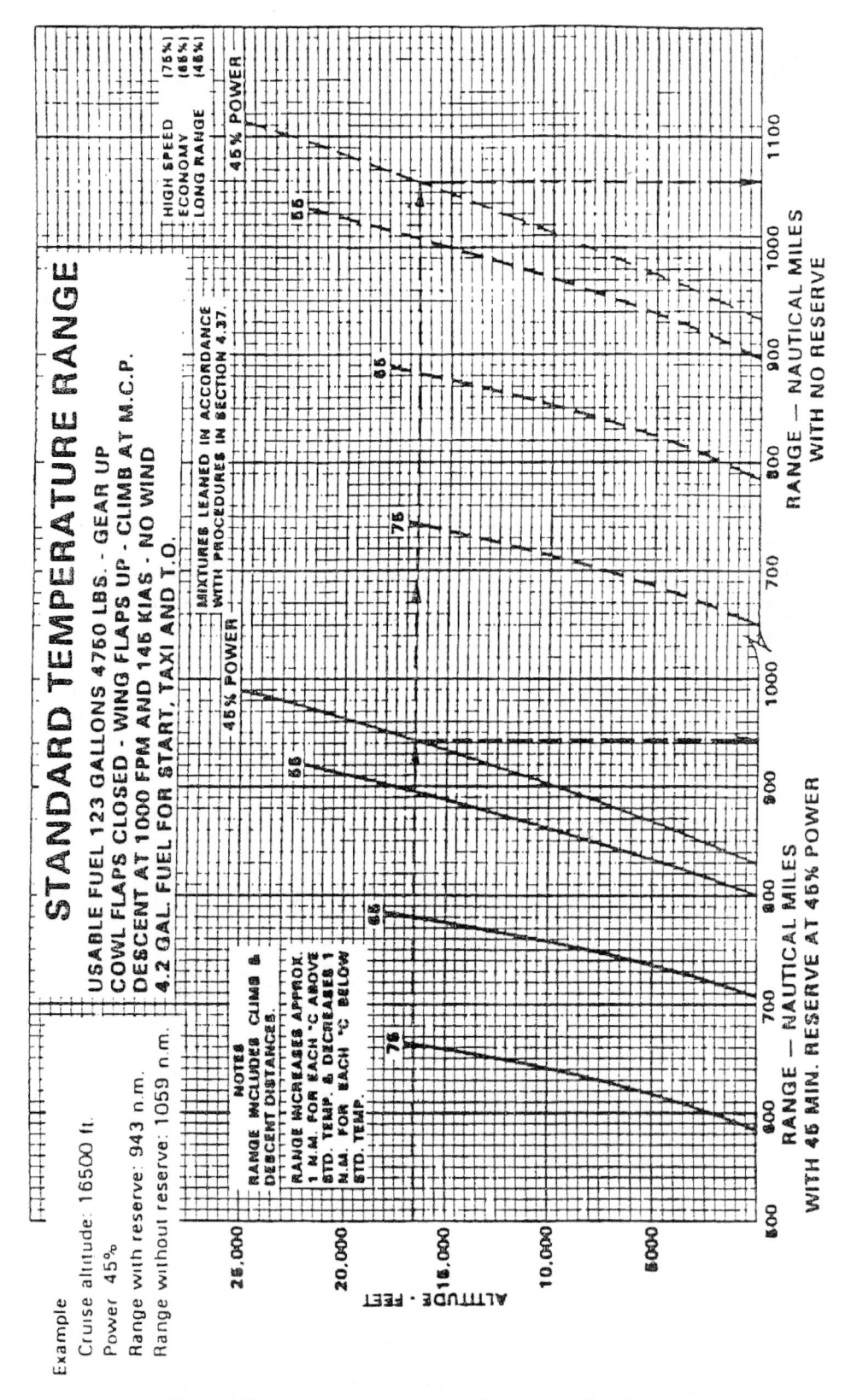

STANDARD TEMPERATURE RANGE

USABLE FUEL 123 GALLONS 4750 LBS. - GEAR UP
COWL FLAPS CLOSED - WING FLAPS UP - CLIMB AT M.C.P.
DESCENT AT 1000 FPM AND 145 KIAS - NO WIND
4.2 GAL FUEL FOR START, TAXI AND T.O.

Example
Cruise altitude: 16500 ft
Power 45%
Range with reserve: 943 n.m.
Range without reserve: 1059 n.m.

NOTES
RANGE INCLUDES CLIMB & DESCENT DISTANCES.
RANGE INCREASES APPROX. 1 N.M. FOR EACH °C ABOVE STD. TEMP. & DECREASES 1 N.M. FOR EACH °C BELOW STD. TEMP.

MIXTURES LEANED IN ACCORDANCE WITH PROCEDURES IN SECTION 4.37.

HIGH SPEED (75%)
ECONOMY (65%)
LONG RANGE (45%)

TEMPERATURE RANGE - 123 GALLONS USABLE

PERFORMANCE MULTI-ENGINED CLASS B: CLIMB AND CRUISE

Appendix N

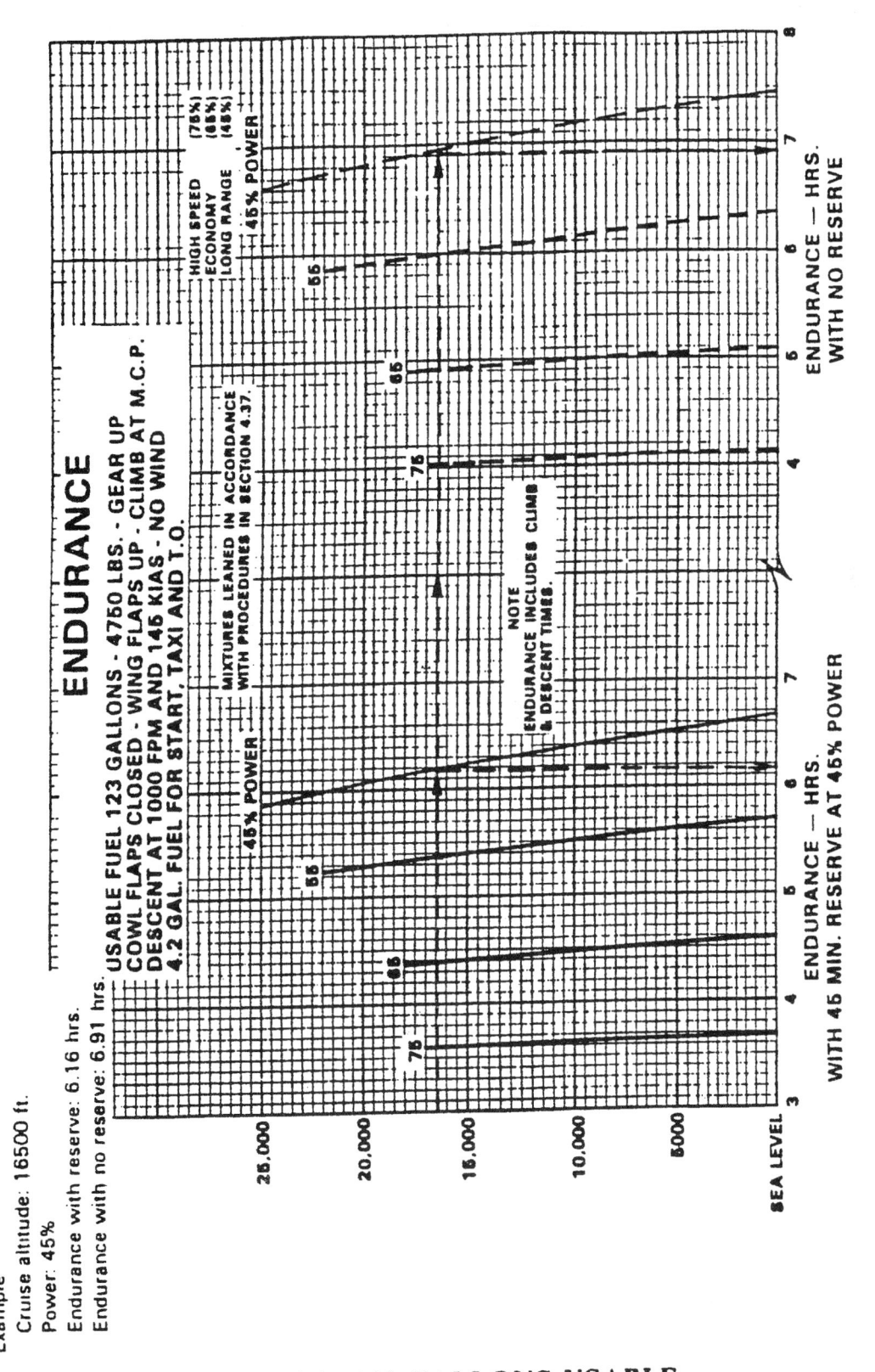

ENDURANCE

USABLE FUEL 123 GALLONS. - 4750 LBS. - GEAR UP
COWL FLAPS CLOSED - WING FLAPS UP - CLIMB AT M.C.P.
DESCENT AT 1000 FPM AND 146 KIAS - NO WIND
4.2 GAL. FUEL FOR START, TAXI AND T.O.

MIXTURES LEANED IN ACCORDANCE
WITH PROCEDURES IN SECTION 4.37.

NOTE
ENDURANCE INCLUDES CLIMB
& DESCENT TIMES.

HIGH SPEED (75%)
ECONOMY (65%)
LONG RANGE (45%)
45% POWER

Example
Cruise altitude: 16500 ft.
Power: 45%
Endurance with reserve: 6.16 hrs.
Endurance with no reserve: 6.91 hrs.

ENDURANCE - 123 GALLONS USABLE

Appendix O

Press Alt Feet	Std Alt Temp C°	45% Power Approx. Fuel 16 G.P.H. RPM AND MAN. PRESS					55% Power Approx. Fuel 18.7 G.P.H. RPM AND MAN. PRESS.						65% Power Approx. Fuel 23.2 G.P.H. RPM AND MAN PRESS			75% Power Approx. Fuel 29.0 G.P.H. RPM AND MAN PRESS	
		2100	2200	2300	2400	2500	2600	2100	2200	2300	2400	2500	2600	2400	2500	2600	2500	2600
S.L.	15	27.1	26.4	25.5	24.3	23.3	22.5	31.2	30.3	29.4	28.2	27.2	26.3	33.8	32.0	31.0	34.0	33.0
2000	11	26.4	25.8	24.6	23.7	22.8	22.1	30.5	29.7	28.8	27.8	26.8	26.0	33.2	31.7	30.7	33.8	32.7
4000	7	25.8	25.0	24.0	23.2	22.3	21.8	30.0	29.2	28.3	27.4	26.4	25.6	32.8	31.5	30.5	33.6	32.4
6000	3	25.3	24.5	23.5	22.8	21.9	21.5	29.7	28.8	28.0	27.0	26.2	25.3	32.5	31.2	30.3	33.4	32.2
8000	-1	24.8	24.0	23.0	22.4	21.6	21.2	29.4	28.4	27.7	26.8	25.7	25.0	32.3	31.0	30.1	33.1	32.0
10000	-5	24.4	23.7	22.0	22.0	21.4	21.0		28.3	27.5	26.5	25.5	24.7	32.0	30.9	30.0	33.0	31.9
12000	-9	24.0	23.3	22.5	21.7	21.2	20.9		28.3	27.2	26.3	25.3	24.6	31.8	30.7	29.8	32.5	31.8
14000	-13		23.0	22.3	21.4	21.1	20.8			27.1	26.1	25.2	24.4		30.5	29.7		31.7
16000	-17			22.0	21.3	21.0	20.6				25.9	25.0	24.3		30.4	29.5		31.6
18000	-21				21.2	20.9	20.5					25.0	24.2			29.4		
20000	-25				21.2	20.8	20.4						24.2			29.3		
22000	-28						20.4						24.1					
24000	-33						20.4	1650°F MAX E.G.T. (See P.O.H. Section 4)									1525°F MAX E.G.T. (See P.O.H. Section 4)	
25000	-34						20.4											

To maintain constant power add approximately 1% for each 6°C above standard. Subtract approximately 1% for each 6°C below standard. Do not exceed 34" MAP in cruise.

PERFORMANCE   **MULTI-ENGINED CLASS B: CLIMB AND CRUISE**

Appendix P

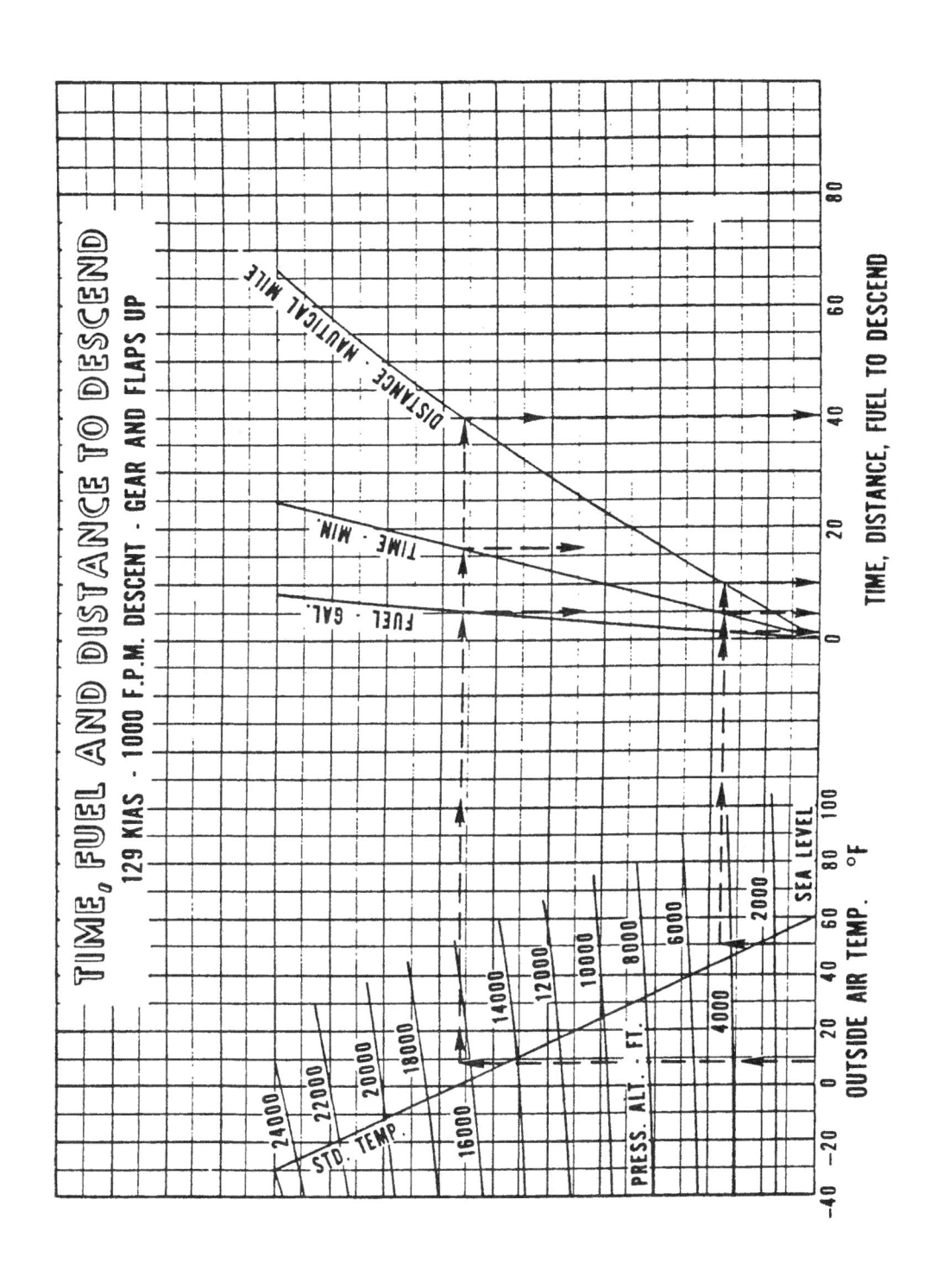

ANSWERS

Questions	Answers
1	C
2	B
3	C
4	A
5	B
6	B
7	C
8	D
9	D
10	C

CHAPTER TWELVE - MULTI-ENGINED CLASS B: LANDING

Contents

		Page
12.1	LANDING REQUIREMENTS	12 - 1
12.2	GRADIENT REQUIREMENT	12 - 1
12.3	LANDING AND APPROACH CLIMB JAR-OPS 1.525 (b)	12 - 1
12.4	LANDING DISTANCE REQUIREMENTS JAR-OPS 1.545	12 - 3
	MULTI-ENGINED CLASS B AIRCRAFT: LANDING	12 - 5

© Oxford Aviation Services Limited

PERFORMANCE   MULTI-ENGINED CLASS B: LANDING

12.1 LANDING REQUIREMENTS

To determine the maximum landing mass it is necessary to consider:

a) the mass for altitude and temperature (gradient requirement)

b) the landing distance requirement

c) the maximum structural landing mass

12.2 GRADIENT REQUIREMENT

Two climb requirements have to be considered:

a) **the landing climb**. This determines an ability to climb at an adequate gradient if a go-around is initiated just before touchdown.

b) **the approach climb**. This determines an ability to climb if a go-around is initiated from the approach phase with an engine inoperative.

12.3 LANDING AND APPROACH CLIMB JAR-OPS 1.525 (b)

i) Landing Climb (All engines operating)

The gradient of climb must be at least 2.5% with:

a) Not more than the power that is available 8 seconds after initiation of movement of the power controls from the minimum flight idle position.

b) The landing gear extended.

c) The wing flaps in the landing position

d) A climb speed equal to V_{REF}

Figure 12.1 shows a typical presentation of the landing climb performance.

PERFORMANCE MULTI-ENGINED CLASS B: LANDING

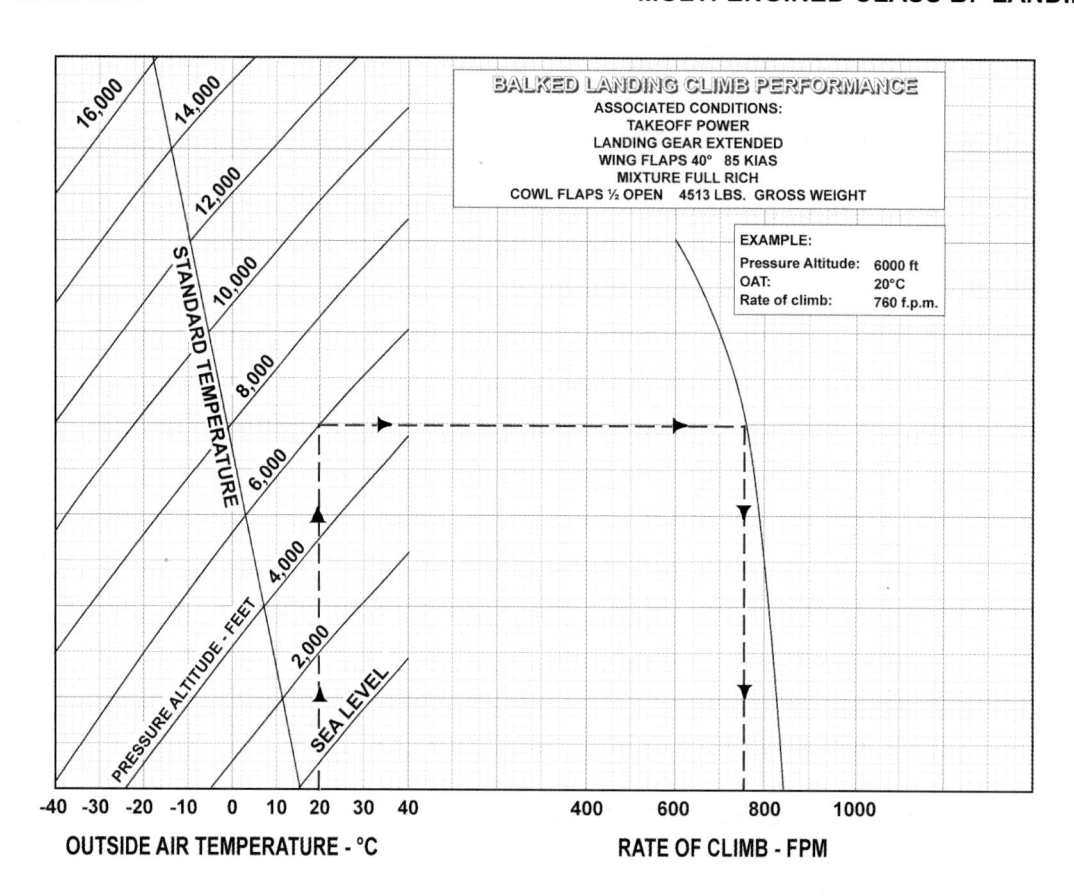

Figure 12.1 Balked Landing Climb Performance

The example shows that at a pressure altitude of 6000 ft, OAT 20°C, at maximum authorised landing weight the rate of climb would be 760 ft/min. at a speed of 85kt IAS. To compare this to the requirement it must be expressed as a gradient.

Climb gradient = $\dfrac{\text{Rate of Climb}}{\text{True Air Speed}}$ in the same units

85kt IAS = 96 kt TAS at 6000ft 20°C

$\therefore$ Gradient = $\dfrac{760}{96} \times \dfrac{60}{6080} \times 100 = 7.8\%$

ii)   The one engine inoperative climb.

The gradient of climb must not be less than 0.75% at an altitude of 1500 ft above the landing surface with:

a)   The critical engine inoperative and its propeller in the minimum drag position.

b) The remaining engine at not more than maximum continuous power.

c) The landing gear retracted.

d) The wing flaps retracted.

e) A climb speed not less than 1.2 V_{S1}.

12.4 LANDING DISTANCE REQUIREMENTS JAR-OPS 1.545

The landing distance requirements for multi-engined Class B aircraft are the same as for single engined aircraft (see Chapter 9).

To summarise these:

a) The aircraft must land from 50 ft to a full stop within 70% of the landing distance available (Lower screen heights may be permitted).

b) The landing distance required must take account of:

  i) The aerodrome altitude.

  ii) Standard day temperature

  iii) Factored winds.

  iv) Runway surface conditions (x 1.15 for grass).

  v) Runway slope (5% distance for 1% slope).

c) Still air and forecast wind situations must be considered.

d) For wet runways increase the distance by 15%.

e) For contaminated runways acceptable data must be used

f) L.D. required ≯ L.D.A.

Figure 12.2 shows a typical presentation of landing distance.

PERFORMANCE

MULTI-ENGINED CLASS B: LANDING

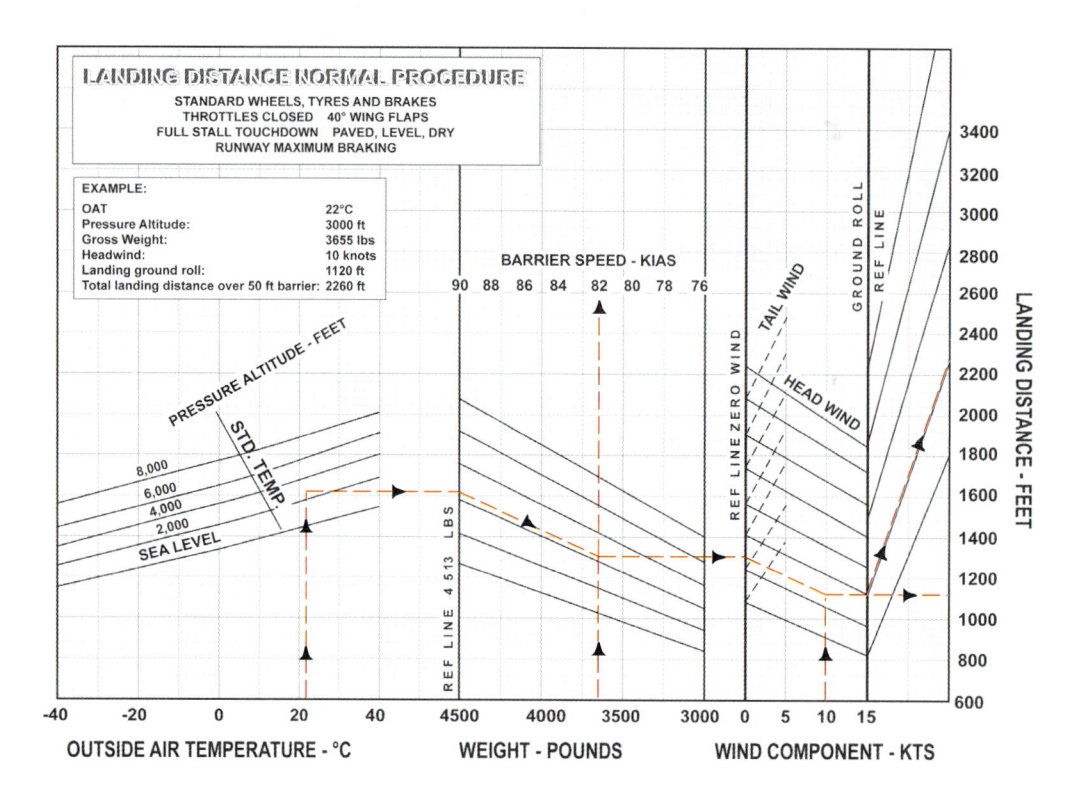

Figure 12.2. Landing Distance Normal Procedure

The example shows that at a pressure altitude of 3000 ft; OAT 22°C, weight 3655 lb, 10kt headwind, level dry runway:

The landing distance from 50ft = 2260 ft

The ground roll = 1120 ft

The threshold speed = 82 kt IAS

Note : 1. For planning purposes the standard day temperature would be used.

2. Both still air and forecast wind would need to be considered.

3. The wind factors have been applied to the data shown, but the 70% rule must be applied (i.e. a factor of 1.43) and if appropriate, the factors for runway surface and slope.

PERFORMANCE   **MULTI-ENGINED CLASS B: LANDING**

PAPER 1

1. The balked landing climb requirement must be met in the following configuration :

   a) Gear down, landing flap, take off power with one engine inoperative.
   b) Gear down, landing flap, take off power on all engines.
   c) Gear down, approach flap, maximum continuous power on all engines.
   d) Gear up, approach flap, maximum continuous power on all engines.

2. For an aircraft flying in conditions that would give the balked landing climb limit gradient of 2.5%, the rate of climb that would be achieved at a TAS of 90 knots would be :

   a) 228 ft./min
   b) 22.5 ft./min.
   c) 3600 ft./min.
   d) 360 ft./min.

Refer to CAP 698 Figure 3.9 for questions 3, 4, 5 and 6

3. Find the gross landing distance required, taking into account the temperature and factored wind, for the following conditions :
   Pressure altitude   2000 ft.
   Temperature   4° C
   Weight   4100 lb.
   Reported wind comp.   5 kt. head
   Runway slope   level

   a) 1220 ft.
   b) 2280 ft.
   c) 2400 ft.
   d) 3600 ft.

4. Find the maximum planned landing weight to comply with the regulations, for the following conditions :

   Pressure altitude   Sea level
   Temperature   30° C
   Forecast wind comp.   10 kt. headwind
   There is a single runway, length 3300 ft. hard, dry, level.

   a) 3500 lb.
   b) 3650 lb.
   c) 4000 lb
   d) 4250 lb.

PERFORMANCE — MULTI-ENGINED CLASS B: LANDING

5. An aerodrome has a single level grass runway, length 4200 ft. If the pressure altitude is 5000 ft. and the temperature 30° C find the maximum planned landing weight if the forecast wind component is 10 kt. headwind. :

    a) 3350 lb.
    b) 3600 lb.
    c) 4250 lb.
    d) 4513 lb.

6. Find the gross landing distance required for the following conditions :
    Pressure altitude   4000 ft.
    Temperature   16° C
    Weight   3800 lb.
    Reported wind comp.   15 kt. Headwind
    Runway slope   1.5% downhill

    a) 2040 ft.
    b) 2200 ft.
    c) 2100 ft.
    d) 2300 ft.

7. An aerodrome has a paved level runway of 5000 ft. What distance should be used in the graph to calculate the maximum planned landing weight if the runway is wet :

    a) 3040 ft.
    b) 3500 ft.
    c) 4025 ft.
    d) 4348 ft.

8. The maximum landing weight may be determined by the landing climb gradient requirement. This is to ensure that :

    a) there is adequate obstacle clearance during approach.
    b) the landing distance is not exceeded.
    c) the climb performance is adequate in the event of a go-around.
    d) manoeuvrability is adequate.

Refer to CAP 698 Figure 3.9 for question 9

9. If the weight for landing is 3600 lb. the speed at the screen should be :

    a) 79.5 kts.
    b) 80.5 kts.
    c) 81.5 kts
    d) 82.5 kts:

10. The speed at the 50 ft. screen (V_{Ref}) should be :

   a) not greater than V_{MC} or 1.3 V_{SO}
   b) not less than the lower of V_{MC} or 1.3 V_{SO}
   c) not less than the greater of 1.1 V_{MC} or 1.2 V_{SO}
   d) not less than the greater of V_{MC} or 1.3 V_{SO}

ANSWERS

Questions	Answers
1	B
2	A
3	C
4	B
5	B
6	D
7	A
8	C
9	C
10	D

CHAPTER THIRTEEN - CLASS A AIRCRAFT: TAKE OFF

Contents

		Page
13.1	PERFORMANCE CLASS A	13 - 1
13.2	ADDITIONAL DEFINITIONS	13 - 1
13.3	TAKE-OFF REQUIREMENTS	13 - 2
13.4	AERODROME DISTANCES	13 - 3
13.5	REQUIRED DISTANCES	13 - 3
13.6	DRY RUNWAY	13 - 3
13.7	WET RUNWAY	13 - 6
13.8	TAKE-OFF SPEED REQUIREMENTS JAR 25.107	13 - 10
13.9	PRESENTATION OF DATA	13 - 13
13.10	BALANCED FIELD	13 - 13
13.11	UNBALANCED FIELD	13 - 14
13.12	TAKE-OFF FROM AN UNBALANCED FIELD	13 - 16
13.13	V_1 RANGE	13 - 16
13.14	CLIMB GRADIENT LIMIT	13 - 17
13.15	TYRE SPEED LIMIT	13 - 19
13.16	BRAKE ENERGY LIMIT	13 - 20
13.17	BRAKE COOLING	13 - 21
13.18	RUNWAY STRENGTH	13 - 21
13.19	MAXIMUM TAKE-OFF MASS	13 - 23
13.20	TAKE OFF SPEEDS AND THRUST SETTINGS	13 - 23

© Oxford Aviation Services Limited

13.21	CORRECTION FOR STOPWAY AND CLEARWAY	13 - 25
13.22	STABILISER TRIM SETTING	13 - 27
	CLASS A AIRCRAFT: TAKE OFF PAPER 1.	13 - 28
	CLASS A AIRCRAFT: TAKE OFF PAPER 2	13 - 30

PERFORMANCE CLASS A AIRCRAFT: TAKE OFF

13.1 PERFORMANCE CLASS A

a) All multi-engined turbojet aircraft

b) Multi-engined turboprop aircraft with a maximum approved passenger seating of more than 9 or a maximum take-off mass exceeding 5700 kg.

13.2 ADDITIONAL DEFINITIONS

Decision speed V_1.

An engine failure having been promptly recognised, V_1 is the speed at which:

a) the continued take-off distance required will not exceed the take-off distance available

b) the continued take-off run required will not exceed the take-off run available

c) The accelerate - stop distance required will not exceed the accelerate- stop distance available

Minimum Unstick Speed V_{MU}

The calibrated airspeed at and above which the aeroplane can safely lift off the ground and continue the take-off. The minimum speed at which the aircraft will lift off the ground will be obtained with the aeroplane in the most nose-up attitude that can be achieved. This may determined by:

a) the aircraft geometry (tail bumper in contact with the runway)

b) elevator power

Lift-off Speed V_{LOF}

The calibrated airspeed at which the aircraft first becomes airborne.

Maximum Brake Energy Speed V_{MBE}

The maximum speed from which a stop can be accomplished within the energy capabilities of the brakes.

Engine Failure Speed V_{EF}

The calibrated airspeed at which the critical engine is assumed to fail.

V_{2min}
The minimum take-off safety speed, with the critical engine inoperative.

V_2
The take-off safety speed, with the critical engine inoperative.

V_3
The steady initial climb speed with all engines operating.

V_{MC}
The minimum control speed. The calibrated airspeed at which, when the critical engine is suddenly made inoperative, it is possible to recover control of the aeroplane, and maintain straight flight either with zero yaw or with an angle of bank of not more than 5° V_{MC} may not exceed 1.2 V_S

V_{MCG}
The minimum control speed on the ground. The calibrated airspeed during take-off, at which, when the critical engine is suddenly made inoperative, it is possible to recover control of the aeroplane with the use of primary aerodynamic controls alone, to enable the take-off to be safely continued.

V_{MCL}
The minimum control speed during landing approach.

13.3 TAKE-OFF REQUIREMENTS

To determine the maximum permissible mass for take-off it is necessary to consider the limits set by:

a) the aerodrome distances available

b) the climb requirements

c) obstacle clearance

d) brake energy limitations

e) tyre speed limitations

f) runway strength limitation

g) maximum structural mass

PERFORMANCE   CLASS A AIRCRAFT: TAKE OFF

13.4 AERODROME DISTANCES (JAR-OPS 1.490)

It must be shown that the distances required by the aeroplane, as defined in JAR, do not exceed the appropriate distances available at the aerodrome of take-off.

a) Take-off Run required must not exceed Take-off Run available.

b) Take-off Distance required must not exceed the Take-off Distance available, with a clearway distance not exceeding half of the Take off Run available.

c) Accelerate-stop Distance required must not exceed the Accelerate-stop Distance available.

13.5 REQUIRED DISTANCES

The distances required are defined in JAR and cover the cases of take-off with all engines operating and take-off with engine failure, for both dry and wet runways.

13.6 DRY RUNWAY

Take-off Distance   JAR 25.113

The take-off distance is the greater of:

a) The distance from the start of take-off to the point at which the aeroplane is 35ft above the take-off surface assuming that the critical engine fails at V_{EF}.

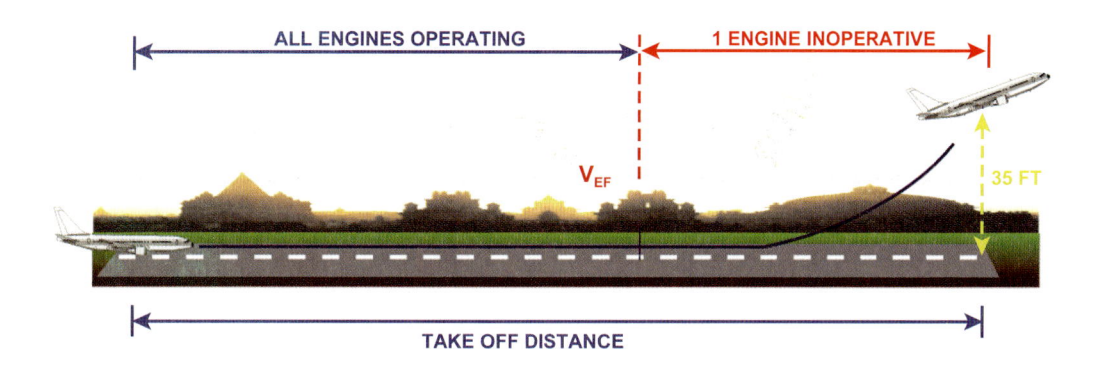

Figure 13.1a

b) 115% of the distance with all engines operating from the start of take-off to the point at which the aircraft is 35ft above the take-off surface.

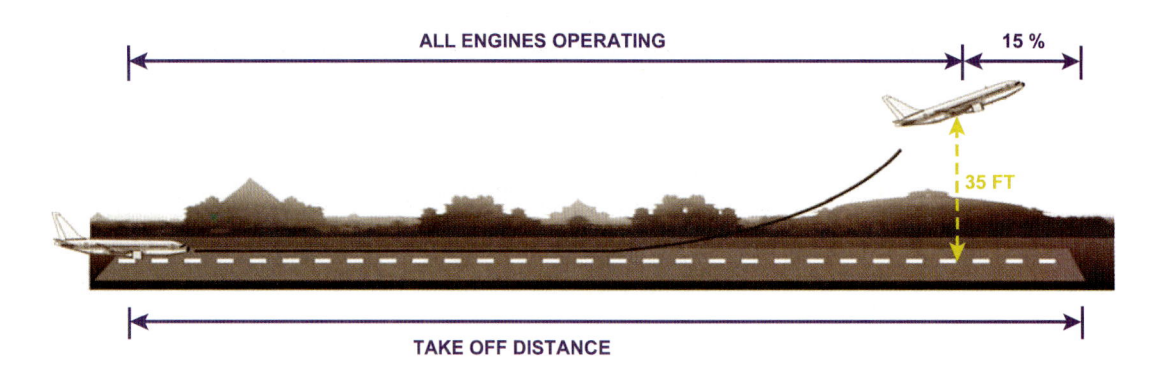

Figure 13.1b

Take-off Run JAR 25.113

If the take-off distance includes a clearway, the take-off run is the greater of:

a) The distance from the start of take-off to a point equidistant between the point at which V_{LOF} is reached and the point at which the aeroplane is 35ft above the take-off surface, the critical engine having failed at V_{EF}.

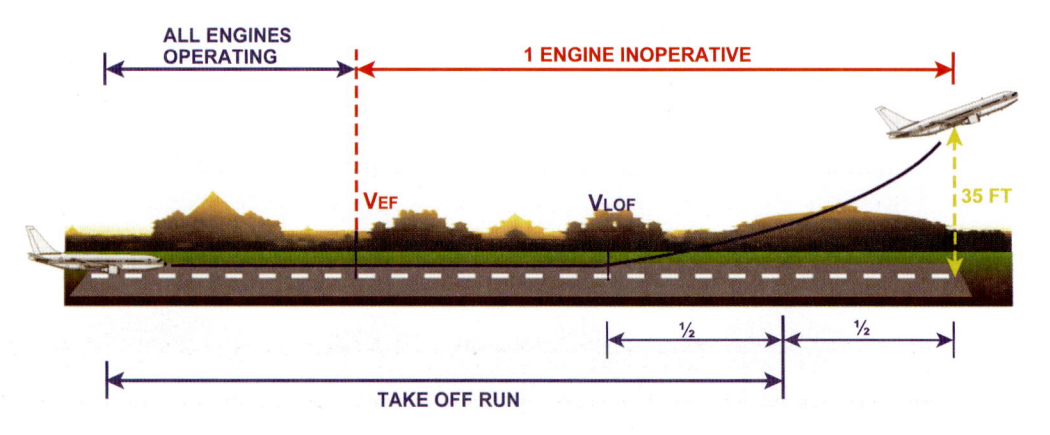

Figure 13.2a

PERFORMANCE   **CLASS A AIRCRAFT: TAKE OFF**

b)   115% of the distance with all engines operating from the start of take-off to a point equidistant between the point at which V_{LOF} is reached and the point at which the aeroplane is 35ft above the take-off surface.

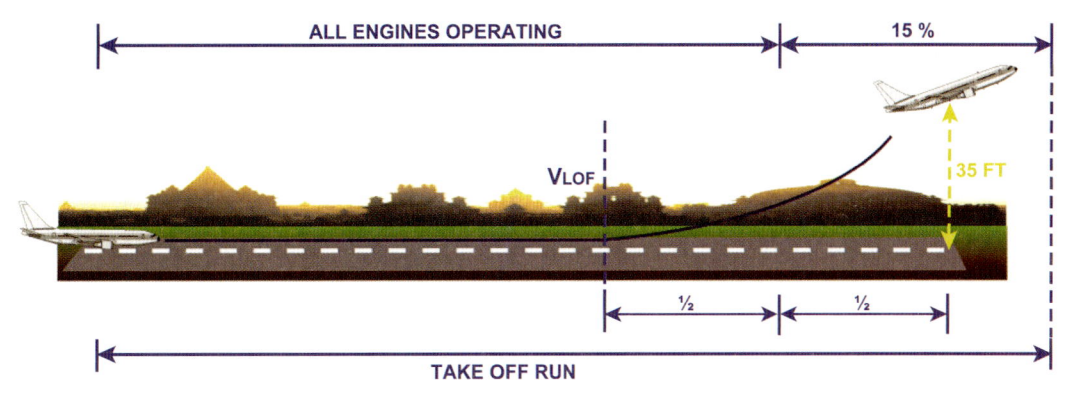

Figure 13.2b

Accelerate-stop distance JAR 25.109

The accelerate-stop distance is the greater of:

a)   (With engine failure)

        The sum of the distances necessary to :

        i)   Accelerate the aeroplane from the start of take-off to V_{EF} with all engines operating.

        ii)   Accelerate from V_{EF} to V_1 and continue the acceleration for 2 seconds after V_1 is reached assuming the critical engine fails at V_{EF}.

        iii)   Come to a full stop from the point reached at the end of the acceleration in (ii) assuming that the pilot does not apply any means of retarding the aeroplane until that point is reached.

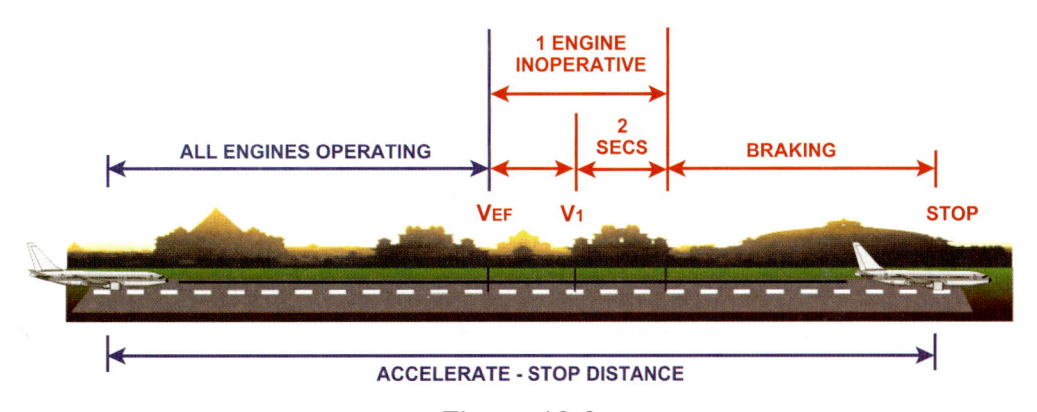

Figure 13.3a

PERFORMANCE CLASS A AIRCRAFT: TAKE OFF

b) (All engines operating)

The sum of the distances necessary to

i) Accelerate the aeroplane from the start of take-off to V_1 and continue the acceleration for 2 seconds after V_1 is reached with all engines operating

ii) Come to a full stop from the point reached at the end of the acceleration in (i) assuming that the pilot does not apply any means of retarding the aeroplane until that point is reached.

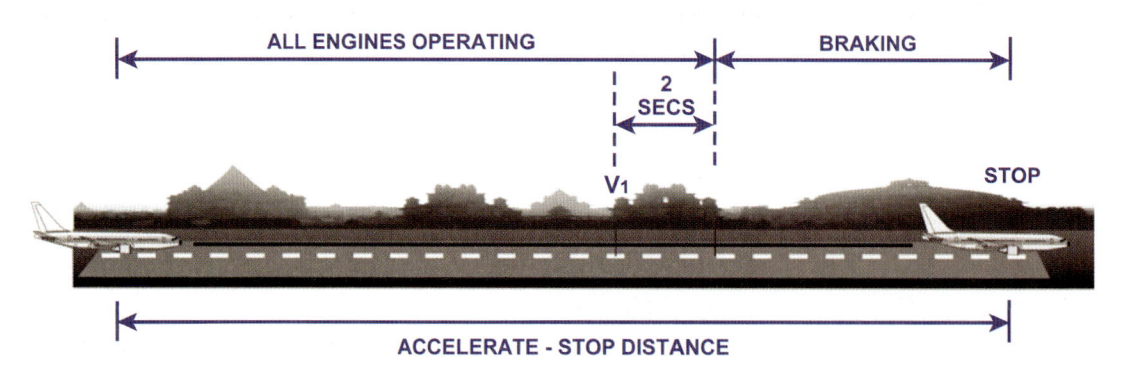

Figure 13.3b.

13.7 WET RUNWAY

JAR 25   AMJ 25X1591

Take-off Distance

The take-off distance required on a wet runway is the greater of :

a) The distance from the start of take-off to the point at which the aeroplane is 15 ft. above the take-off surface, consistent with the achievement of V_2 before reaching 35 ft. Assuming that the critical engine fails at V_{EF} corresponding to V_{GO}

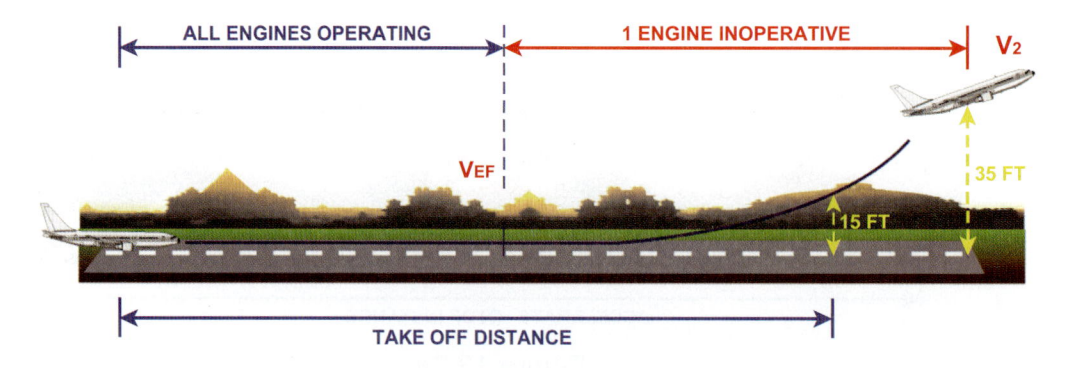

Figure 13.4a

PERFORMANCE — CLASS A AIRCRAFT: TAKE OFF

b) 115% of the distance with all engines operating from the start of take-off to the point at which the aeroplane is 35 ft. above the take-off surface

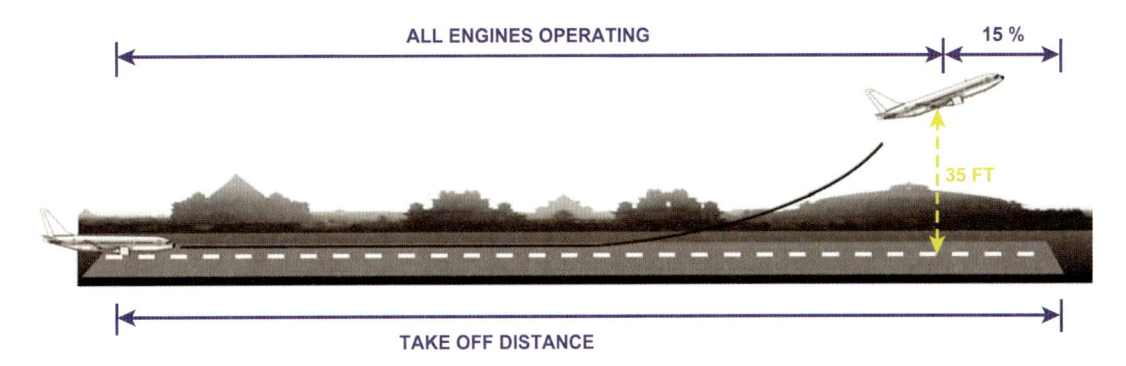

Figure 13.4b

Take-off Run.

The take-off run required on a wet runway is the greater of :

a) The distance from the start of take-off to a point at which V_{LOF} is reached, assuming that the critical engine fails at V_{EF} corresponding to V_{GO}

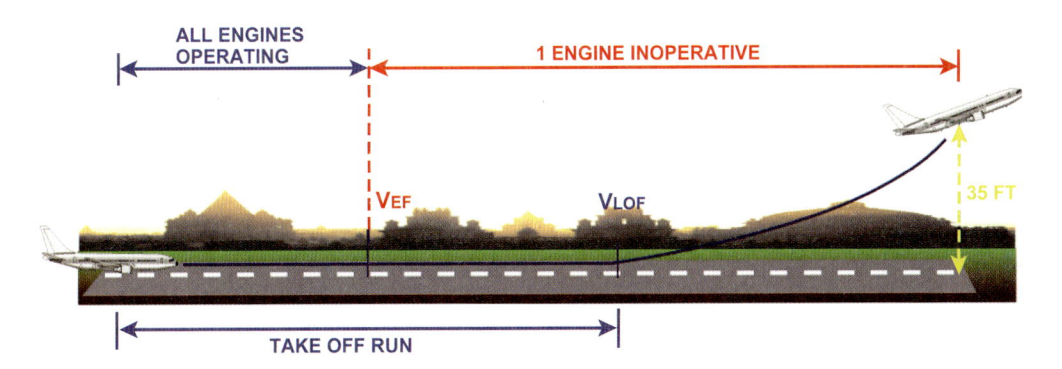

Figure 13.5a

PERFORMANCE

CLASS A AIRCRAFT: TAKE OFF

b) 115% of the distance with all engines operating from the start of take-off to a point equidistant between the point at which V_{LOF} is reached and the point at which the aeroplane is 35 ft. above the take-off surface.

Note : V_{GO} is the lowest decision speed from which a continued take-off is possible within the take-off distance available.

V_{STOP} is the highest decision speed from which the aeroplane can stop within the accelerate-stop distance available.

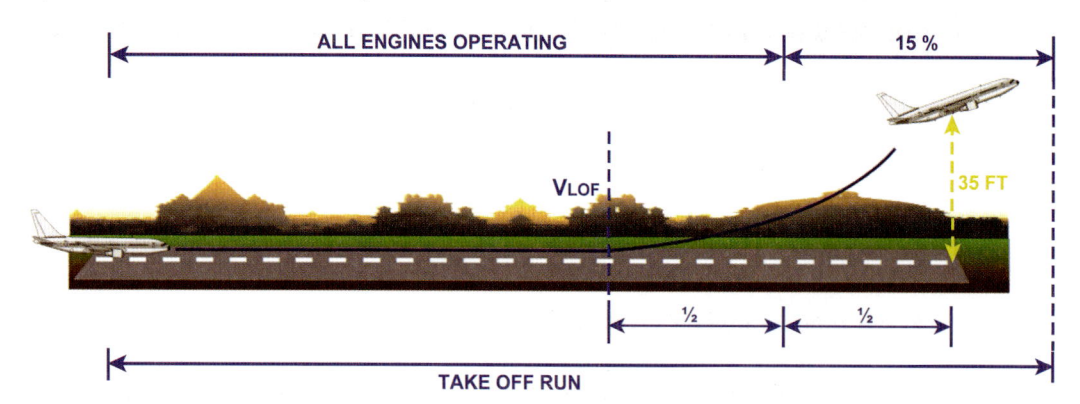

Figure 13.5b

Accelerate-stop distance.

The accelerate- stop distance on a wet runway is the greater of :

a) (With engine failure)

The sum of the distances necessary to :

i) Accelerate the aeroplane with all engines operating to V_{EF} corresponding to V_{STOP}

ii) Accelerate from V_{EF} to V_{STOP} and continue the acceleration for 2 seconds after V_{STOP} is reached, assuming the critical engine fails at VEF

PERFORMANCE   **CLASS A AIRCRAFT: TAKE OFF**

    iii)   Come to a full stop on a wet runway from the point reached at the end of the acceleration period in (ii)

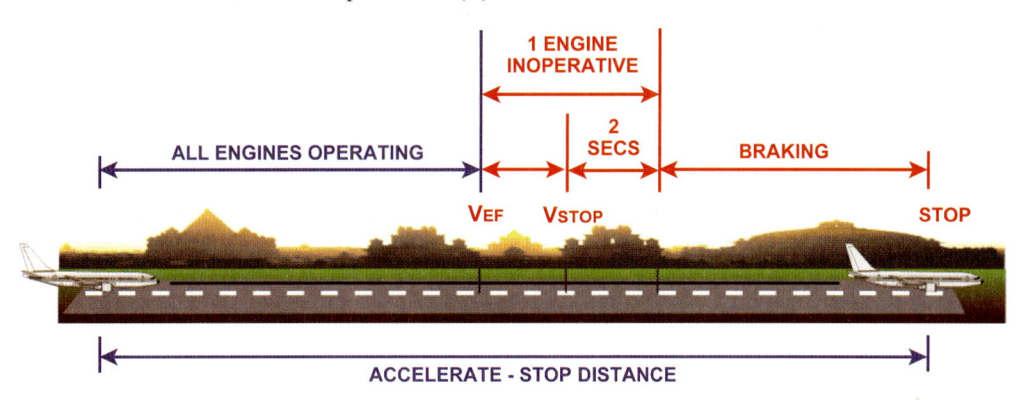

Figure 13.6a

b)   (All engines operating)

        The sum of the distances necessary to :

    i)   Accelerate the aeroplane to V_{STOP} and continue the acceleration for 2 seconds after V_{STOP} is reached, with all engines operating.

    ii)   Come to a full stop on a wet runway from the point reached at the end of the acceleration period in (i)

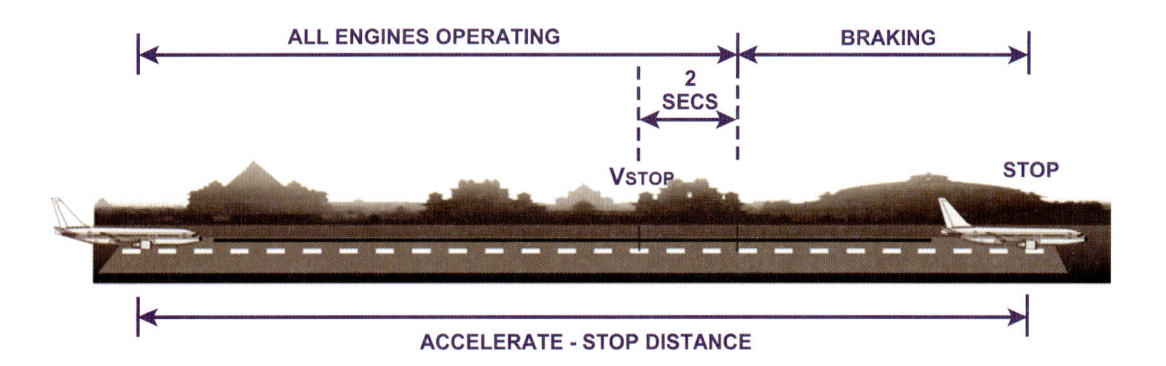

Figure 13.6b.

PERFORMANCE — CLASS A AIRCRAFT: TAKE OFF

13.8 TAKE-OFF SPEED REQUIREMENTS JAR 25.107

V_1:

a) may not be less than V_{EF} plus the speed gained with the critical engine inoperative for the time between engine failure and the point at which the pilot applies the first means of retardation.

b) must not exceed V_R

c) must not exceed V_{MBE}

d) must not be less than V_{MCG}

V_R:

May not be less than

a) V_1

b) $1.05\ V_{MC}$

c) a speed such that V_2 may be attained before 35ft.

d) a speed such that if the aeroplane is rotated at its maximum practicable rate will result in a V_{LOF} of not less than $1.1\ V_{MU}$ (all engines operating) or $1.05 V_{MU}$ (engine inoperative) [if the aeroplane is geometry limited or elevator power limited these margins are $1.08\ V_{MU}$ (all engines) and $1.04\ V_{MU}$ (engine inoperative)]

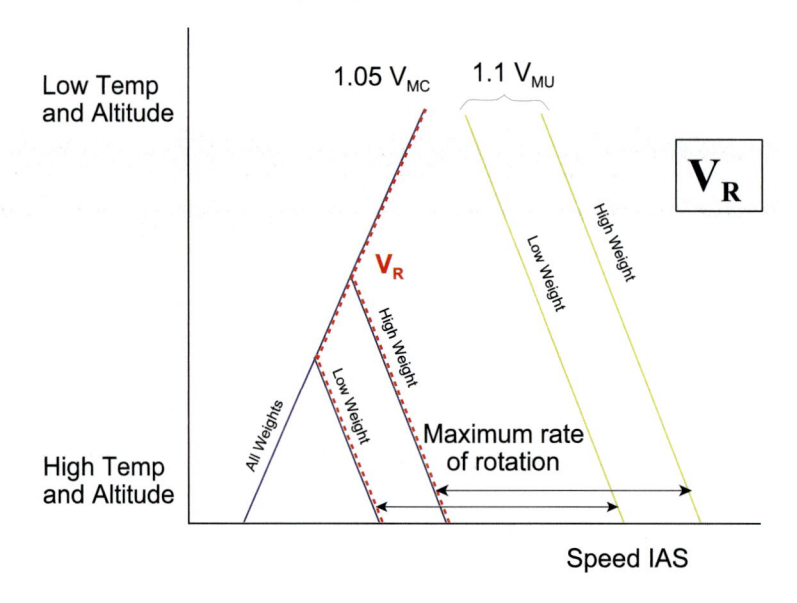

Figure 13.7

PERFORMANCE   **CLASS A AIRCRAFT: TAKE OFF**

V_2 min:

May not be less than:

a) $1.13V_{SR}$ for 2 and 3 engine turboprops and turbojets without provision for obtaining a significant reduction in the one engine inoperative power-on stalling speed.

b) $1.08V_{SR}$ for turboprops with more than 3 engines and turbojets with provision etc.

c) $1.1V_{MC}$

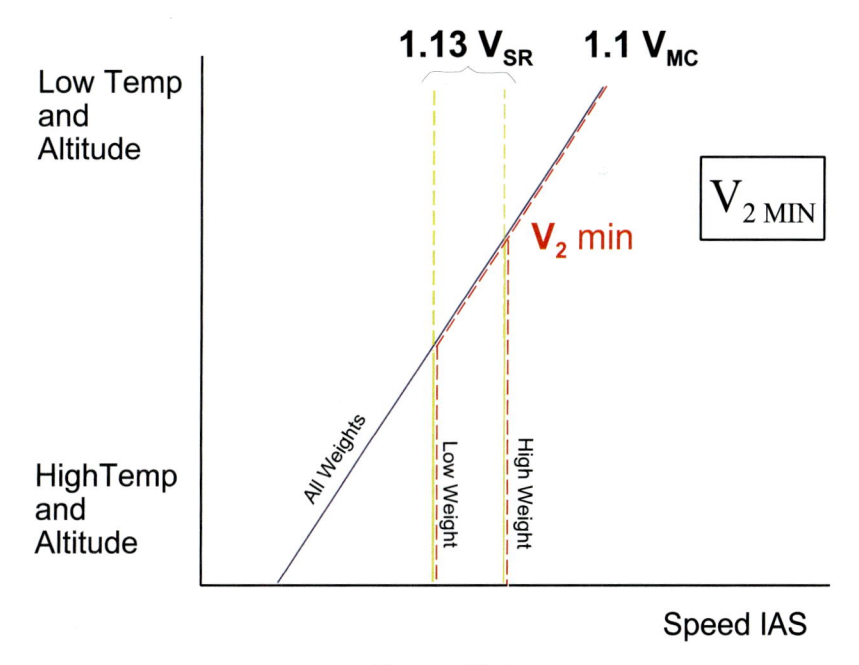

Figure 13.8

V_2 :

May not be less than:

a) V_{2min}

b) V_R plus the speed increment attained up to 35ft.

PERFORMANCE   **CLASS A AIRCRAFT: TAKE OFF**

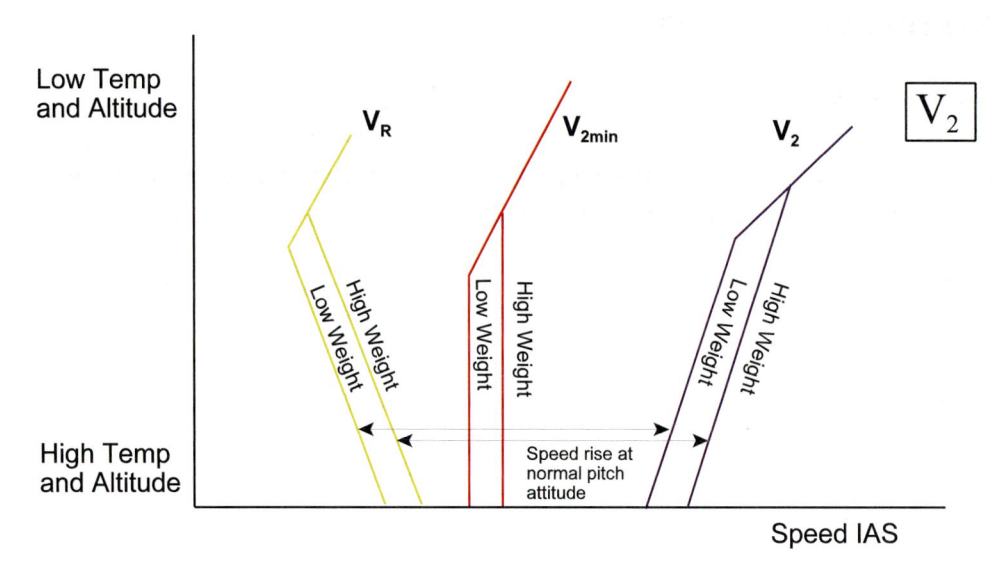

Figure 13.9.

Effects of early and over-rotation

If the aircraft is rotated to the correct attitude but at too low a speed, lift off will not occur until the normal V_{LOF}, but there will be higher drag during the increased time in the rotated attitude, giving increased distance to lift off. Rotation to an attitude greater than the normal lift off attitude could bring the wing close to its ground stalling angle. Ground stall should not be possible with leading edge devices correctly set, so it is of extreme importance that these devices are set to the take-off position.

JAR 25.107 requires:

a)   the take-off distance using a rotation speed of 5 knots less than V_R shall not exceed the take-off distance using the established V_R

b)   reasonable variations in procedures such as over rotation and out of trim conditions must not result in marked increases in take-off distance.

       Note :   The expression 'marked increase' in the take-off distance is defined as any amount in excess of 1% of the scheduled distance.
       (ACJ No.2 to JAR 25.107e4)

PERFORMANCE   CLASS A AIRCRAFT: TAKE OFF

13.9 PRESENTATION OF DATA

A complete analysis of take-off performance requires account to be taken of any stopway and clearway available. As this is time consuming and will often give a maximum permissible take-off mass in excess of that required, simplified data is often presented to permit a rapid assessment of the take-off mass. One method of doing this is to use balance field data.

13.10 BALANCED FIELD

A balanced field exists if the Take-off Distance is equal to the Accelerate-stop Distance. An aerodrome which has no stopway or clearway has a balanced field. For an aeroplane taking off if an engine failure occurs, the later the engine fails, the greater will be the accelerate-stop distance required but the less will be the take-off distance required At some speed the two distances will be equal. Figure 13.10 shows the variation of these distances graphically.

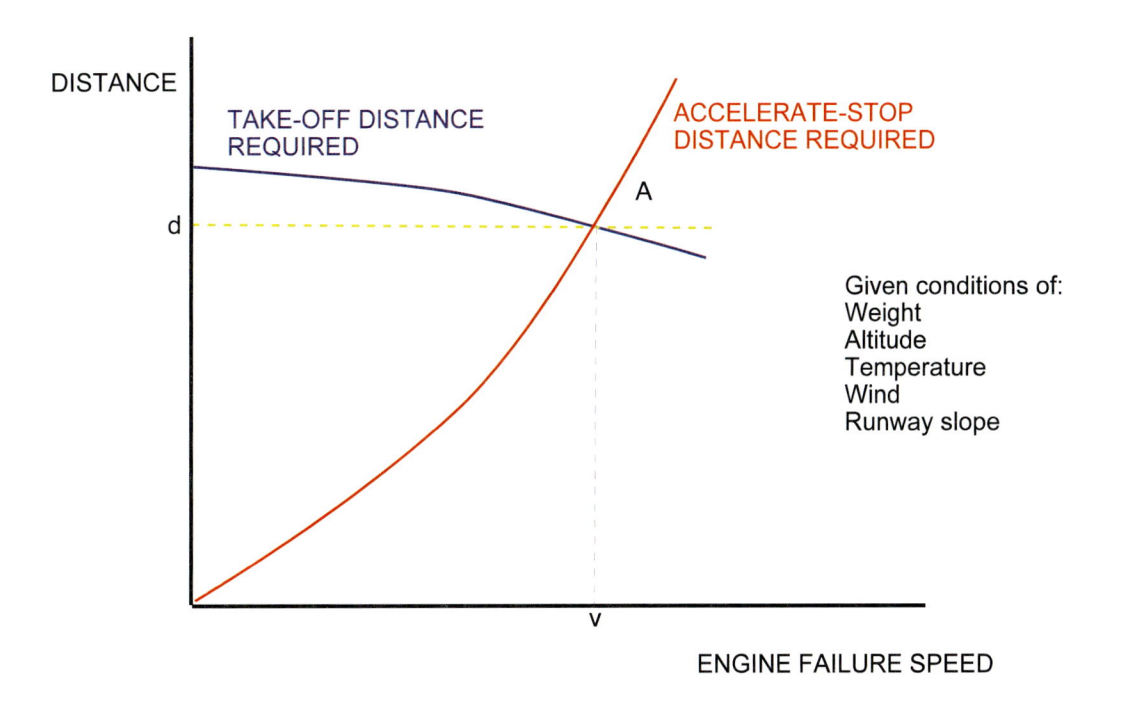

Figure 13.10

The distance, point A in Figure 13.10 is the balanced field length required for the prevailing conditions. It represents the maximum distance required for those conditions, because at whatever speed the engine fails, the distance is adequate, either to stop, if the failure occurs before V_1 or to complete the take-off if the failure occurs after V_1.

PERFORMANCE   **CLASS A AIRCRAFT: TAKE OFF**

13.11 UNBALANCED FIELD.

For a given weight and conditions, the balanced field V_1 will give the optimum performance, since the TODR and the ASDR are equal. In some circumstances however this V_1 will not be acceptable, as V_1 must lie within the limits of V_{MCG}, V_R and V_{MBE}. The following situations will give an unbalanced field :

1. V_1 less than V_{MCG}

    At low weights and altitudes V_1 for the balanced field may be less than V_{MCG}. In this case V_1 would have to be increased to V_{MCG} and so the TODR would be less, and the ASDR would be greater than the balanced field length. The field length required would be equal to the ASDR at V_{MCG}

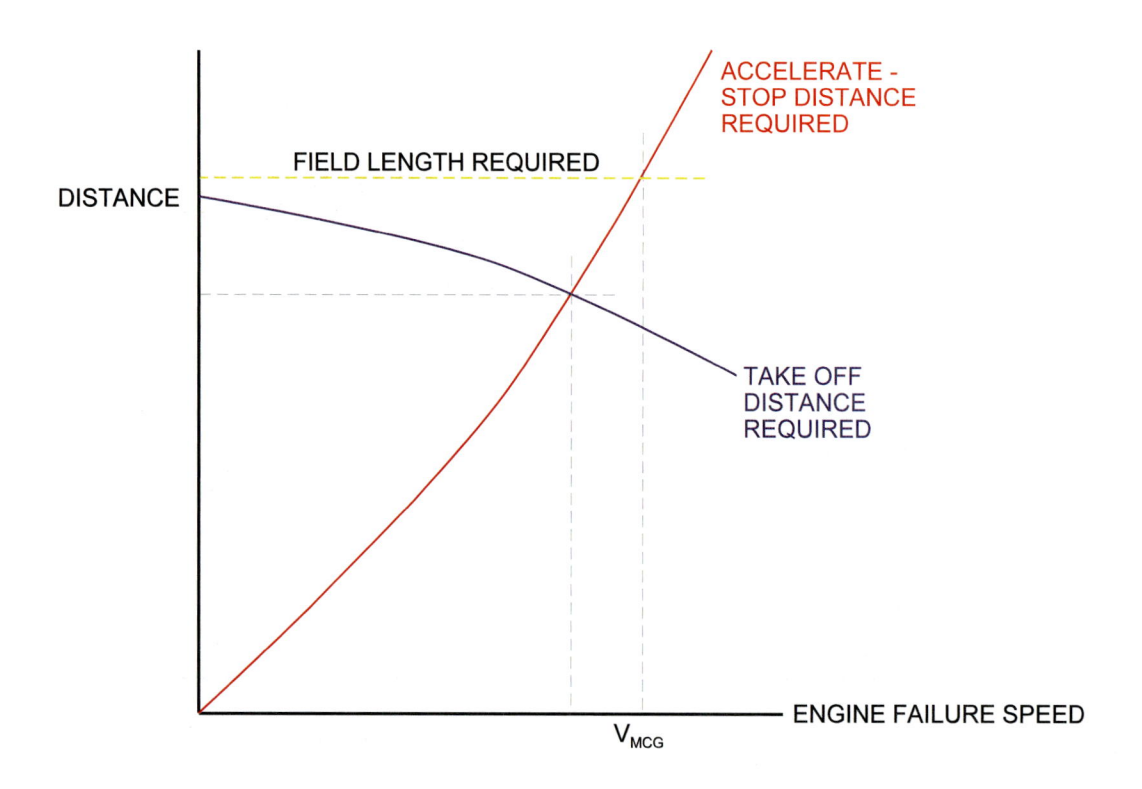

Figure 13.11

2. V_1 greater than V_{MBE}

    At high weight, altitude and temperature, the balanced field V_1 may exceed the V_{MBE}. V_1 would have to be reduced to V_{MBE} giving a TODR greater, and an ASDR which is less, than the balanced field length. The field length required would be equal to the TODR at V_{MBE}.

PERFORMANCE — CLASS A AIRCRAFT: TAKE OFF

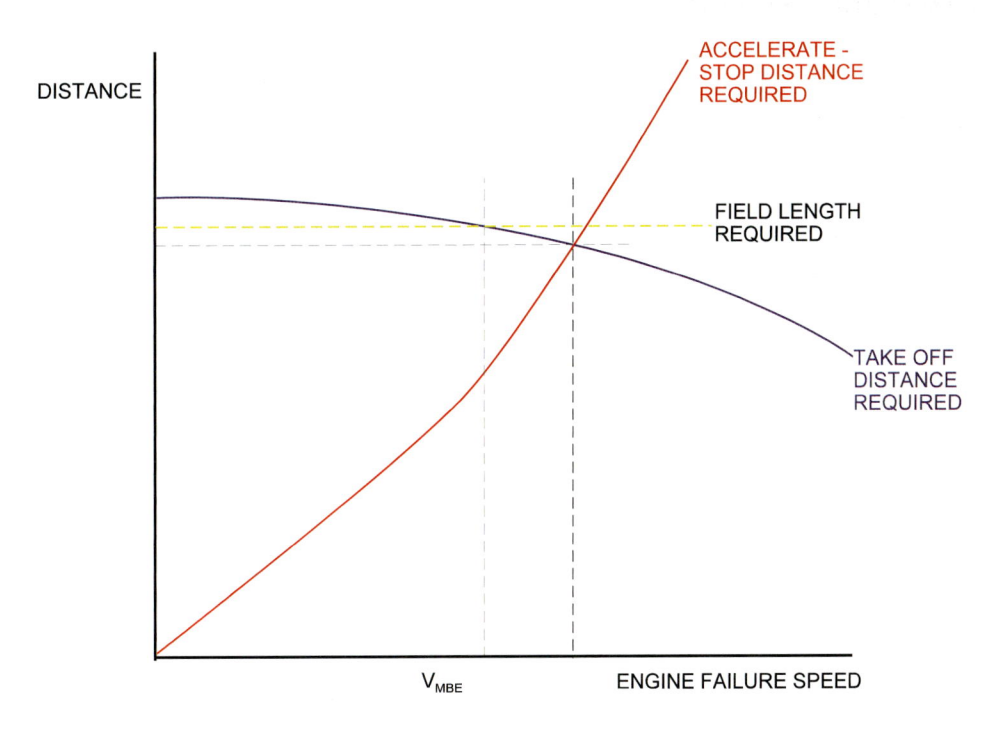

Figure 13.12

3. V_1 greater than V_R

   For aircraft with good braking capabilities, the stopping distance will be short, giving a high balanced field V_1 speed. If this exceeds V_R for the weight, V_1 will have to be reduced to V_R and the field length required will be equal to the TODR at V_R.

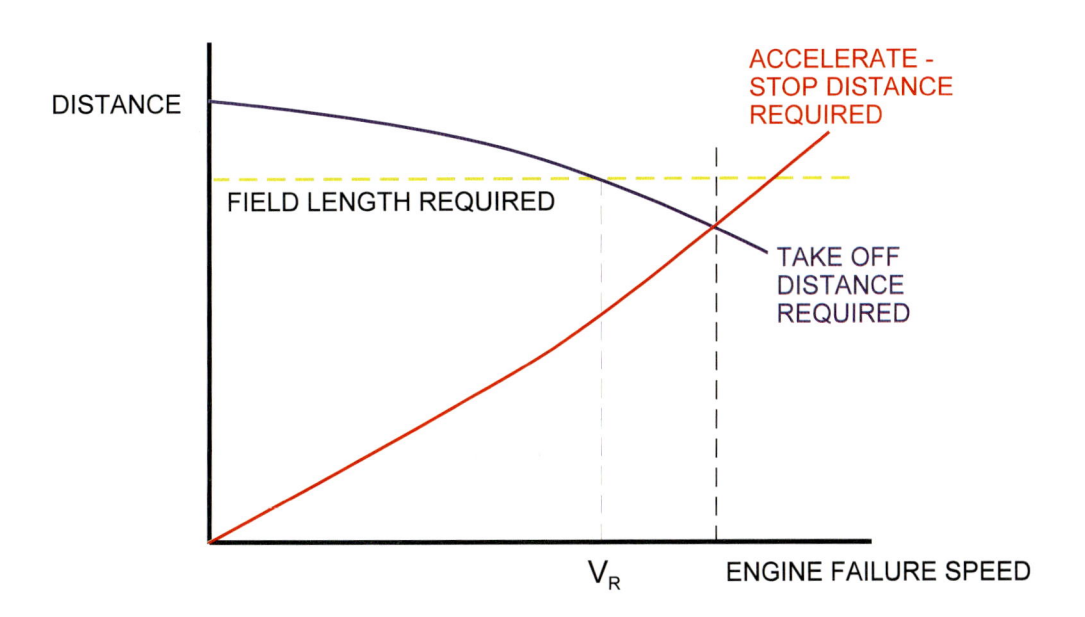

Figure 13.13.

PERFORMANCE   CLASS A AIRCRAFT: TAKE OFF

13.12 TAKE-OFF FROM AN UNBALANCED FIELD

If the take-off aerodrome is not a balanced field, the balanced field data can be used by assuming a balanced field equal to the lesser of the Take-off Distance Available and the Accelerate-stop Distance Available. This distance may exceed the Take Off Run Available unless the TORA becomes limiting. The take-off mass obtained will of course be less than that which could have been obtained by taking account of stopway and clearway, but if the mass is sufficient for the flight, it will not be necessary to go into a more detailed analysis.

Figure 13.15 shows a typical presentation of the balanced field length, with an illustrative example.

13.13 V_1 RANGE

If the balanced field available is greater than the balanced field required for the required take-off mass and conditions there will be a range of speed within which V_1 can be chosen. This situation is illustrated in Figure 13.14.

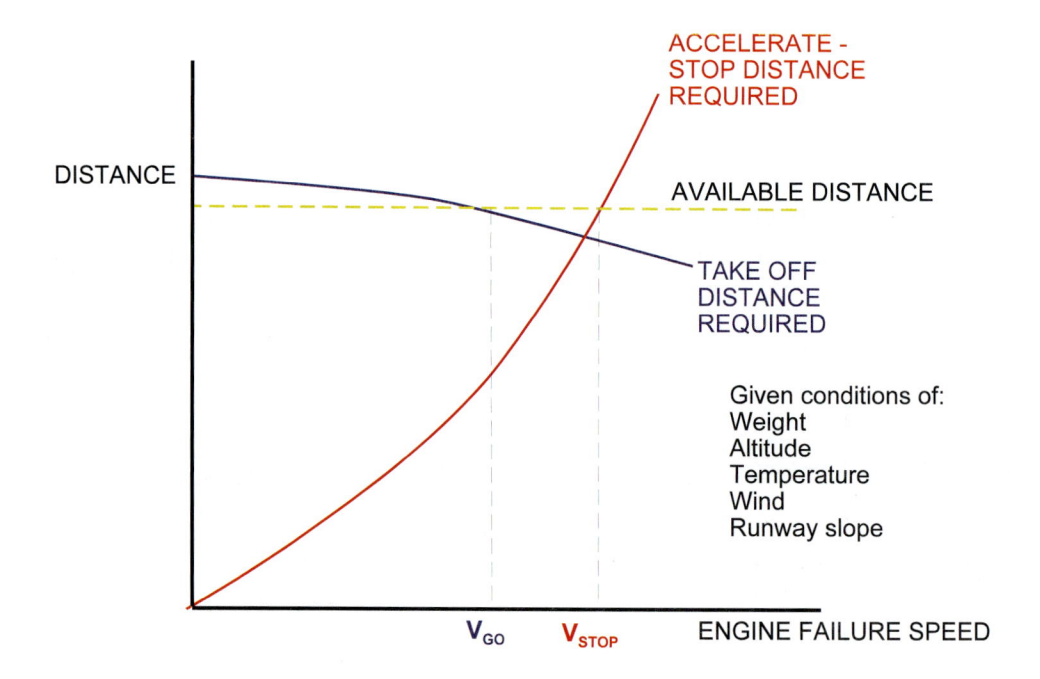

Figure 13.14.

V_{GO} is the first speed at which the take-off can be completed within the distance available, and V_{STOP} is the last speed at which the accelerate-stop could be completed within the distance. The V_1 speed can therefore be chosen anywhere between V_{GO} and V_{STOP}.

PERFORMANCE   **CLASS A AIRCRAFT: TAKE OFF**

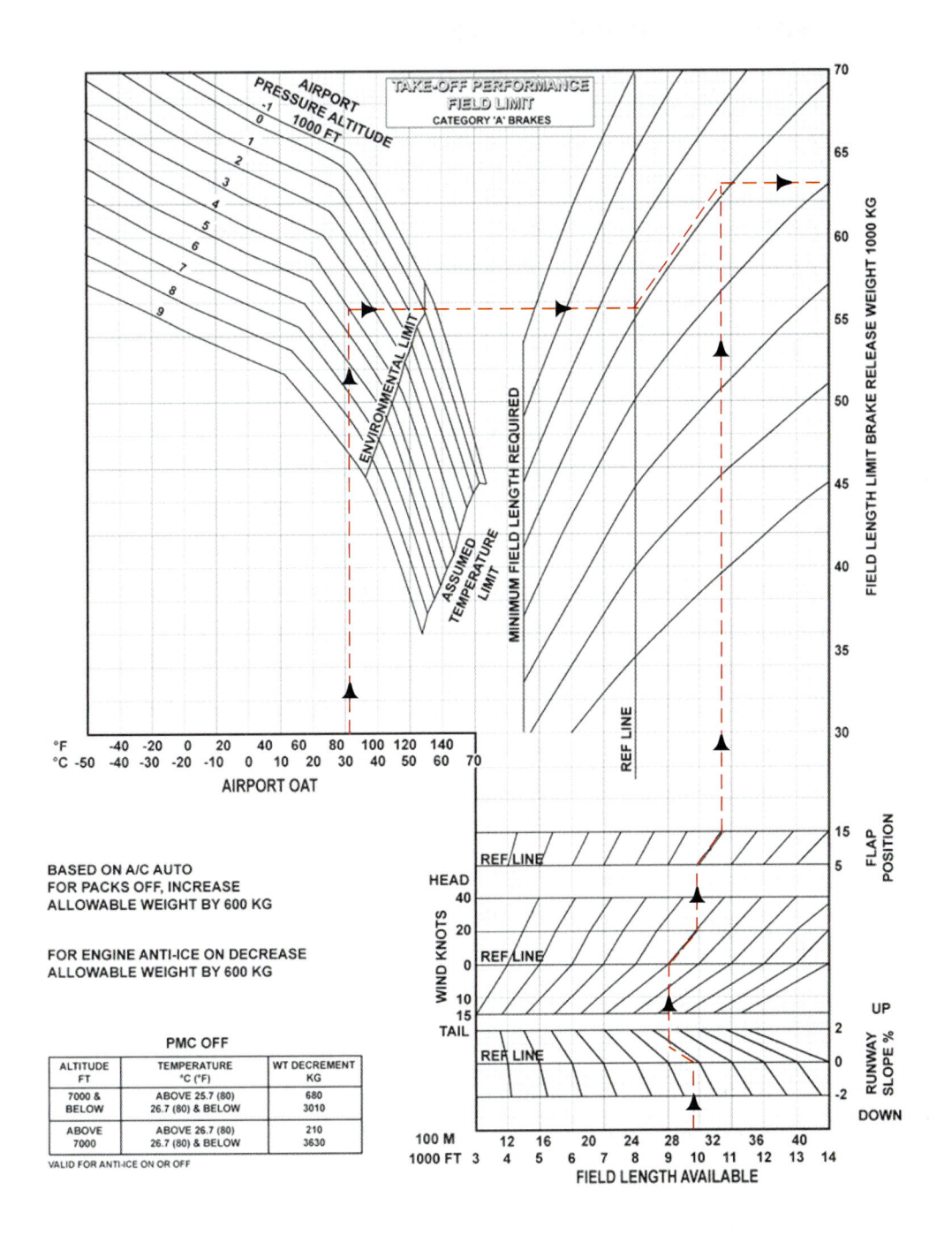

Figure 13.15

13.14 CLIMB GRADIENT LIMIT

The regulations lay down minimum values for climb gradient for each of the climb segments after take-off. Chapter 15 gives details of these requirements, and the most limiting will determine the Maximum Take-off Mass for altitude and temperature. Figure 13.16 shows a typical presentation of the climb limited take-off mass.

PERFORMANCE   **CLASS A AIRCRAFT: TAKE OFF**

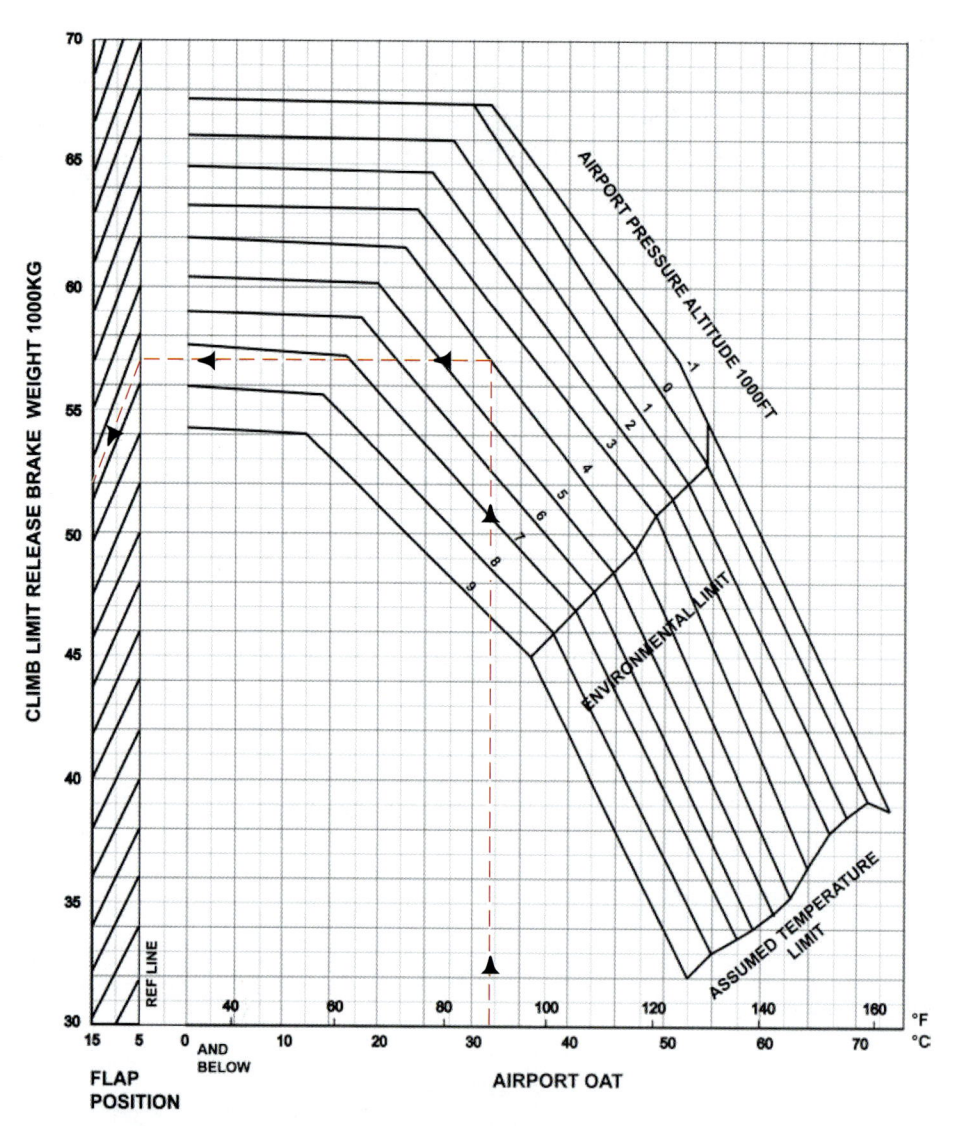

BASED ON A/C AUTO WITH APU ON OR OFF. FOR PACKS OFF, INCREASE ALLOWABLE WEIGHT BY 1000 KG.
FOR OPERATION WITH ENGINE ANTI-ICE ON SUBTRACT 140 KG WHEN AIRPORT PRESSURE ALTITUDE IS AT OR BELOW 8000FT OR 1280 KG WHEN AIRPORT PRESSURE ALTITUDE IS ABOVE 8000FT.

PMC OFF CORRECTION

ALTITUDE FT	TEMPERATURE °C	WT DECREMENT KG
7000 & BELOW	26.7 & ABOVE	790
	BELOW 26.7	4340
ABOVE 7000	26.7 & ABOVE	790
	BELOW 26.7	5110

Figure 13.16 Takeoff Performance Climb Limit

PERFORMANCE   **CLASS A AIRCRAFT: TAKE OFF**

13.15 TYRE SPEED LIMIT

The tyres are rated with a maximum permissible speed rating. This limiting speed is a True Ground Speed, and so the limiting IAS and consequent take-off mass will vary with altitude, temperature, wind speed, and take-off flap setting but will not be affected by runway slope. Figure 13.17 shows a typical presentation of the tyre speed limited take-off mass.

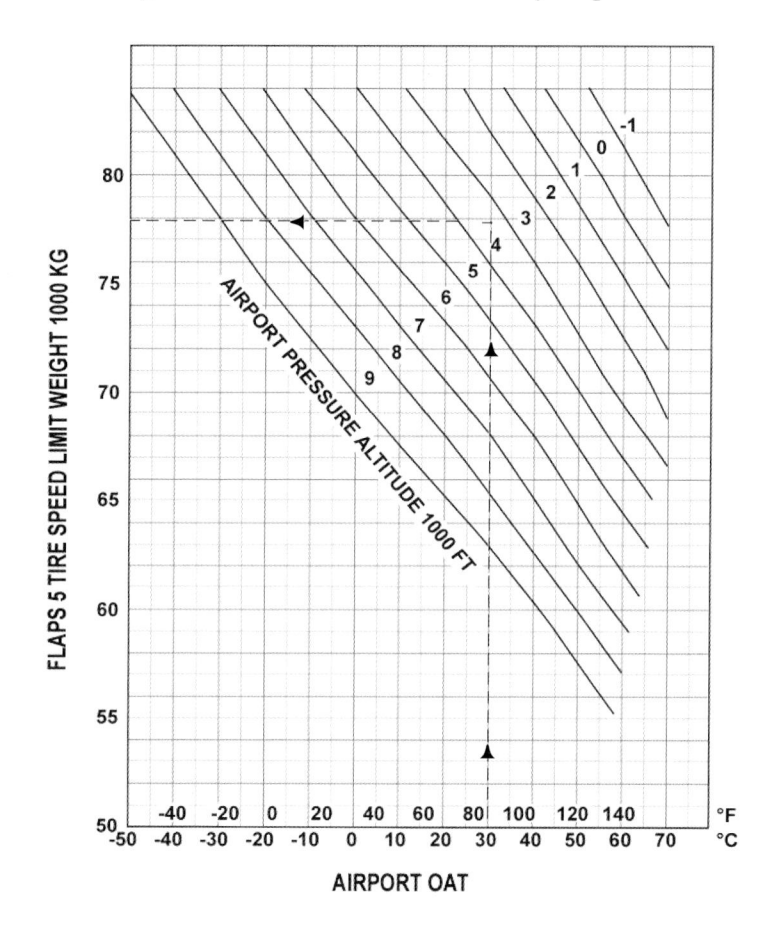

Figure 13.17 Takeoff Performance Tyre Speed Limit

Increase 'flaps 5' tyre speed limit weight by 6600 kg for flaps 15°
Decrease 'flaps 5' tyre speed limit weight by 9600 kg for 210 mph tyres and flaps 5.
Decrease 'flaps 5' tyre speed limit weight by 1500 kg for 210 mph tyres and flaps 15.
Increase tyre speed limit weight by 400 kg per knot of headwind.
Decrease tyre speed limit weight by 650 kg per knot of tailwind.

PMC OFF CORRECTION

ALTITUDE FT	TEMPERATURE °C	WT. DECREMENT KG
Below 5000	Above 21	250
	21 & Below	210
5000 & above	Above 21	200
	21 & Below	270

PERFORMANCE   CLASS A AIRCRAFT: TAKE OFF

13.16 BRAKE ENERGY LIMIT

For an aircraft of mass M, travelling at a true speed of V, the kinetic energy is ½ MV². If the aircraft is braked to a stop from this speed, a large proportion of this energy will go into the brakes as heat. The energy capacity of the brakes is limited and so for a given mass there will be a limiting speed from which a stop can be made. This will be a True Ground Speed, and so the corresponding IAS will vary with altitude, temperature and wind. Runway slope will also affect the speed, as a change in height involves a change in potential energy.

The brake energy limit speed V_{MBE} must not be less than the V_1 speed. If it is, the mass must be reduced until V_1 and V_{MBE} are the same. The Flight Manual will give the amount of weight to be deducted for each knot that V_1 exceeds V_{MBE}

For most aircraft, V_{MBE} will only be limiting in extremely adverse conditions of altitude, temperature wind and runway slope. Figure 13.18 shows a typical presentation for V_{MBE}.

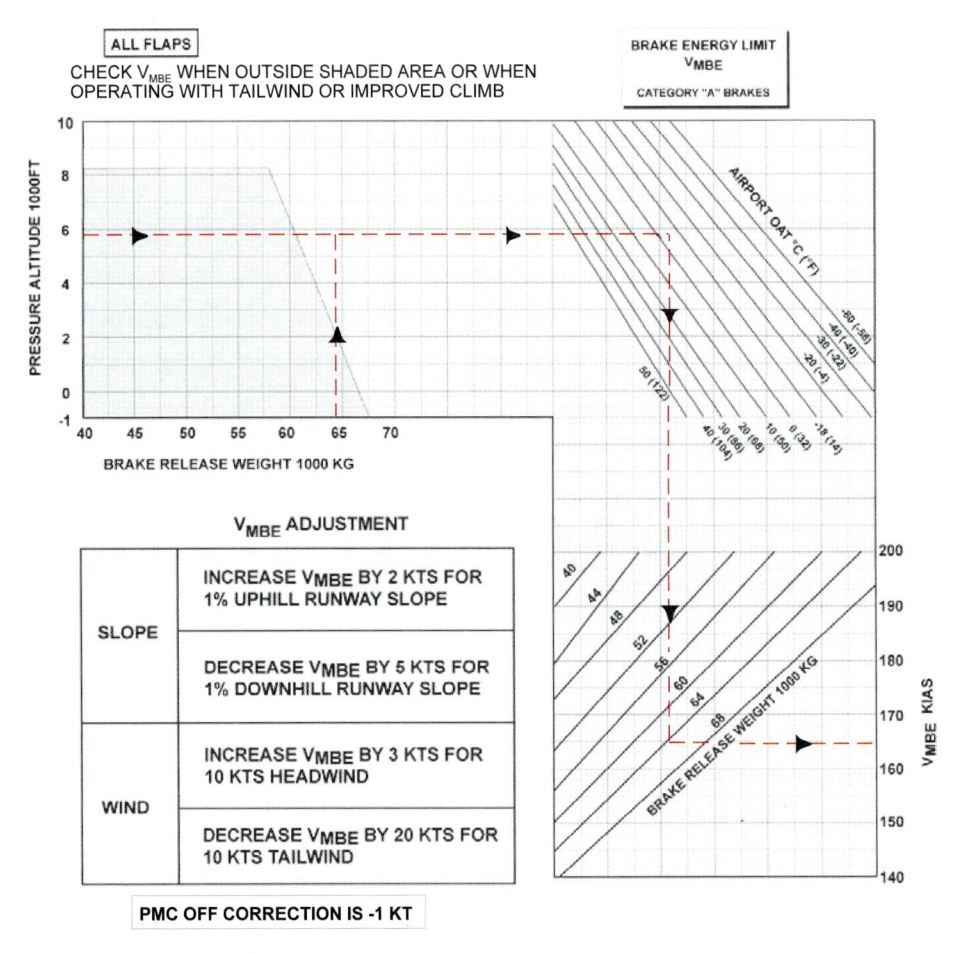

DECREASE BRAKE RELEASE WEIGHT BY 300 KG FOR EACH KNOT V_1 EXCEEDS V_{MBE}. DETERMINE NORMAL V_1, V_R, V_2 SPEEDS FOR LOWER BRAKE RELEASE WEIGHT.

Figure 13.18 Brake Energy Limit

13 - 20   © Oxford Aviation Services Limited

13.17 BRAKE COOLING

The value of V_{MBE} obtained from the data assumes that the brakes are at ambient temperature before the start of take-off. If a take-off is abandoned following a recent landing, or after prolonged taxying the brakes will already be at a fairly high temperature, and their ability to absorb further energy will be reduced. Data is given in the Manual to show the time to be allowed for the brakes to cool.

Figure 13.19 shows a typical brake cooling schedule.

13.18 RUNWAY STRENGTH

The operating mass of the aircraft may be limited by runway strength considerations. The bearing strength of a pavement is expressed by a PCN (Pavement Classification Number) and this is compared to the ACN (Aircraft Classification Number). The UK system of classification is the LCN (Load Classification Number) but this can be converted into the PCN system. The PCN is compared to the ACN. Operation on the pavement is permissible if the ACN is less than or equal to the PCN. Because the PCN includes a safety factor, a 10% increase of ACN over PCN is generally acceptable for pavements that are in good condition, and occasional use by aircraft with ACN's up to 50% greater than the PCN may be permitted.

PERFORMANCE   **CLASS A AIRCRAFT: TAKE OFF**

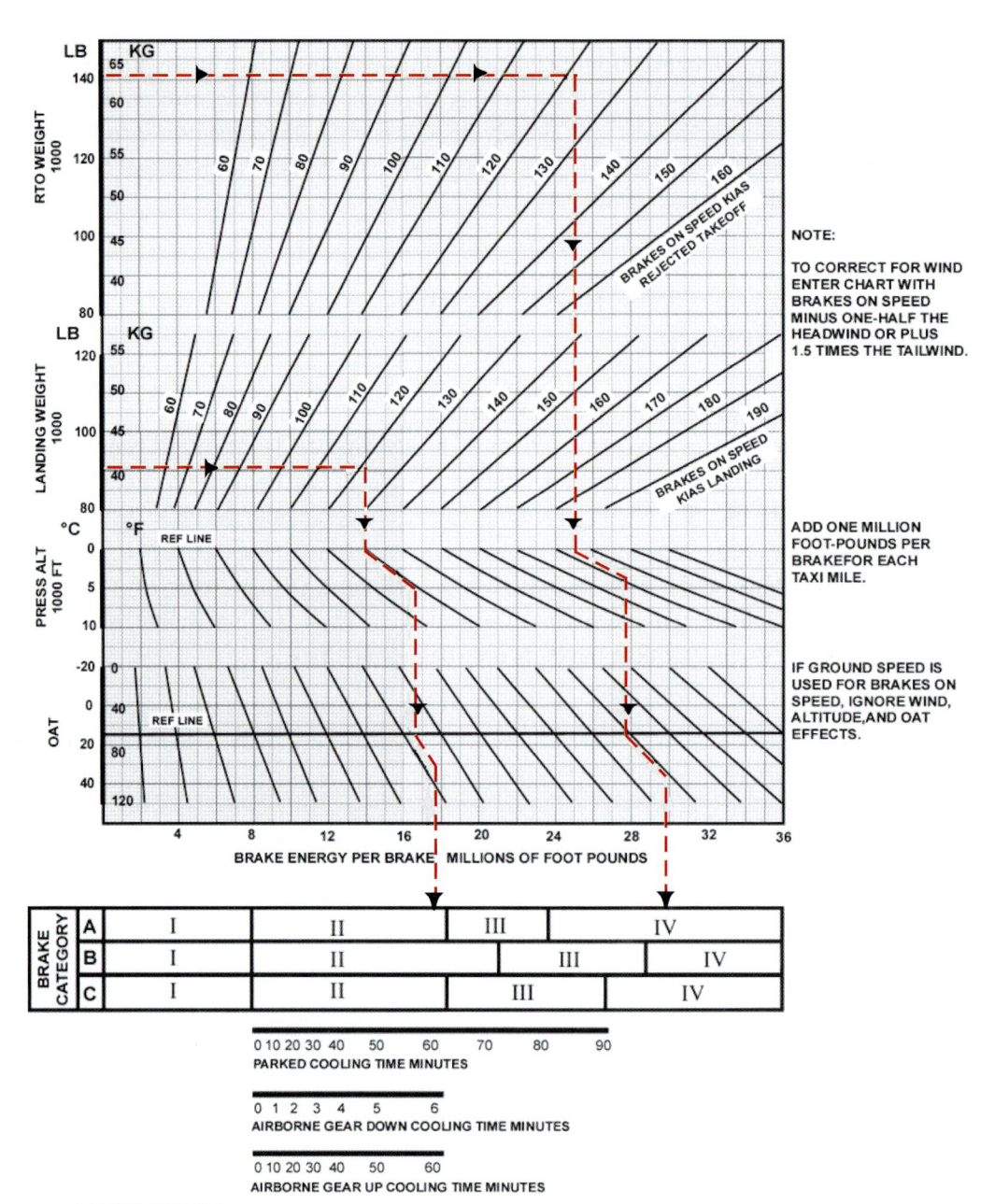

Figure 13.19

PERFORMANCE CLASS A AIRCRAFT: TAKE OFF

13.19 MAXIMUM TAKE-OFF MASS

Consideration of the mass determined by the field length available, the climb requirement, the tyre speed limit, the brake energy limit, and the maximum structural limit, will determine the maximum mass for take-off. It will be the lowest of the masses given by the above limitations. If there are obstacles to be considered on the take-off flight path this may determine a further limitation on take-off mass. Analysis of obstacle clearance limited mass is examined in Chapter 15.

13.20 TAKE OFF SPEEDS AND THRUST SETTINGS

When the maximum permissible take-off mass has been determined it is necessary to find the corresponding take-off speeds and thrust settings. Figures 13.20 and 13.21. show the presentation of the take-off speeds V_1, V_R and V_2 and the % N_1 for take off.

Take off speeds

The relevant column is determined from the "column reference" graph and the speeds read off for the take-off mass. Slope and wind corrections to V_1 can be determined from the V_1 adjustment table. V_1 must not exceed V_R, or be less than V_{MCG}.

If V_1 is less than V_{MCG} set V_1 equal to V_{MCG}.
If V_R is less than V_{MCG} set V_R equal to V_{MCG} and determine a new V_2 by adding the difference between the normal V_R and V_2 to V_{MCG}.
No take off weight adjustment is necessary provided that the field length exceeds the minimum field length required shown on the Field Limit performance chart.

PERFORMANCE   CLASS A AIRCRAFT: TAKE OFF

PMC ON

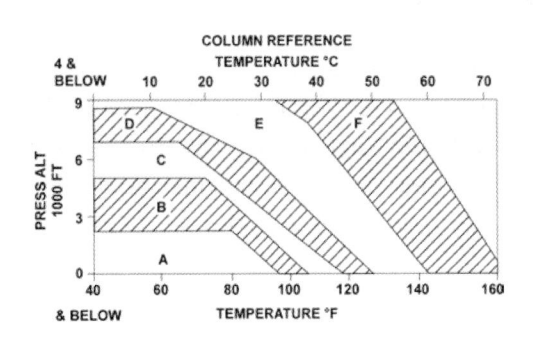

TAKEOFF SPEEDS

SLOPE/WIND V₁ ADJUSTMENT					
WEIGHT 1000	SLOPE %		WIND		KTS
	DN	UP	TAIL		HEAD
KG	-2	0	2	-15	0	40
70	-3	0	4	-3	0	1
60	-2	0	3	-3	0	1
50	-2	0	2	-3	0	2
40	-2	0	2	-3	0	2

*V₁ NOT TO EXCEED V_R

FLAPS 15

WT 1000 KG	A			B			C		
	V₁	V_R	V₂	V₁	V_R	V₂	V₁	V_R	V₂
70	148	151	157						
65	142	145	152	143	146	151	145	147	152
60	135	138	146	136	139	146	138	140	146
55	128	131	141	129	132	140	131	133	140
50	121	123	135	122	124	134	123	125	134
45	113	116	129	113	116	128	115	118	128
40	105	107	122	105	108	122	107	110	122

WT 1000 KG	D			E			F		
	V₁	V_R	V₂	V₁	V_R	V₂	V₁	V_R	V₂
70									
65									
60	139	140	146						
55	132	133	140	134	134	139			
50	124	126	134	126	127	133			
45	116	118	128	118	119	127	120	120	127
40	107	110	121	109	111	120	112	113	120

IN BOXED AREA CHECK MINIMUM V₁ (MCG) FOR ACTUAL TEMP.

MINIMUM V₁ (MCG)

ACTUAL OAT		PRESS ALT FT					
°C	°F	-1000	0	2000	4000	6000	8000
50	122	111	109	104			
40	104	116	114	110	105	100	96
30	86	120	120	115	110	106	102
20	68	120	120	117	114	110	106
10	50	120	120	117	114	111	108
-50	-58	121	121	118	115	112	109

FOR A/C OFF INCREASE V₁ (MCG) BY 3 KNOTS

Figure 13.20

PERFORMANCE   **CLASS A AIRCRAFT: TAKE OFF**

13.21 CORRECTION FOR STOPWAY AND CLEARWAY

The speeds shown in the table are based on a balanced field length (TORA = TODA = ASDA) and are not valid if the take-off mass has been derived using stopway or clearway. Where this is the case the V_1 may be adjusted for the effects of stopway or clearway from the table below.

V_1 ADJUSTMENTS

CLEARWAY MINUS STOPWAY FT	NORMAL V_1 KIAS			
	100	120	140	160
800	-	-	-3	-2
600	-	-3	-3	-2
400	-4	-3	-2	-1
200	-2	-2	-1	0
0	0	0	0	0
-400	2	2	2	1
-800	2	2	2	1

The adjusted V_1 must not exceed V_R

The maximum allowable clearway for a given field length is shown below.

Field Length ft	Max Allowable Clearway for V_1 Reduction Ft
4000	450
6000	550
8000	600
10000	650
12000	700
14000	750

The adjusted V_1 must be compared to V_{MCG} which can be read from the Minimum V_1 table.

If V_1 is less than V_{MCG}, set V_1 = VMCG

PERFORMANCE

CLASS A AIRCRAFT: TAKE OFF

Thrust setting

MAX TAKE-OFF %N1

VALID FOR 2 PACKS ON (AUTO) ENGINE A / I ON OR OFF											
AIRPORT OAT		AIRPORT PRESSURE ALTITUDE FT									
°C	°F	-1000	0	1000	2000	3000	4000	5000	6000	7000	8000
55	131	93.6	93.5	93.0	92.5						
50	122	94.8	85.1	94.9	94.8	94.4	93.3				
45	113	95.4	95.9	95.9	96.0	95.8	95.4	95.3	94.7	94.1	
40	104	96.1	96.6	96.6	96.6	96.5	96.4	96.4	96.4	96.4	95.9
35	93	96.6	97.3	97.2	97.2	97.1	97.1	97.2	97.1	97.1	97.1
30	86	96.3	97.8	97.8	97.9	97.8	97.8	98.9	97.9	98.0	97.9
25	77	95.7	97.0	97.6	98.3	98.6	98.6	98.7	98.7	98.8	98.7
20	68	94.9	96.2	96.7	97.5	98.1	98.7	99.3	99.4	99.5	99.4
15	59	94.1	95.4	95.9	96.6	97.2	97.8	98.4	99.1	99.7	99.9
10	50	93.3	94.6	95.1	95.8	96.4	97.0	97.6	98.2	98.8	99.3
5	41	92.5	93.7	94.2	94.9	95.5	96.1	96.7	97.3	97.9	98.4
0	32	91.6	92.9	93.4	94.1	94.7	95.2	95.8	96,5	97.0	97.5
-10	14	89.9	91.1	91.6	92.3	92.9	93.5	94.1	94.7	95.3	95.7
-20	-4	88.2	89.4	89.9	90.6	91.1	91.7	92.3	92.9	93.4	93.9
-30	-22	86.4	87.6	88.1	88.8	89.3	89.9	90.4	91.0	91.6	92.0
-40	-40	84.7	85.8	86.3	86.9	87.4	88.0	88.5	89.1	89.7	90.1
-50	-58	82.8	83.9	84.4	85.0	85.6	86.1	86.6	87.2	87.7	88.1

PMC ON

% N1 BLEED ADJUSTMENTS	
CONFIGURATION	
A / C PACKS OFF	+1.0

DO NOT OPERATE ENGINE ANTI-ICE "ON" AT AIRPORT OAT ABOVE 10°C (50°F)

Figure 13.21

13.22 STABILISER TRIM SETTING

The stabiliser trim setting appropriate to the C.G. position and take-off mass can be read from the table below.

FLAPS 15 STABILISER TRIM SETTING

C.G. % MAC	6	10	14	18	22	26	30
STAB TRIM	5	4¼	3¾	3	2½	1¾	1

FOR WEIGHTS AT OR BELOW 45,400 KG SUBTRACT ½ UNIT.
FOR WEIGHTS AT OR ABOVE 61,300 KG ADD ½ UNIT.

PERFORMANCE — CLASS A AIRCRAFT: TAKE OFF

PAPER 1

1. The speed V_{MU} is :

    a) the maximum speed for flight with the undercarriage extended.
    b) the maximum speed at which the aircraft should become airborne.
    c) the minimum speed at which the aircraft can safely lift off the ground.
    d) the minimum speed at which the elevators can rotate the aircraft until the tail bumper is in contact with the runway.

2. The speed V_1 is :

    a) the stalling speed with the flaps in a prescribed position.
    b) the critical speed for engine failure during take off.
    c) the speed at which, with the critical engine inoperative, the TODR will not exceed the TODA, the TORR will not exceed the TORA. And the ASDR will not exceed the ASDA.
    d) the speed at which rotation to the unstick attitude is initiated.

3. The Take off Distance Required on a dry runway is :

    a) 115% of the distance to 35 ft. the critical engine having failed at V_{EF}
    b) the distance to reach 35 ft. with all engines operating.
    c) the distance to reach 15 ft. with the critical engine having failed at V_{EF}
    d) the greater of the distance to reach 35 ft. with the critical engine having failed at V_{EF}, and 115% of the distance to reach 35 ft. with all engines operating.

4. When calculating the Accelerate-Stop Distance Required, braking is assumed to begin

    a) at the speed at which the engine fails.
    b) at the speed V_1
    c) at the speed reached 2 seconds after the engine failure has occured
    d) at the speed reached after 2 seconds after V_1, with all engines operating, or with one engine inoperative, whichever is limiting.

5. Select the correct sequence of speeds :

    a) V_{MCG} V_{EF} V_1 V_R
    b) V_{MCG} V_{EF} V_R V_1
    c) V_{EF} V_1 V_{MCG} V_R
    d) V_{MCG} V_1 V_{EF} V_R

PERFORMANCE   CLASS A AIRCRAFT: TAKE OFF

6. The minimum value of V_2 (V_{2min}) :

   a) must not be less than 1.1 V_{MC} or 1.13 V_{SR}
   b) must not be less than 1.1 V_{SR} or 1.2 V_{MC}
   c) must not be less than 1.1 V_{MU} or 1.13 V_{SR}
   d) must not be less than 1.1 V_{LOF} or 1.13 V_{SR}

7. Which of the following statements is correct :

   a) If the aircraft is rotated before V_R to the normal attitude, the take off distance required will be reduced.
   b) If the aircraft is rotated before V_R to a greater than normal attitude, the take off distance required will be reduced.
   c) If the aircraft is rotated after V_R to the normal attitude, the take off distance required will be increased.
   d) If the aircraft is rotated after V_R to a greater than normal attitude, the take off distance will be reduced.

8. The Balanced Field Length for an aircraft is when, in the event of an engine failure during take off :

   a) the distance to accelerate is equal to the distance to stop.
   b) the take off distance required is equal to the accelerate- stop distance required
   c) the take off run required is equal to the accelerate-stop distance required
   d) the take off distance required is equal to the take off run required.

9. If the balanced field V_1 exceeds V_{MBE}

   a) V_{MBE} must be increased to V_1 and the field length required will be greater than the balanced field length.
   b) V_1 must be reduced to V_{MBE} and the field length required will be greater than the balanced field length.
   c) V_1 must be reduced to V_{MBE} and the field length required will be less than the balanced field length
   d) V_{MBE} must be increased to V_1 and the field length required will be less than the balanced field length.

10. If the balanced field length required for a given weight is less than the balanced field length available :

    a) Take off may not be made at that weight.
    b) the V_1 must be increased above the balanced field V_1
    c) the V_1 must be decreased below the balanced field V_1
    d) there will be a range of V_1 speed available.

PERFORMANCE   CLASS A AIRCRAFT: TAKE OFF

PAPER 2

Refer to CAP 698 Fig. 4.4 for questions 1, 2 and 3.

1. Find the field length limited take off mass, given the following data :

    Pressure altitude   3000 ft.   Take off flap   15°
    Temperature   15°C   A/C   Auto
    B.F.L. Available   2000 m.   PMC   On
    Runway slope   1.0% Downhill
    Reported wind comp.   20 kt. Headwind

    a)   52200 kg.
    b)   58200 kg.
    c)   53000 kg.
    d)   61500 kg.

2. Find the field length limited take off mass given the following data :

    Pressure altitude   2500 ft.   Take off flap   5°
    Temperature   30°C   A/C   Off
    B.F.L. Available   6000 ft.   PMC   On
    Runway slope   level
    Reported wind comp.   25 kt. Headwind

    a)   53500 kg.
    b)   53750 kg.
    c)   55000 kg.
    d)   55820 kg.

3. What minimum balanced field length must be available for take off at a mass of 56000 kg. For the following conditions :

    Pressure altitude   4000 ft.   Take off flap   15°
    Temperature   20°C   A/C   Auto
    Reported wind comp.   20 kt. Headwind   PMC   On
    Runway slope   1.0% Downhill

    a)   6300 ft.
    b)   8000 ft.
    c)   8500 ft.
    d)   4500 ft.

PERFORMANCE   **CLASS A AIRCRAFT: TAKE OFF**

Refer to CAP 698 Fig. 4.5 for questions 4 and 5

4. Find the maximum take off mass determined by the climb requirement for the following take off conditions :

Pressure altitude	3500 ft.	Take off flap	15°
Temperature	30°C	A/C	Auto
		PMC	On

   a)   60200 kg.
   b)   53500 kg.
   c)   56000 kg.
   d)   51600 kg.

5. Find the maximum take off mass determined by the climb requirement for the following take off conditions. :

Pressure altitude	500 ft.	Take off flap	15°
Temperature	10°C	A/C packs	Off
		PMC	On

   a)   59700 kg.
   b)   58800 kg.
   c)   53000 kg.
   d)   63400 kg.

Refer to CAP 698 Fig. 4.6 for questions 6 and 7

6. Find the tyre speed limited take off mass, given :

Pressure altitude	4000 ft.	Take off flap	5°
Temperature	15°C	A/C	Auto
Reported wind comp.	10 kt. Tailwind	PMC	Off
		Tyre rating	225 mph

   a)   79500 kg.
   b)   73000 kg.
   c)   78850 kg.
   d)   74000 kg.

PERFORMANCE   CLASS A AIRCRAFT: TAKE OFF

7. When considering the tyre speed limit, up to what temperature would a take off mass of 64000 kg. be permissible in the following conditions, with 225 mph. rated tyres :

   Pressure altitude   2000 ft.   Take off flap   5°
   Reported wind comp.   10 kt. Tailwind   PMC   On
                                                             A/C   Auto

   a)   50°C
   b)   67°C
   c)   70°C
   d)   In excess of 70°C

8. **Refer to CAP 698 Fig. 4.7 for question 8**

   Find the brake energy limit speed for the following conditions :

   Pressure altitude   4000 ft.   Take off flap   5°
   Temperature   25° C   A/C   Auto
   Take off mass   65000 kg.   PMC   Off
   Reported wind comp.   15 kt. Headwind
   Runway slope   1.5% downhill

   a)   153.5 kt.
   b)   157.5 kt.
   c)   160.5 kt.
   d)   162.5 kt.

9. **Refer to CAP 698 Fig. 4.9 for question 9**

   For the following aerodrome conditions, a take off mass of 62000 kg. has been determined.
   Pressure altitude   3000 ft.   Take off flap   15°
   Temperature   20° C   A/C   Auto
   Reported wind comp.   20 kt. Headwind   PMC   On
   Runway slope   1.0% Uphill

   Find the speeds V_1 (i) V_R (ii) V_2 (iii)

	(i)	(ii)	(iii)
a)	136 kt.	139 kt.	146 kt.
b)	139 kt.	142 kt.	148 kt.
c)	145 kt.	143.5 kt.	149.5 kt.
d)	142.5 kt.	142.5 kt.	148 kt.

PERFORMANCE   CLASS A AIRCRAFT: TAKE OFF

10. For the conditions in Question 9, find the take off % N_1 (i) and the stabilator trim setting,(ii) if the C.G. position is 22% MAC.

	(i)	(ii)
a)	96.9%	2½ units
b)	95.9%	3 units
c)	91.8%	3½ units
d)	91.5%	3 units

PERFORMANCE **CLASS A AIRCRAFT: TAKE OFF**

ANSWERS

PAPER 1

Questions	Answers
1	C
2	C
3	D
4	D
5	D
6	A
7	C
8	B
9	B
10	D

PAPER 2

Questions	Answers
1	B
2	B
3	A
4	D
5	A
6	B
7	B
8	A
9	D
10	B

CHAPTER FOURTEEN - CLASS A: ADDITIONAL TAKE OFF PROCEDURES

Contents

                                                      Page

14.1   NON STANDARD TAKE-OFF PROCEDURES . . . . 14 - 1

14.2   CONTAMINATED RUNWAYS . . . . 14 - 1

14.3   TAKE-OFF WITH INCREASED V_2 SPEED . . . . 14 - 5

14.4   TAKE-OFF WITH REDUCED THRUST . . . . 14 - 8

14.5   TAKE-OFF WITH ANTI-SKID INOPERATIVE . . . . 14 - 11

         CLASS A AIRCRAFT: ADDITIONAL TAKE OFF PROCEDURES . . . . . . 14 - 13

© Oxford Aviation Services Limited

© Oxford Aviation Services Limited

PERFORMANCE — CLASS A: ADDITIONAL TAKE OFF PROCEDURES

14.1 NON STANDARD TAKE-OFF PROCEDURES

The procedure to determine the take-off mass, take-off speeds, and thrust settings for the normal take-off procedure is given in the previous chapter. Chapter 14 gives additional procedures to cover:

a) take-off with contaminated runway

b) take-off with increased V_2 speed

c) take-off with reduced thrust

d) Take-off with anti-skid inoperative

14.2 CONTAMINATED RUNWAYS

A runway is considered to be contaminated when more than 25% of the runway surface area (whether in isolated areas or not) within the required length and width being used, is covered by surface water, more than 3mm. deep, or by slush or loose snow, equivalent to more than 3mm. of water. (AMJ 25X1591)

Slush, loose snow or standing water on the runway will affect both the Take-off distance Required and the Accelerate-stop Distance Required. The Take-off Distance Required will increase because of the additional wheel drag and impingement drag. The accelerate-stop distance will increase because of the increased distance to accelerate and the increased distance to stop resulting from the reduced runway coefficient of braking friction. For given distances available, the maximum take-off mass and the V_1 will therefore be reduced compared to the dry runway.

The supplementary performance information required by JAR25X1591 should include accelerate-stop distance, take-off distance and take-off run appropriate to the relevant contaminant, derived in a similar manner to those distances with a wet runway.

The acceleration distance should take account of the additional drag due to the gear displacement drag, and the spray impingement drag, and the decrease of drag which occurs above the aquaplaning speed. The aquaplaning speed V_p may be taken as :

$$V_p = 9 \sqrt{\frac{p}{\sigma}}$$

Where p = tyre pressure in lb/sq.in.

σ = specific gravity of precip.

V_p = speed in knots

The stopping distance for runways contaminated by standing water, slush or loose snow, should assume a coefficient of friction of 0.25 x the dry runway coefficient, at or below 0.9 V_p and 0.05 above this speed.

PERFORMANCE   **CLASS A: ADDITIONAL TAKE OFF PROCEDURES**

Typical presentation with instructions for use are shown in Figure 14.1.

PROCEDURE.

1. Determine the normal maximum take-off mass (Field length or obstacle limit)

2. Enter the left hand table for the appropriate depth and pressure altitude (interpolating if necessary) with the mass from (1) and read the weight reduction.

3. If in the shaded area enter the right hand table with the available field length and pressure altitude and read the limiting mass with contamination for the available field length and $V_1 = V_{MCG}$.

4. The maximum permissible mass is the less of those obtained from (2) or (3)

Take off speeds with contamination

1. Determine V_1 V_R and V_2 for the actual take off weight from the take off speed table (Figure 13.30)

2. If V_{MCG} limited, set $V_1 = V_{MCG}$
If not V_{MCG} limited, re-enter the left hand table with the actual weight to determine the V_1 reduction to apply to V_1 .
If the adjusted V_1 is less than V_{MCG}, set $V_1 = V_{MCG}$

PERFORMANCE **CLASS A: ADDITIONAL TAKE OFF PROCEDURES**

ADVISORY INFORMATION
ONE ENGINE INOPERATIVE

SLUSH/ STANDING WATER TAKEOFF

0.5 INCH (13mm) SLUSH/STANDING WATER DEPTH

GROSS WEIGHT 1000 KG	WEIGHT AND V_1 REDUCTIONS			
	PRESS ALT	S.L.	4000 FT	8000 FT
40	1000 KG KIAS	4.1 14	5.2 9	6.6 1
44	1000 KG KIAS	5.4 12	6.6 6	8.4 0
52	1000KG KIAS	7.9 4	9.9 0	11.7 0
56	1000 KG KIAS	9.4 0	11.4 0	12.6 0
60	1000 KG KIAS	10.8 0	12.4 0	13.0 1
64	1000KG KIAS	11.8 0	12.9 0	12.8
68	1000 KG KIAS	12.7 0	13.2 0	12.3 3

FIELD LENGTH AVAILABLE FT	$V_1 = V_{1(MCG)}$ LIMIT WEIGHT 1000 KG		
	PRESSURE ALTITUDE		
	SEA LEVEL	4000 FT	8000 FT
5200	39		
5400	42		
5600	45	38	
5800	48	40	
6000	51	43	
6200	54	45	38
6400	57	48	40
6600	60	50	42
6800	64	53	44
7000	67	55	47
7200		58	49
7400		61	51
7600		63	53
7800		66	55
8000		69	57

INSTRUCTIONS:

-T.O. WEIGHT
1. ENTER WEIGHT AND V_1 REDUCTIONS TABLE FOR SLUSH DEPTH WITH DRY FIELD/OBSTACLE LIMIT TO OBTAIN SLUSH WEIGHT DECREMENT.

2. IF IN SHADED AREA FIND $V_{1\,(MCG)}$ LIMITED WEIGHT FOR AVAILABLE FIELD LENGTH AND PRESSURE ALTITUDE.

3. MAX ALLOWABLE SLUSH WEIGHT IS LESSOR OF WEIGHTS FROM 1 AND 2.

-T.O. SPEEDS
1. OBTAIN V_1, V_R, V_2 FOR ACTUAL WEIGHT FROM QRH.

b) IF $V_{1\,(MCG)}$ LIMITED, SET V_1 $V_{1\,(MCG)}$
IF NOT $V_{1(MCG)}$ LIMITED, ENTER WEIGHT AND V_1 REDUCTIONS TABLE WITH ACTUAL WEIGHT TO OBTAIN V_1 SPEED ADJUSTMENT.
IF ADJUSTED V_1 IS LESS THAN $V_{1(MCG)}$ SET V_1 $V_{1\,(MCG)}$

Figure 14.1

PERFORMANCE — CLASS A: ADDITIONAL TAKE OFF PROCEDURES

The procedure is illustrated by the following example:

Given: Pressure altitude: 4000ft
        Temperature: 0°C
        Flap setting: 5°
        TORA: 7000ft
        R/W slope: 2% uphill
        Wind: Still Air
        Contamination: 0.5in standing water
        PMC and ACS on

Solution: Using the data in CAP 698

Figure 4.4   Field length limit weight = 51800 kg

Figure 4.14   (0.5 in slush/standing water)
                from left hand table:
                Weight reduction 10100 kg,
                ∴T.O. Wt = 41700 kg

Data is in the shaded area, therefore enter right-hand table with field length and press alt.
$V_1 = V_{MCG}$ limited weight = 57000 kg
∴ Max take-off weight = 41700 kg (lower weight)

Figure 4.8   Take-off speeds
                Column reference: zone B
                For 41700 kg, 2% uphill slope,
                V_1 = 118 kt VR = 118 kt V_2 = 131 kt
                V_{MCG} = 111 kt ∴not V_{MCG} limited.

Figure 4.14   Re-enter left hand table with 41700 kg
                V_1 reduction = 4 kt
                ∴V_1 = 114 kt (not less than V_{MCG})
                ∴V_1 = 114 kt

PERFORMANCE   CLASS A: ADDITIONAL TAKE OFF PROCEDURES

14.3 TAKE-OFF WITH INCREASED V_2 SPEED

V_2 is the speed achieved at the screen after engine failure during take-off, and determines both the Take-off Distance Required and the climb gradient achieved after take-off. As V_2 is not the speed for best climb angle, an increased climb gradient could be achieved by increasing the V_2 speed. However, this would increase the Take-off Distance Required.

If the maximum take-off mass is limited by the climb requirement, take-off at this mass will use less than the Take-off Distance Available. Increasing the V_2 speed will therefore increase the climb limited mass and decrease the field length limited mass, as shown in figure 14.2.

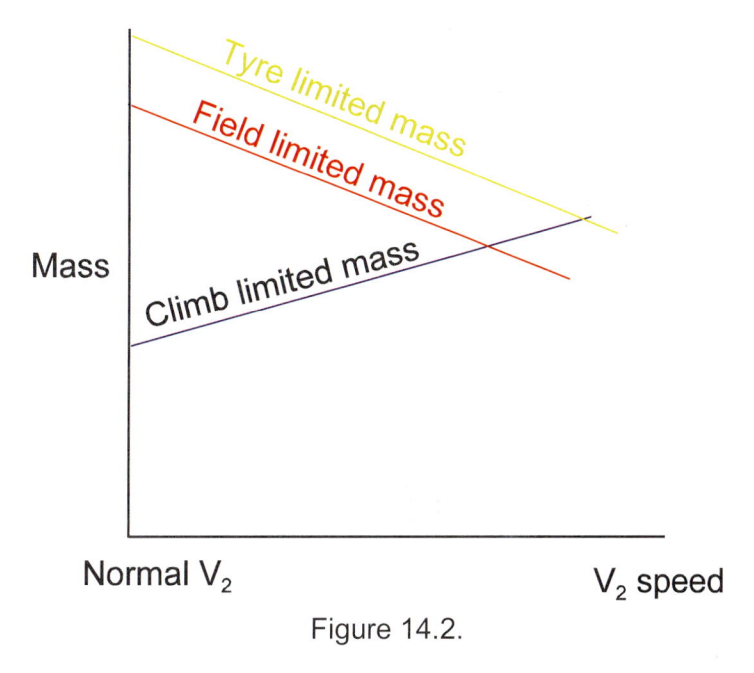

Figure 14.2.

It can be seen that there will be a value of speed increase that will cause the gradient and field length to be equally limiting. This will be the optimum increase for V_2. A lesser increase would give an improved climb gradient mass, but there would still be a surplus Take-off Distance Available.

To achieve an increased V_2 speed, the V_R must be increased, and it will be necessary to ensure that the tyre limit speed is not exceeded. This is usually done by finding the V_2 increase that would give equal mass limits from the tyre speed limit and the climb gradient limit. The lesser mass improvement is then selected and added to the original climb limited mass to obtain the improved take-off mass. The speed increments are then added to the normal speeds for this improved take-off mass. Figure 14.3 shows a typical presentation for the increased V_2 data.

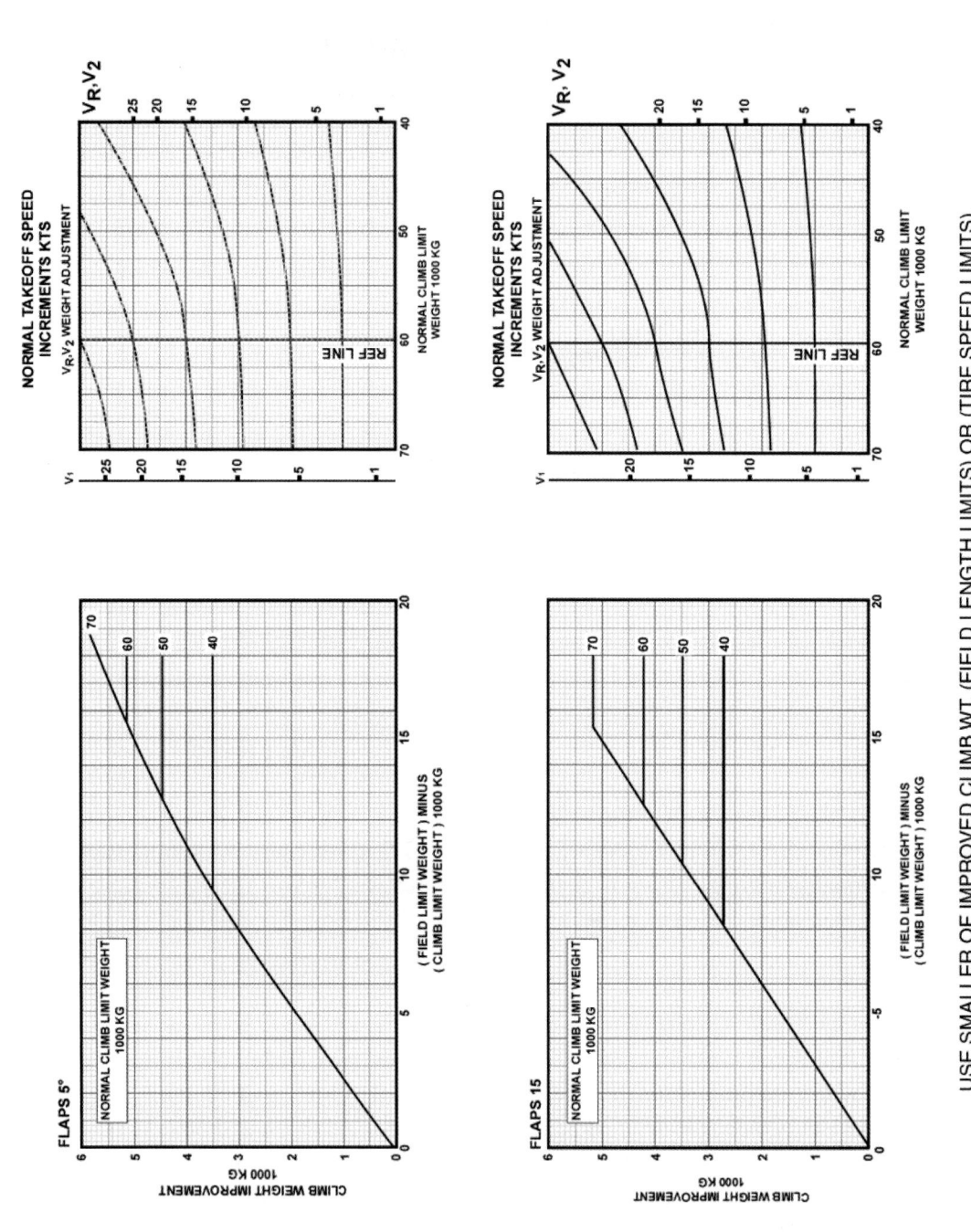

Figure 14.3 Improved Climb Performance Field Length Limit

PERFORMANCE   CLASS A: ADDITIONAL TAKE OFF PROCEDURES

PROCEDURE

1. Find the difference between the field length limit mass and the climb limit mass.

2. Enter the left hand graph with this mass difference, and read the mass increment on the left hand scale.

3. Move horizontally to the V_1 scale and read the V_1 increment.

4. Move horizontally to the V_R / V_2 reference line, correct for climb limit mass, and read the V_R and V_2 increments.

5. Add the mass increment to the climb limit mass.

6. Repeat the procedure on the tyre limit graph, using the difference between the tyre limit mass and the climb limit mass.

7. The maximum improved mass is the less of the field length and the tyre speed improved masses.

8. Find the normal speeds for the selected improved take-off mass.

9. Add the speed increments from (3) and (4) to the speeds in (8)

The use of the graphs is illustrated by the following example:

Given: Aerodrome pressure altitude: 6000ft
        Temperature: 20°C
        Flap setting: 15°
        Take-off Runway: Level, 9000ft
        Wind: Still air
        PMC and AC On

Solution: Using the data in CAP 698
    Figure 4.4 Field Length limit weight 58000 kg
    Figure 4.5 Climb limit weight 50800 kg
    Figure 4.6 Tyre limit weight 79400 kg
    Figure 4.15 (Flaps 15°)

Enter left-hand graph with Field length weight - Climb weight
        58000 kg - 50800 kg = 7200 kg
Read weight improvement on left-hand scale = 2450 kg
Move horizontally to V_1 scale V_1 increase = 10.5 kt
Move horizontally to Ref. Line, correct for climb limit weight
Read V_R, V_2 increase = 13 kt

PERFORMANCE   **CLASS A: ADDITIONAL TAKE OFF PROCEDURES**

Figure 4.16   Enter with Tyre limit weight - Climb weight
79400 - 50800 = 28600 kg

Weight improvement = 3500 kg

Revised Take-off weight = 50800 kg + 2450 kg = 53250 kg.
(Note: If the tyre limit gives a lower improved weight, the lower weight must be used)

Figure 4.8   Check column reference graph - Zone D

Read normal speeds for 53250 kg:	V_1 131kt,	V_R 131kt	V_2 138 kt
Add speed increments	10.5 kt	13 kt	13 kt
Increased speeds:	V_1, 141.5 kt	V_R 144 kt	V_2 151 kt

14.4 TAKE-OFF WITH REDUCED THRUST

For many aerodromes, if take-off is made with the maximum permissible thrust, in high performance conditions such as low temperature, low pressure altitude, low aeroplane mass, the take-off will be completed well within the aerodrome distances available. In these conditions it would be possible to take off with reduced thrust and use part or all of the surplus distances. Using reduced thrust for take-off results in increased engine life, and consequently reduced operating costs.

Take-off with reduced thrust is not permitted with:

a)   icy or very slippery runways

b)   contaminated runways

c)   ant-skid inoperative

d)   reverse thrust inoperative

e)   increased V_2 procedure

f)   PMC off

Reduced thrust take off procedure is not recommended if potential windshear conditions exist.

PERFORMANCE — CLASS A: ADDITIONAL TAKE OFF PROCEDURES

PROCEDURE

For take-off at a given pressure altitude, the maximum % N_1 is determined by the OAT. To find the reduced rpm to be used it is necessary to find the OAT at which the aircraft would become performance limited at the pressure altitude. This is done by entering the graphs for:

a) field length
b) climb limit
c) tyre speed limit
d) obstacle limit

with the **required** take-off mass and pressure altitude and extracting the limiting OAT for each case. The most limiting OAT is the lowest of these. The maximum thrust reduction permitted is 25% and so the assumed OAT must not exceed that which would give a 25% reduction. The take-off rpm is then based on the assumed OAT. The procedure is sometimes known as the assumed temperature take-off.

The take-off speeds are determined for the assumed temperature, apart from the V_{MCG} which must be determined for the actual aerodrome temperature. This is to allow for the situation where the thrust of the operating engine is increased to the normal maximum take-off thrust after engine failure, and directional control must be maintained with this asymmetric thrust.

Figure 14.4 shows a typical presentation of the reduced thrust data.

CHART READING PROCEDURE

a) For the required take-off mass determine the limiting OAT for the field length, climb, tyre limit, obstacles.

b) From the top table (figure 14.4) read the maximum assumed OAT (25% thrust reduction).

c) Take the lower temperature from (a) and (b).

d) With this temperature in the centre table, read the maximum take-off %N_1.

e) Do not use an assumed temperature less than the Minimum assumed Temperature shown in the bottom line of the centre table.

f) Subtract the actual OAT from the assumed OAT and enter the bottom table with this difference and read the %N_1 adjustment to be subtracted from the %N_1 from (d).

g) Determine the take-off speeds using the assumed temperature (except for V_{MCG} use actual OAT).

PERFORMANCE CLASS A: ADDITIONAL TAKE OFF PROCEDURES

-400 23.5K ASSUMED TEMPERATURE REDUCED THRUST

[PMC ON]

ASSUMED TEMP %N_1 = MAX %N_1 MINUS %N_1 ADJUSTMENT

MAXIMUM ASSUMED TEMPERATURE *

OAT °C	PRESS ALT 1000 FT									OAT °F	PRESS ALT 1000 FT								
	-1	0	1	2	3	4	5	6	7	8		-1	0	1	2	3	4	5	6	7	8
55	70	71	69								130	158	158	156							
50	67	67	67	67	65	63					120	152	152	151	151	148	146				
45	65	65	64	64	62	61	60	59	57		110	149	148	147	145	143	140	139	137	135	
40	64	63	62	62	60	59	58	57	56	55	100	145	144	143	142	139	136	134	133	131	129
35	62	61	61	60	58	57	56	55	54	53	90	142	140	139	138	136	132	130	129	127	126
30	61	59	59	57	57	55	54	53	52	51	80	142	139	136	134	132	129	127	125	123	122
25	61	59	58	56	55	53	52	51	50	49	70	142	139	136	133	130	126	124	122	120	118
20	61	59	58	56	55	52	51	50	49	48	60 AND BELOW	142	139	136	133	130	126	123	121	118	116
15 AND BELOW	61	59	58	56	55	52	51	49	48	46											

*BASED ON 25% TAKEOFF THRUST REDUCTION

MAX TAKEOFF %N_1

VALID FOR 2 PACKS ON (AUTO) ENGINE A/I ON OR OFF	FOR A/C OFF ADD 1.0 %N_1									
ASSUMED TEMP	AIRPORT PRESSURE ALTITUDE FT									
°C / °F	-1000	0	1000	2000	3000	4000	5000	6000	7000	8000
75 / 167	83.0	86.2	87.2							
70 / 158	85.7	86.3	86.5	87.3	87.3					
65 / 149	88.5	88.8	88.4	88.0	86.8	86.8	87.4	88.2	88.9	89.5
60 / 140	91.2	91.2	90.8	90.3	89.4	88.2	87.6	87.5	88.2	88.8
55 / 131	93.5	93.6	93.1	92.6	91.9	90.8	90.2	89.6	89.0	88.3
50 / 122	94.7	95.1	94.9	94.8	94.4	93.3	92.2	92.2	91.6	90.9
45 / 113	95.3	95.9	95.9	96.0	95.8	95.4	95.3	94.8	94.1	93.4
40 / 104	96.0	96.7	96.6	96.6	96.5	96.4	96.4	96.4	96.4	95.9
35 / 95	96.5	97.3	97.3	97.2	97.2	97.1	97.2	97.2	97.1	97.1
30 / 86	96.4	97.8	97.8	97.9	97.8	97.8	98.0	97.9	98.0	97.9
25 / 77	95.6	97.0	97.6	98.3	98.6	98.6	98.7	98.7	98.8	98.7
20 / 68	94.8	96.2	96.8	97.5	98.1	98.7	99.3	99.4	99.5	99.4
15 / 59	94.0	95.4	95.9	96.6	97.2	97.8	98.4	99.1	99.7	99.9
MINIMUM ASSUMED TEMP °C (°F)	32 (90)	30 (86)	28 (82)	26 (79)	24 (75)	22 (72)	20 (68)	18 (64)	16 (61)	14 (57)

%N_1 ADJUSTMENT FOR TEMPERATURE DIFFERENCE

ASSUMED TEMP MINUS OAT	OUTSIDE AIR TEMPERATURE													
	°C	-40	-20	0	5	10	15	20	25	30	35	40	45	50	55
°C / °F	°F	-40	-4	32	41	50	59	68	77	86	95	104	113	122	131
10 / 18						1.6	1.6	1.6	1.5	1.5	1.5	1.5	1.4	1.4	1.3
20 / 36				3.3	3.2	3.2	3.1	3.0	3.0	2.9	2.8	2.8	2.7	2.5	2.3
30 / 54				4.8	4.8	4.6	4.5	4.4	4.3	4.0	3.8	3.6	3.6	3.6	3.6
40 / 72			6.0	6.2	6.1	6.0	5.8	5.7	5..2	5.0	5.0				
50 / 90			8.2	7.5	7.3	7.2	6.6								
60 / 108		10.4	9.5	8.7	8.1	7.9									
70 / 126		11.8	10.7	9.3											
80 / 144		13.0	11.8	10.1											
90 / 162		14.8	12.4												
100 / 180		15.0	12.8												
110 / 198		15.4													

Figure 14.4

PERFORMANCE **CLASS A: ADDITIONAL TAKE OFF PROCEDURES**

The procedure is illustrated by the following example:

Given: Pressure altitude: 2000ft
Temperature: 20°C
Wind: Still air
TORA: Level, 6500ft
Flap setting: 5°

Weight required for take-off = 52000 kg

Solution: Using the data in CAP 698

Figure 4.4 Field length limiting OAT = 35°C
Figure 4.5 Climb limit OAT = 46°C
Figure 4.6 Tyre limit OAT = in excess of 70°C
Figure 4.17 From the first table, max assumed OAT for 25% thrust reduction = 58°C
Lowest assumed OAT = 35°C

Figure 4.17 From the centre table for 35°, N_1 = 95.6%
(Minimum assumed temp. = 26°C)
Assumed OAT - Actual OAT = 35° -20° = 15°C

Figure 4.17 From the bottom table, N_1 adjustment = 2.3%
∴Take-off N_1 = 95.6 - 2.3 = 93.3%

Figure 4.8 Take-off speeds: Column reference (2000 ft 35°C) zone C
V_1 = 134 kt V_R = 136kt V_2 = 144kt
V_{MCG} (20°C) = 113 kt

14.5 TAKE-OFF WITH ANTI-SKID INOPERATIVE

If the anti-skid system is inoperative, the stopping distance required will be increased for a given aeroplane mass and V_1 speed. As the available accelerate-stop distance is fixed, the V_1 speed must be decreased, and the mass decreased such that the take-off distance available is not exceeded.

A simplified method of determining the take-off mass and speeds with anti-skid inoperative is given below.

Procedure.

a) Determine the normal Field length / obstacle limited Take-off Mass.
b) Reduce this mass by 7700 kg.
c) Find the V_1 for this mass.

PERFORMANCE   CLASS A: ADDITIONAL TAKE OFF PROCEDURES

- d) Reduce this V_1 by the decrement shown in table 1.
- e) Compare the V_1 to V_{MCG} from table 2.
- f) If V_1 less than V_{MCG}, V_1 may be set = V_{MCG} provided the Accelerate - Stop Distance available exceeds approx. 7900 ft.

Table 1

ANTI-SKID INOPERATIVE V_1 DECREMENTS	
FIELD LENGTH FT	V_1 REDUCTION KTS
6000	28
8000	22
10000	18
12000	14
14000	13

Table 2
Minimum V_{1MCG}

OAT		PRESSURE ALTITUDE FT					
°C	°F	0	2000	4000	6000	8000	9000
55	131	105					
50	122	109	104				
45	113	111	107	102			
40	104	114	110	105	100	96	
35	95	117	112	108	104	100	97
30	86	120	115	110	106	102	99
25	77	120	117	114	110	106	102
20	68	120	117	114	110	106	103
15	59	120	117	114	111	108	105
-50	-58	121	118	115	112	109	107

FOR PACKS OFF ADD 2 KNOTS

Example.

Given: Pressure altitude: 3000ft
        Temperature: 25°C
        Wind: Still air
        Flap setting   5°
        Runway   9000ft
        Slope   1% Uphill
        ACS and PMC on.   No obstacles.

Solution: Using CAP 698

         Figure 4.4   Field Length limit =   59000 kg
                       Reduction   7700 kg
                       Revised mass   51300 kg

         Figure 4.8   Zone B
                       For 51300 kg V_1 =132.5 kt V_R =134 kt V_2 = 144 kt
         Figure 4.18   From table 1, V_1 decrement = 19 kt
                       ∴ Adjusted V_1 = 113.5 kt
         Figure 4.19   From table 2, V_{MCG} = 111 kt
                       ∴ V_1 = 113.5 kt

PERFORMANCE — CLASS A: ADDITIONAL TAKE OFF PROCEDURES

PAPER 1

1. A runway is considered to be contaminated when :

   a) more than 50% of the runway surface is covered by more than a 5 mm. depth of water.
   b) more than 25% of the runway surface is covered by more than a 2.5 mm. depth of water.
   c) more than 25% of the runway surface is covered by more than a 3 mm. depth of water.
   d) more than 30% of the runway surface is covered by more than a 3 mm. depth of water.

2. For an aircraft with a tyre pressure of 130 lb/sq. in., the aquaplaning speed on a runway contaminated by 5 mm. of standing water, would be :

   a) 46 knots
   b) 80 knots
   c) 103 knots
   d) 130 knots

3. The aquaplaning speed :

   a) is not affected by changes of contaminant density, but increases if tyre pressure increases.
   b) is not affected by changes of contaminant density, but increases if tyre pressure decreases.
   c) increases if contaminant density increases, and tyre pressure decreases.
   d) decreases if contaminant density increases, and tyre pressure decreases.

4. **Refer to CAP 698 fig. 4.14**

   Take off is to be made on a runway with 13 mm. (0.5 in.) of standing water.
   The pressure altitude is 2000 ft. and the temperature is 10°C
   The field length limited mass for the 6200 ft. runway is 55000 kg.
   The maximum permissible mass for take off is :

   a) 54870 kg.
   b) 49220 kg.
   c) 51500 kg.
   d) 44800 kg.

5. The effect of increasing the V_2 speed for a given take off mass and aerodrome conditions is :

   a) the climb gradient increases and the take off distance decreases.
   b) the climb gradient decreases and the take off distance increases.
   c) the climb gradient increases and the take off distance increases.
   d) the climb gradient decreases and the take off distance decreases.

PERFORMANCE　　　　　　　　　　　　　CLASS A: ADDITIONAL TAKE OFF PROCEDURES

6. **Refer to CAP 698 Fig. 4.15 and 4.16 for question 6**

    For the ambient aerodrome conditions the following mass limits have been determined for take off with 5° of flap.

    Field length limit mass   61000 kg.
    Climb limit mass   55000 kg.
    Tyre limit mass   75000 kg.

    The maximum possible mass at which take off could be made using the increased V_2 procedure would be :

    a)   56750 kg.
    b)   58500 kg.
    c)   57400 kg.
    d)   59300 kg.

7. Which of the following statements is correct :

    a)   The increased V_2 procedure may be used with a reduced thrust take off.
    b)   Reduced thrust may be used for take off if the runway is contaminated.
    c)   The increased V_2 procedure would only be used if the climb limited take off mass is less than the field length limited mass.
    d)   Reduced thrust may be used for take off if the anti skid system is inoperative.

    Refer to CAP 698 Fig. 4.17 for questions 8 and 9

8. Take off is to be made using the reduced thrust procedure at a mass of 56000 kg.
    Pressure altitude is 1000 ft. and the aerodrome temperature is 20°C
    BFL = 7500 ft. level runway, zero wind component. Flaps 5°
    The assumed temperature which should be used to determine the $\%N_1$ for take off is :

    a)   28°C
    b)   43°C
    c)   47°C
    d)   70°C

9. An assumed temperature of 45°C has been determined for take off with reduced thrust at an aerodrome where the pressure altitude is 2000 ft. and the temperature is 15°C
    A/C is on.
    The $\%N_1$ for take off would be :

    a)   94.6%
    b)   90.1%
    c)   91.5%
    d)   88.8%

PERFORMANCE   **CLASS A: ADDITIONAL TAKE OFF PROCEDURES**

10. Take off is to be made with the anti skid system inoperative. Compared to normal operation from the same aerodrome in the same ambient conditions, the maximum take off mass (i) the V_1 speed (ii) and the accelerate-stop distance (iii) would :

	(i)	(ii)	(iii)
a)	decrease	increase	increase
b)	decrease	decrease	remain the same
c)	increase	decrease	decrease
d)	decrease	decrease	decrease

ANSWERS

Questions	Answers
1	C
2	C
3	D
4	D
5	C
6	C
7	C
8	B
9	B
10	B

CHAPTER FIFTEEN - CLASS A: TAKE OFF CLIMB

Contents

		Page
15.1	TAKE-OFF CLIMB	15 - 1
15.2	GRADIENT REQUIREMENTS	15 - 2
15.3	OBSTACLE CLEARANCE (Net take off flight path)	15 - 2
15.4	TURNS ON THE FLIGHT PATH	15 - 3
15.5	ALLOWANCE FOR TURNS	15 - 3
15.6	PRESENTATION OF DATA	15 - 4
15.7	NOISE ABATEMENT PROCEDURES	15 - 6
15.8	NOISE ABATEMENT CLIMB - EXAMPLE OF A PROCEDURE ALLEVIATING NOISE CLOSE TO THE AERODROME (NADP 1)	15 - 7
15.9	NOISE ABATEMENT CLIMB - EXAMPLE OF A PROCEDURE ALLEVIATING NOISE DISTANT FROM THE AERODROME (NADP 2)	15 - 7
	CLASS A AIRCRAFT: TAKE OFF CLIMB	15 - 9

© Oxford Aviation Services Limited

© Oxford Aviation Services Limited

PERFORMANCE   CLASS A: TAKE OFF CLIMB

15.1 TAKE-OFF CLIMB

The take-off climb from the screen to a height of 1,500 ft above aerodrome level is divided into segments, the end of a segment being determined by a change of configuration or engine power. The critical engine is assumed to have failed at V_{EF}.

Segment	Gear	Flap	Power	Speed
First	Down	Take-off	Take-off	V_2
Second	Up	Take-off	Take-off	V_2
Third	Up	From take-off to full up	Take-off	Increasing from V_2 to final segment speed
Final (fourth)	Up	Up	Max Continuous	Final segment speed

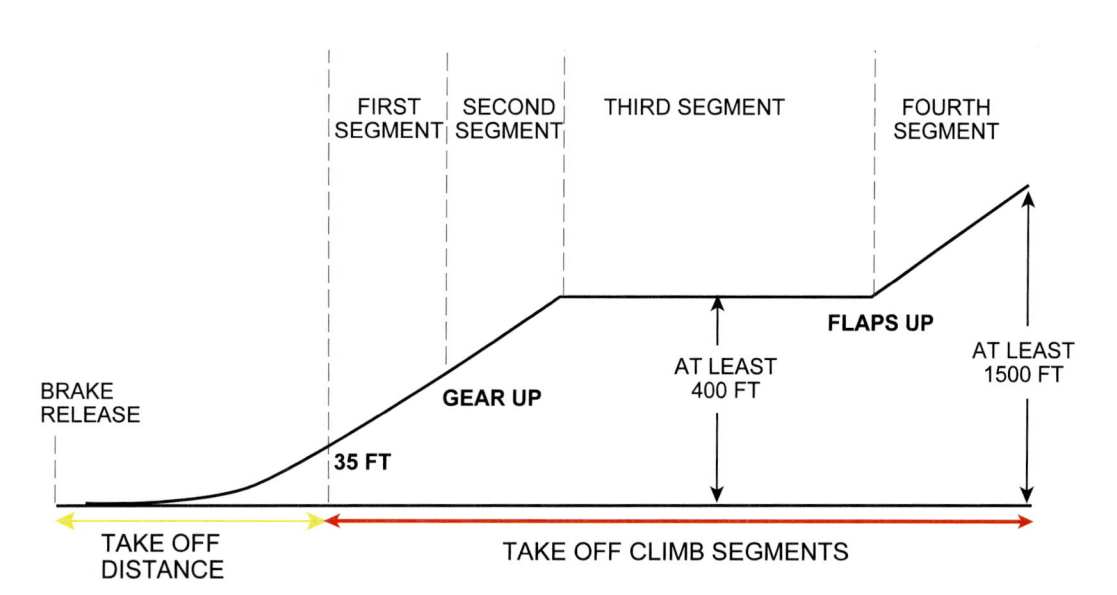

Figure 15.1

The take off climb will determine:

a)   the take-off mass limit for altitude and temperature
b)   the clearance of obstacles on the net take-off flight path

15.2 GRADIENT REQUIREMENTS

Minimum gradient requirement are laid down for each of the following climb segments. Each of these will determine a mass for the pressure altitude and ambient temperature, and the lowest of these will determine the M-A-T limit for take-off. The percentage gradient required varies with the number of engines fitted to the aircraft.

Segment	Number of Engines		
	2	3	4
First	$\not< 0$	0.3%	0.5%
Second	2.4%	2.7%	3.0%
Third and Final	1.2%	1.5%	1.7%

Note: The third segment is a level segment, but the equivalent gradient must be available for acceleration.

15.3 OBSTACLE CLEARANCE (Net take off flight path)

The net gradient is the gross gradient reduced by:

- 0.8% for 2 engined aircraft
- 0.9% for 3 engined aircraft
- 1.0% for 4 engined aircraft

The net take-off flight path to the point at which the aeroplane is 1,500 ft above the take-off surface or at which transition to the en-route configuration is completed, whichever is higher, must clear relevant obstacles by 35 ft, or by 50 ft during a turn in which the aeroplane is banked by more than 15°, or by a lateral distance of:

90m + 0.125 D, where D is the distance from the end of take-off distance available, or for aeroplanes with a wing span of less than 60m,

60m + ½ wing span + 0.125D

but not greater than:

a) If the change of track is less than 15°, 300m if sufficient navigational accuracy can be maintained through the obstacle accountability area, or 600m for flights under all other conditions.

b) If the change of track is greater than 15°, 600m if sufficient navigational accuracy or 900m for all other flights

PERFORMANCE   **CLASS A: TAKE OFF CLIMB**

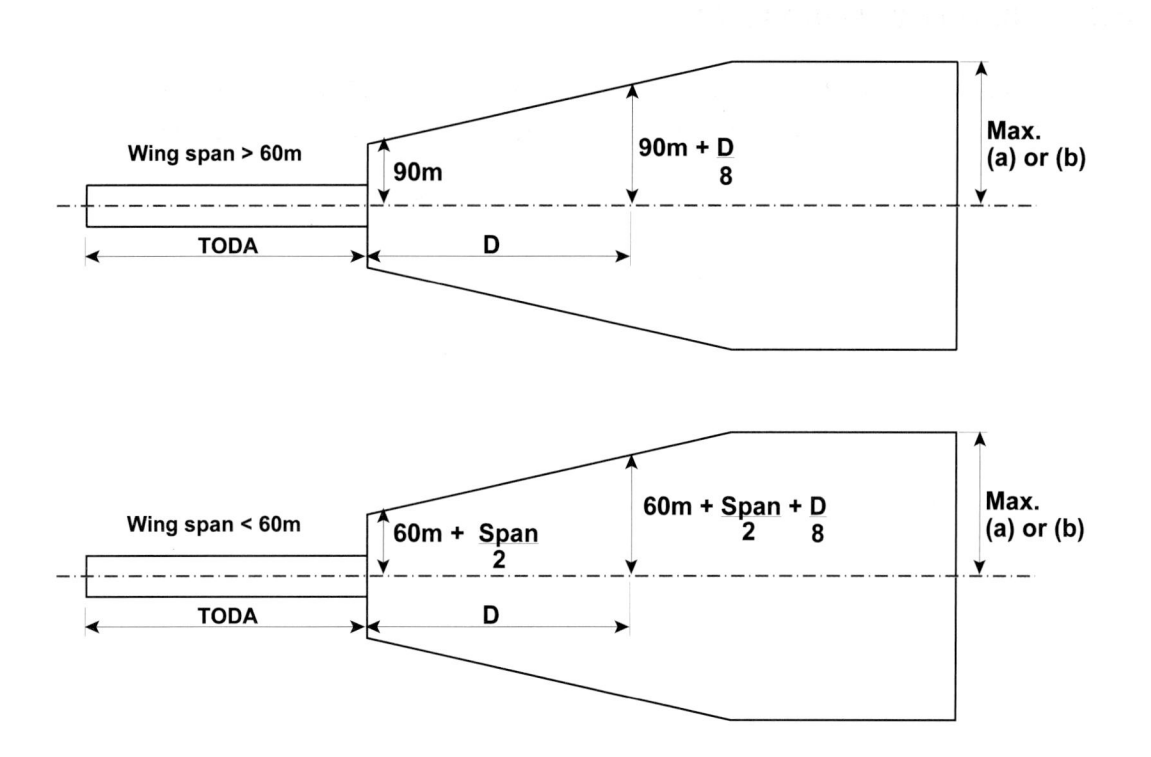

Figure 15.2 Obstacle Clearance

15.4 TURNS ON THE FLIGHT PATH

a)   Turns are not allowed below a height of half the wingspan or 50ft whichever is greater.

b)   Up to 400ft, bank angle may not be more than 15°.

c)   Above 400 ft, bank angle may not be more than 25°.

15.5 ALLOWANCE FOR TURNS

Allowance must be made for the effect of the turn on the climb gradient and speed. The flight manual usually gives a gradient decrement for a 15° banked turn at V_2. For greater bank angles:

     For 20° bank, use 2 x gradient decrement and $V_2 + 5$ kt
     For 25° bank, use 3 x gradient decrement and $V_2 + 10$ kt

Distances in turns should be increased in proportion to the increased speeds.

PERFORMANCE   CLASS A: TAKE OFF CLIMB

15.6 PRESENTATION OF DATA

Figure 15.3 shows a simplified presentation of the flight path data. The Flight Manual will contain data for a full flight path analysis, but the Operation Manual may contain simplified data for rapid planning.

The data shown is for the air conditioning packs on. To obtain the performance with packs off it would be necessary to refer to the full Flight Manual data.

The data is not valid for take-offs using the improved climb technique.

The obstacle height should be calculated from the lowest point on the runway to account conservatively for the runway slope.

The broken arrowed line in Figure 15.3. illustrates the use of the graph for the following example:

Obstacle: Height above lowest point on runway: 360 ft
            Distance from brake release point: 18,000 ft

            Aerodrome Pressure Altitude: 1,000 ft

            OAT: 37°C

            Reported Wind: 20 kt headwind

            Flap: 5°

            From Fig. 15.3 Maximum Obstacle Clearance Weight = 53,700 kg.

PERFORMANCE CLASS A: TAKE OFF CLIMB

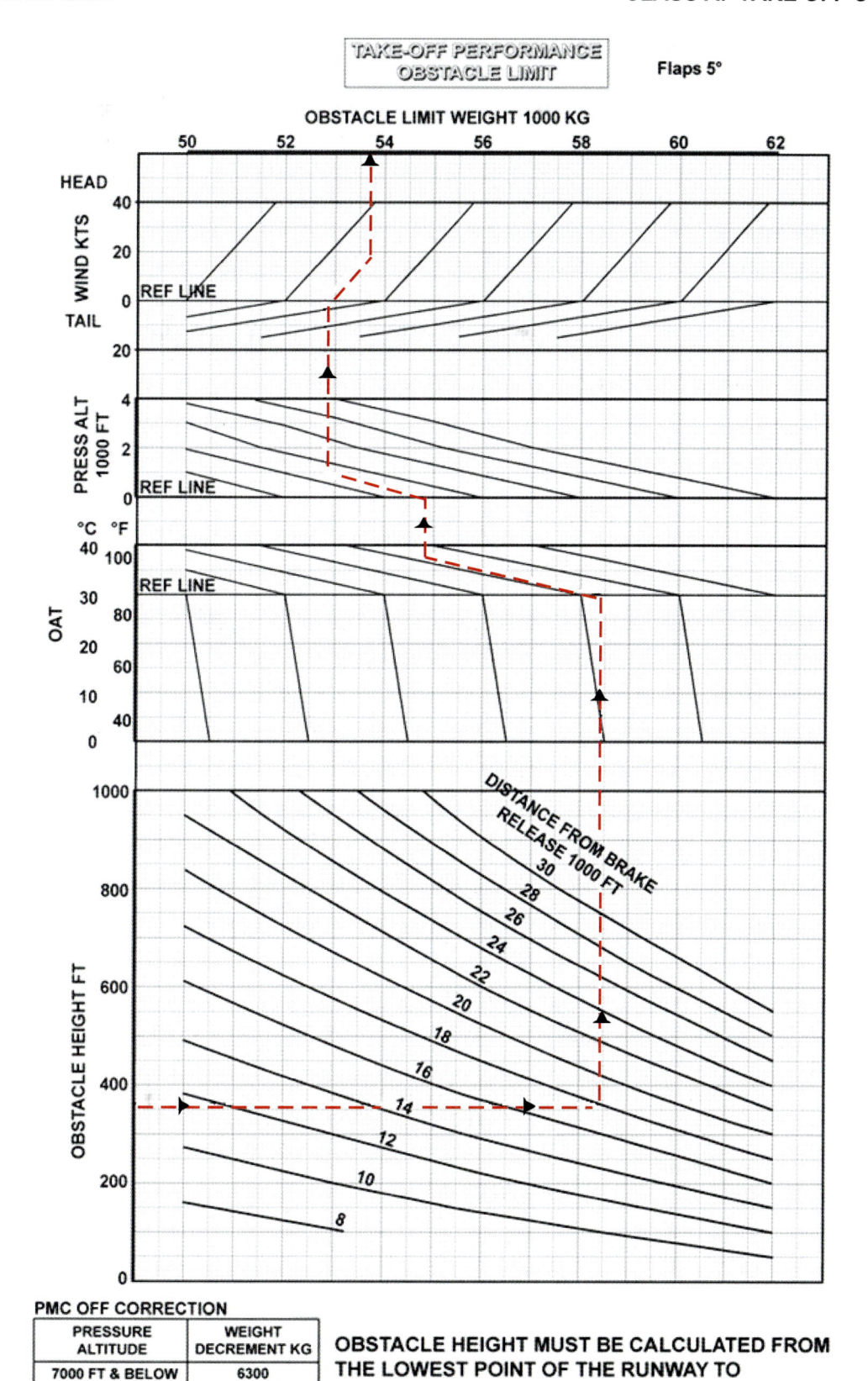

TAKE-OFF PERFORMANCE OBSTACLE LIMIT Flaps 5°

Figure 15.3

PMC OFF CORRECTION

PRESSURE ALTITUDE	WEIGHT DECREMENT KG
7000 FT & BELOW	6300
ABOVE 7000 FT	7100

OBSTACLE HEIGHT MUST BE CALCULATED FROM THE LOWEST POINT OF THE RUNWAY TO CONSERVATIVELY ACCOUNT FOR RUNWAY SLOPE.

PERFORMANCE   CLASS A: TAKE OFF CLIMB

15.7 NOISE ABATEMENT PROCEDURES

Aeroplane operating procedures for the take-off climb shall ensure that the necessary safety of flight operations is maintained whilst minimizing exposure to noise on the ground.

The first procedure (NADP 1) is intended to provide noise reduction for noise-sensitive areas in close proximity to the departure end of the runway (see Figure 15.4). The second procedure (NADP 2) provides nosie reduction to areas more distant from the runway end (see Figure 15.5).

The two procedures differ in that the acceleration segment for flap/slat retraction is either initiated prior to reaching the maximum prescribed height or *at* the maximum prescribed height. To ensure optimum acceleration performance, thrust reduction may be initiated at an intermediate flap setting.

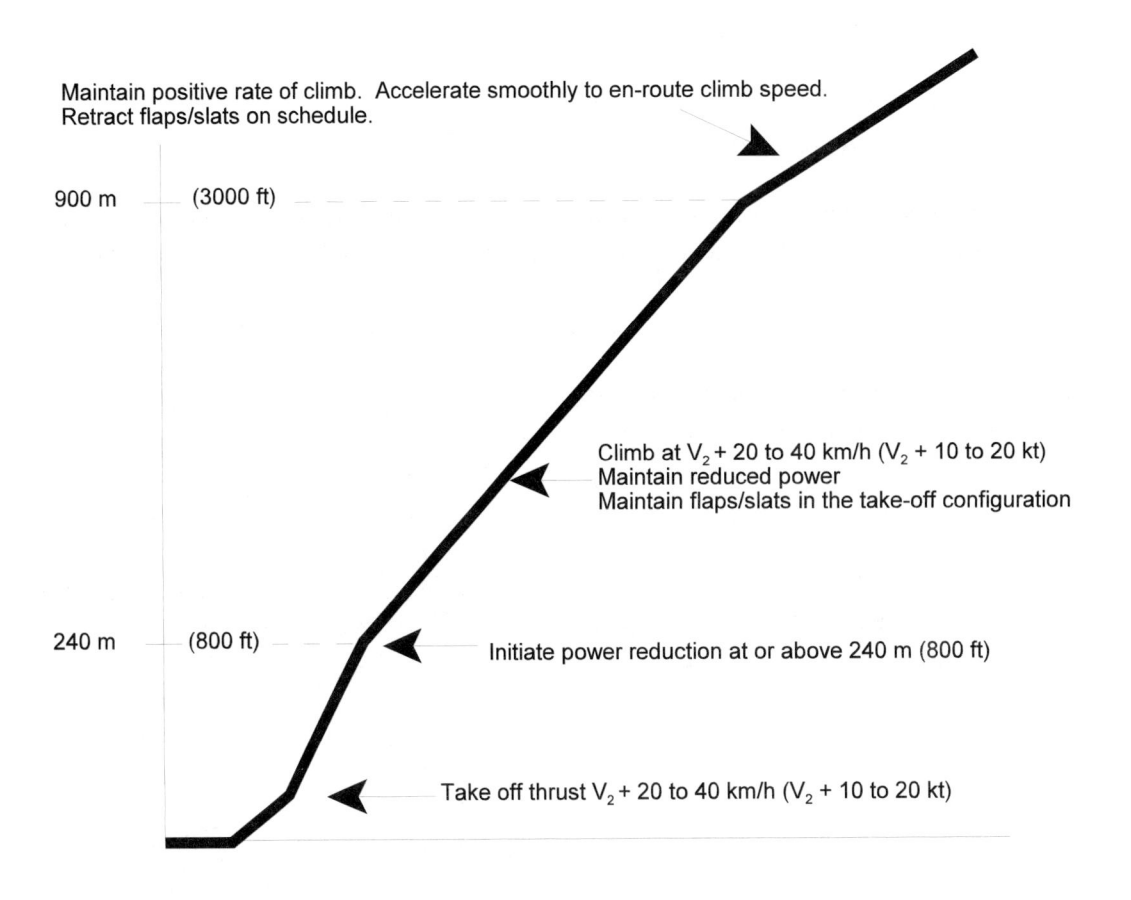

Figure 15.4 Noise Abatement Take-Off Climb - Example of a Procedure Alleviating Noise Close to the Aerodrome (NADP 1)

PERFORMANCE   CLASS A: TAKE OFF CLIMB

15.8 NOISE ABATEMENT CLIMB - EXAMPLE OF A PROCEDURE ALLEVIATING NOISE CLOSE TO THE AERODROME (NADP 1)

This procedure involves a power reduction at or above the prescribed minimum altitude and the delay of flap/slat retraction until the prescribed maximum altitude is attained. At the prescribed maximum altitude, accelerate and retract flaps/slats on schedule whilst maintaining a positive rate of climb, and complete the transition to normal en-route climb speed.

The noise abatement procedure is not to be initiated at less than 240 m (800 ft) above aerodrome elevation.

The initial climbing speed to the noise abatement initiation point shall not be less than V_2 plus 20 km/h (10 kt).

On reaching an altitude at or above 240 m (800 ft) above aerodrome elevation, adjust and maintain engine power/thrust in accordance with the nosie abatement power/thrust schedule provided in the aircraft operating manual. Maintain a climb speed of V_2 plus 20 to 40 km/h (10 to 20 kt) with flaps and slats in the take-off configuration.

At no more than an altitude equivalent to 900 m (3,000 ft) above aerodrome elevation, whilst maintaining a positive rate of climb, accelerate and retract flaps/slats on schedule.

At 900 m (3,000 ft) above aerodrome elevation, accelerate to en-route climb speed.

15.9 NOISE ABATEMENT CLIMB - EXAMPLE OF A PROCEDURE ALLEVIATING NOISE DISTANT FROM THE AERODROME (NADP 2)

This procedure involves initiation of flap/slat retraction on reaching the minimum prescribed altitude. The flaps/slats are to be retracted on schedule whilst maintaining a positive rate of climb. The power reduction is to be performed with the initiation of the first flap/slat retraction **or** when the zero flap/slat configuration is attained. At the prescribed altitude, complete the transition to normal en-route climb procedures.

The noise abatement procedure is not to be initiated at less than 240 m (800 ft) above aerodrome elevation.

The noise abatement procedure is not to be initiated at less than 240 m (800 ft) above aerodrome elevation.

The initial climbing speed to the noise abatement initiation point is V_2 plus 20 to 40 km/h (10 to 20 kt).

On reaching an altitude equivalent to at least 240 m (800 ft) above aerodrome elevation, decrease aircraft body angle/angle of pitch whilst maintaining a positive rate of climb, accelerate towards V_{ZF} and either:

PERFORMANCE   CLASS A: TAKE OFF CLIMB

a)   reduce power with the initiation of the first flap/slat retraction; or

b)   reduce power after flap/slat retraction.

Maintain a positive rate of climb, and accelerate to and maintain a climb speed of V_{ZF} + 20 to 40 km/h (10 to 20 kt) to 900 m (3,000 ft) above aerodrome elevation.

On reaching 900 m (3,000 ft) above aerodrome elevation, transition to normal en-route climb speed.

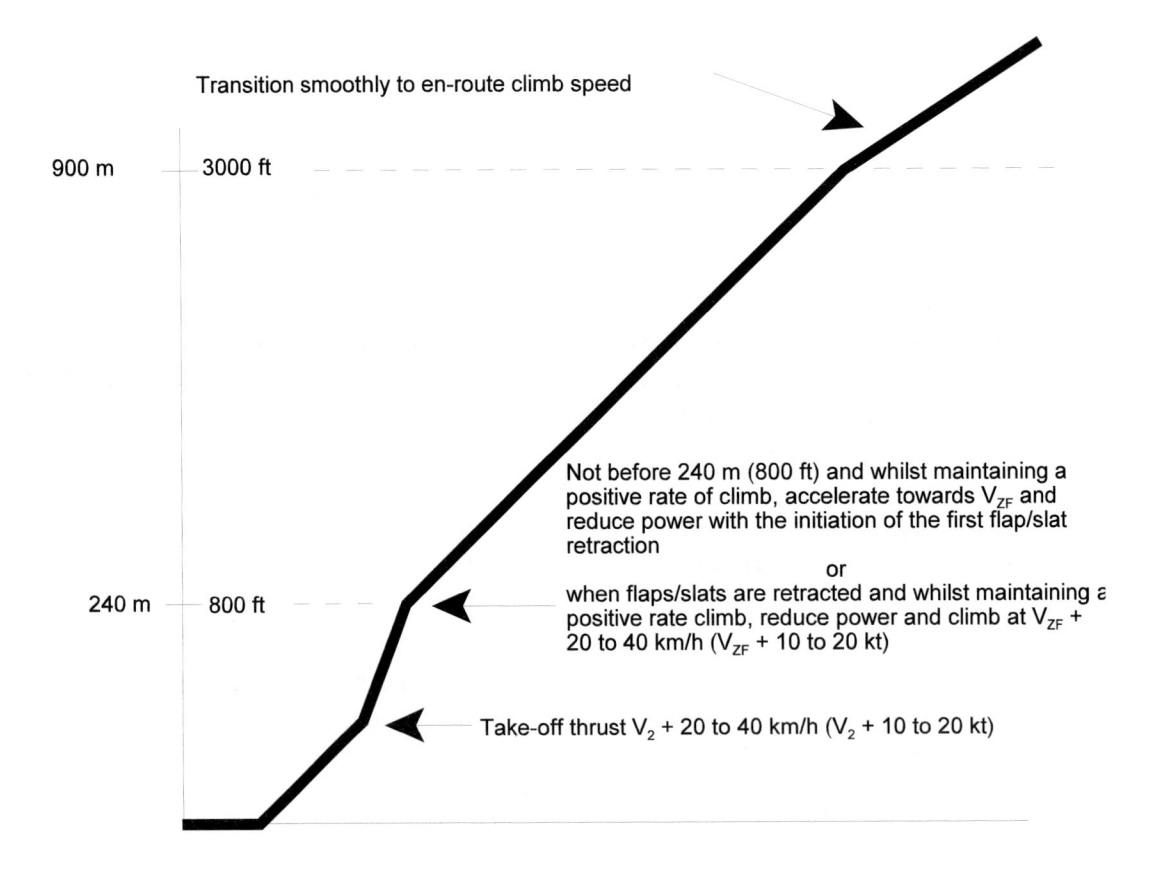

Figure 15.5 Noise Abatement Take-Off Climb - Example of a Procedure Alleviating Noise Distant from the Aerodrome (NADP 2)

PERFORMANCE   CLASS A: TAKE OFF CLIMB

PAPER 1

1. The first segment of the take off flight path ends :

   a) when the flaps and landing gear are fully retracted.
   b) when the landing gear is retracted and the power reduced to maximum continuous.
   c) when the landing gear is fully retracted.
   d) when the aircraft reaches a height of 400 ft.

2. For the third segment of the take off flight path, which of the following combinations of speed, power, flap setting, and landing gear position is correct ?

	Speed	Power	Flap	Gear
a)	V_2	Take off	Take off	Retracted
b)	Accelerating	Take off	Being retracted	Retracted
c)	V_2	Max. Continuous	Being retracted	Retracted
d)	Accelerating	Take off	Take off	Down

3. The second segment gradient requirement for a two engined aircraft is 2.4 %
   If the V_2 speed is 150 knots IAS the rate of climb at sea level ISA when operating at the W.A.T. limit would be *approximately* :

   a) 150 ft./min.
   b) 240 ft./min.
   c) 360 ft./min.
   d) 625 ft./min.

4. The climb gradient requirement determines a maximum weight for take off. This ensures that:

   a) the take off distance available will not be exceeded.
   b) there will be adequate obstacle clearance after take off.
   c) in the event of an engine failure on take off a minimum climb gradient will be achievable.
   d) in the event of an engine failure on take off adequate control will be available.

5. The second segment gradient requirement for a two engined aircraft is 2.4 %
   If the total drag of the aircraft at V_2 is 18500 lb., and the thrust per engine is 22000 lb., the maximum permissible climb limited take off weight would be :

   a) 145800 lb.
   b) 1062500 lb.
   c) 91700 lb.
   d) 90300 lb.

PERFORMANCE

CLASS A: TAKE OFF CLIMB

6. The vertical clearance of obstacles required on the take off flight path is :

   a) 35 ft. for all flight paths
   b) 35 ft. for the net flight path, and 50 ft. for the gross flight path.
   c) 50 ft. for all flight paths.
   d) 35 ft., or 50 ft. during a turn in which the bank exceeds 15°

7. A Boeing 747-400 has a wing span of 64.4 m. When considering relevant obstacles for the take off flight path what is the semi-width of the zone at a distance of 1500 m. from the TODA. :

   a) 198.7 m.
   b) 277.5 m.
   c) 279.7 m.
   d) 300 m.

8. Which of the following statements is correct with regard to turns on the take off flight path ?

   a) Turns may not be assumed to take place below 400 ft.
   b) For an aircraft with a wing span of 120 ft. turns may not be assumed to take place below 60 ft.
   c) Above 400 ft. a turn may not be assumed with a bank angle greater than 15°
   d) During a turn the climb gradient may be assumed to be the same as on the straight flight path.

9. **Refer to CAP 698 Fig. 4.20 for question 9**

   Find the maximum weight to clear the obstacle for the following data :

Pressure altitude	2000 ft.	Take off flap	5°
Temperature	15°C	PMC	On
Reported wind comp.	25 kt. Headwind		
Obstacle	Height 300 ft. at a distance of 14000 ft. from brake release		

   a) 48800 kg.
   b) 50700 kg.
   c) 52300 kg.
   d) 53500 kg.

PERFORMANCE   **CLASS A: TAKE OFF CLIMB**

10. Runway 09 / 27 has a length of 2000 m., the elevation of the 09 threshold is 250 ft. and the elevation of 27 threshold is 240 ft. The elevation of the obstacle is 360 ft.
The height of the obstacle which should be assumed to determine the maximum mass for take off is :

    a)   110 ft.
    b)   120 ft.
    c)   360 ft.
    d)   370 ft.

ANSWERS

Questions	Answers
1	C
2	B
3	C
4	C
5	A
6	D
7	B
8	B
9	B
10	B

CHAPTER SIXTEEN - CLASS A: EN ROUTE

Contents

		Page
16.1	CLIMB TO CRUISE ALTITUDE	16 - 1
16.2	CRUISE	16 - 2
16.3	SPECIFIC RANGE IN TERMS OF MACH NUMBER	16 - 4
16.4	TYPICAL CRUISE DATA	16 - 5
16.5	WIND-ALTITUDE TRADE	16 - 6
16.6	LIMITATIONS ON CRUISE SPEED AND ALTITUDE	16 - 7
16.7	THRUST LIMIT	16 - 7
16.8	BUFFET LIMIT	16 - 8
16.9	CRUISE FOR MAXIMUM ECONOMY	16 - 10
16.10	CRUISE WITH ONE ENGINE INOPERATIVE	16 - 10
16.11	RANGE LIMIT	16 - 12
16.12	EXTENDED RANGE TWIN OPERATIONS (ETOPS)	16 - 12
16.13	ETOPS DATA	16 - 12
16.14	AREA OF OPERATION	16 - 12
16.15	CRITICAL FUEL RESERVES.	16 - 14
16.16	NET LEVEL OFF ALTITUDE	16 - 16
16.17	ENDURANCE.	16 - 17
16.18	EN-ROUTE REQUIREMENTS	16 - 18
16.19	EN ROUTE GRADIENT MARGINS	16 - 18
16.20	OBSTACLE CLEARANCE REQUIREMENTS EN-ROUTE JAR-OPS 1.500	16 - 18

© Oxford Aviation Services Limited

16.21	PRESENTATION OF DATA	16 - 19
16.22	PROCEDURE	16 - 19
16.23	TWO ENGINES INOPERATIVE	16 - 21
	CLASS A AIRCRAFT: EN - ROUTE	16 - 23

PERFORMANCE   CLASS A: EN ROUTE

16.1   CLIMB TO CRUISE ALTITUDE

The recommended climb procedure is usually a constant IAS at low altitudes, changing to a constant Mach number a high altitudes. Climbing at a constant IAS will give:

a)   increasing TAS

b)   increasing Mach number

When the scheduled Mach number is reached, the climb is continued with this Mach number constant and will give:

a)   decreasing IAS

b)   decreasing TAS

A typical presentation of data is shown in figures 16.1 and 16.2. Figure 16.1 shows the rpm setting to be used, and Figure 16.2 shows the time, distance, fuel used and TAS on the climb.
MAX CLIMB % N1 250/280/.74M

VALID FOR 2 PACKS ON (AUTO) ENGINE A / I OFF									
TAT °C		PRESSURE ALTITUDE 1000 FT							
		0	5	10	15	20	25	30	35	37
	50	93.7	93.9	94.1						
	40	93.9	94.0	94.3	95.0	95.2				
	30	93.8	94.8	95.1	95.8	96.0	96.0			
	20	92.2	94.4	96.1	96.8	97.1	97.1	97.1		
	10	90.6	92.8	95.2	97.3	98.3	98.5	98.6	98.6	98.6
	0	89.0	91.1	93.5	95.5	98.2	100.1	100.3	100.3	100.3
	-10	87.4	89.4	91.8	93.8	96.4	98.8	101.4	102.1	102.1
	-20	85.7	87.7	90.0	92.0	94.5	96.9	99.4	102.3	102.5
	-30	84.0	86.0	88.3	90.1	92.6	95.0	97.4	100.3	100.5
	-40	82.3	84.2	86.4	88.3	90.7	93.0	95.4	98.2	98.4
	-50	80.5	82.3	84.5	86.4	88.8	91.0	93.4	96.1	96.3
% N₁ CORR	A/C PACKS OFF	+1.0	+0.9	+0.9	+1.0	+1.0	+1.0	+1.1	+1.1	+1.1
	A/C PACKS HIGH	-0.2	-0.2	-0.3	-0.3	-0.4	-0.4	-0.4	-0.4	-0.4
	ENGINE A/I ON	-0.6	-0.7	-0.7	-0.8	-0.9	-0.9	-0.9	-0.9	-0.9
	WING A/I ON	-0.9	-0.9	-1.0	-1.2	-1.4	-1.7	-1.9	-2.2	-2.2

PERFORMANCE   CLASS A: EN ROUTE

```
EN-ROUTE CLIMB
    280/.74
     ISA
```

PRESSURE ALTITUDE FEET	UNITS MIN/KG NAM/KNOTS	BRAKE RELEASE WEIGHT KG									
		68000	66000	64000	62000	60000	58000	56000	52000	48000	44000	40000
37000	TIME/FUEL DIST/TAS			31/2300 196/401	25/1950 150/396	22/1750 130/393	20/1650 116/391	18/1500 106/390	16/1350 91/388	14/1200 79/386	12/1050 69/385	11/950 60/384
36000	TIME/FUEL DIST/TAS		27/2150 164/396	23/1900 140/393	21/1750 124/390	19/1650 113/389	18/1550 104/388	17/1450 97/387	15/1300 84/385	13/1150 74/384	12/1050 65/383	10/900 57/382
35000	TIME/FUEL DIST/TAS	25/2100 152/393	23/1900 134/390	21/1800 122/388	19/1650 112/387	18/1550 103/386	17/1500 96/385	16/1400 90/384	14/1250 79/382	13/1100 69/381	11/1000 58/378	10/900 51/378
34000	TIME/FUEL DIST/TAS	22/1950 131/387	21/1800 119/386	19/1700 110/384	18/1600 102/383	17/1500 95/383	16/1400 89/382	15/1350 84/381	13/1200 74/380	12/1100 65/379	11/1000 58/378	10/900 51/378
33000	TIME/FUEL DIST/TAS	20/1800 118/383	19/1700 109/382	18/1600 101/381	17/1550 95/380	16/1450 89/379	15/1350 83/379	14/1300 78/378	13/1200 70/377	11/1050 62/376	10/950 55/376	9/850 49/375
32000	TIME/FUEL DIST/TAS	19/1750 108/379	18/1650 101/378	17/1550 94/378	16/1450 88/377	15/1400 83/376	14/1350 78/375	14/1250 74/376	12/1150 66/374	11/1050 58/374	10/950 52/373	9/850 46/373
31000	TIME/FUEL DIST/TAS	18/1650 100/376	17/1600 94/375	16/1500 88/374	15/1400 83/374	14/1350 78/373	14/1300 74/373	13/1250 69/372	12/1100 62/371	11/1000 55/371	10/900 49/370	9/800 44/370
30000	TIME/FUEL DIST/TAS	17/1600 93/372	16/1500 87/371	15/1450 82/371	15/1400 78/370	14/1300 73/370	13/1250 69/369	12/1200 69/369	11/1100 59/368	10/1000 52/368	9/900 47/367	8/800 42/367

FUEL ADJUSTMENT FOR HIGH ELEVATION AIRPORTS EFFECT ON TIME AND DISTANCE IS NEGLIGIBLE	AIRPORT ELEVATION	2000	4000	6000	8000	10000	12000
	FUEL ADJUSTMENT	-50	-100	-150	-250	-300	-350

Figure 16.2

16.2 CRUISE

The choice of cruise procedure depends on whether it is required to obtain the maximum possible range or the maximum possible speed.

To obtain the maximum possible range it is necessary to fly at the optimum altitude and optimum speed for the aeroplane mass. However, small deviations from these optimum conditions do not make a very great difference to the range, and the Long Range Cruise procedure scheduled may not correspond exactly to the optimum conditions and the cruising altitude may differ from the optimum altitude by small amounts without great loss of range. A typical Long Range Cruise procedure is at a speed which gives 99% of the optimum range, and typical loss of range for deviations from the optimum altitude are shown below for the Long Range Cruise.

Deviation from optimum	% loss in range
2000 ft above	1
Optimum	0
2000 ft below	1
4000 ft below	4
8000 ft below	10
12000 ft below	15

If maximum range is not essential the cruise may be done at higher speeds, up to the limit imposed by the thrust available or the onset of buffet.

Figure 16.3 shows the relationship between these procedures

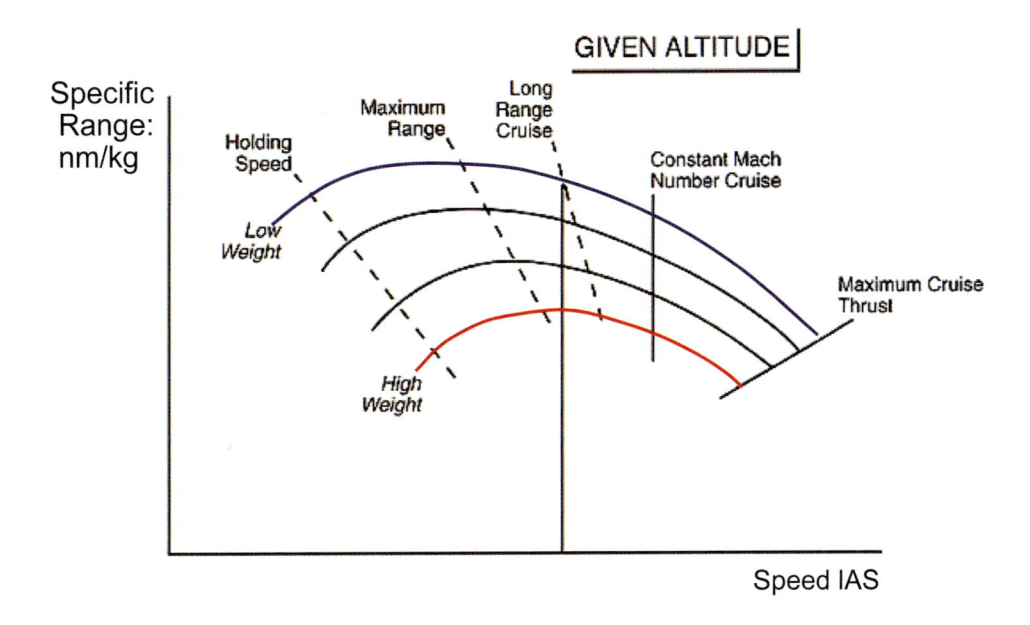

Figure 16.3 Relationship of Cruise Procedures

PERFORMANCE — CLASS A: EN ROUTE

16.3 SPECIFIC RANGE IN TERMS OF MACH NUMBER

The specific range $\dfrac{V}{F}$ may be expressed also in terms of Mach number.

$$\frac{V}{F} = \frac{V}{D} \times \frac{1}{sfc}$$

Since $V = aM$, and $L = W$

$$\frac{V}{F} = \frac{aM}{D} \times \frac{1}{sfc} \times \frac{L}{W}$$

$$= aM \frac{L}{D} \times \frac{1}{sfc} \times \frac{1}{W}$$

This shows that for a given weight the specific range is a maximum when the product of $M \times \dfrac{L}{D}$ is a maximum.

Although $\dfrac{L}{D}$ decreases with increasing Mach number, the product $M \times \dfrac{L}{D}$ increases initially with Mach number and then decreases again. (Figure 16.4)

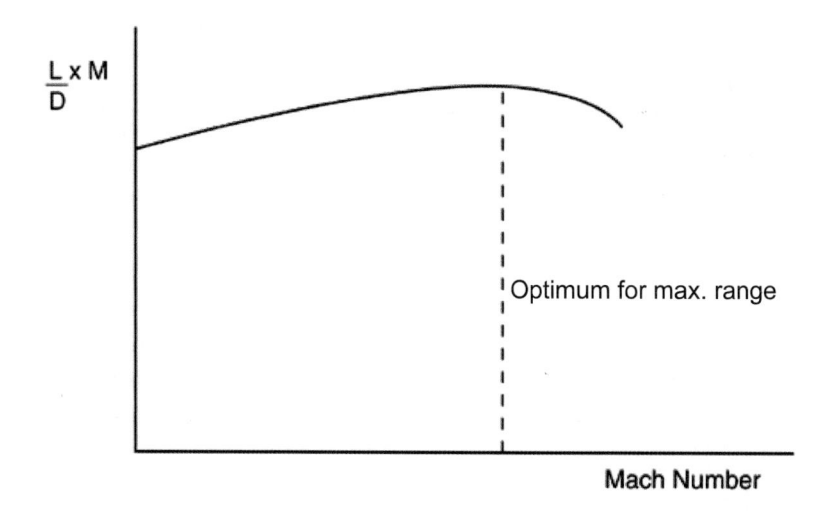

Figure 16.4 Mach Number for Maximum Range

PERFORMANCE

CLASS A: EN ROUTE

16.4 TYPICAL CRUISE DATA

Figure 16.5 shows a typical presentation of cruise data.

% N₁ REQUIRED
MAX TAT FOR THRUST RATING
IAS KNOTS
MACH NUMBER
ISA FUEL FLOW PER ENG
ISA TAS KTS

LONG RANGE CRUISE
ALL ENGINE A/C AUTO
MAX CRUISE THRUST LIMITS

PRESSURE ALTITUDE   **37000FT TO 32000 FT**

PRESS ALT 1000 FEET (STD TAT)	GROSS WEIGHT 1000 KG													
	66	64	62	60	58	56	54	52	50	48	46	44	42	40	38
37 (-32)					92.2 -18 239 .742 1301 426	90.4 -9 240 .744 1232 427	88.8 -3 240 .745 1174 427	87.6 240 .745 1127 427	86.5 240 .745 1084 427	85.5 240 .745 1046 427	84.6 240 .745 1012 427	83.8 240 .745 979 427	83.0 240 .745 949 427	82.2 240 .744 921 427	81.5 239 .742 891 426
36 (-32)				91.4 -14 245 .743 1238 426	89.7 -6 245 .744 1263 427	88.4 246 .745 1210 427	87.3 246 .745 1165 428	86.2 246 .745 1123 428	85.3 246 .745 1086 428	84.4 246 .745 1051 427	83.7 246 .745 1019 427	82.9 246 .745 989 427	82.2 245 .744 961 427	81.5 245 .742 932 426	80.6 242 .736 897 422
35 (-30)		92.7 -18 250 .742 1434 428	90.9 -9 251 .744 1363 429	89.5 -3 251 .745 1303 429	88.3 251 .745 1253 429	87.3 252 .745 1208 430	86.3 251 .745 1168 429	85.4 251 .745 1132 429	84.6 251 .745 1097 429	83.9 251 .745 1065 429	83.2 251 .742 1035 429	82.5 251 .744 1006 429	81.8 250 .743 977 428	81.0 248 .737 943 425	80.1 245 .728 908 420
34 (-28)	92.1 -13 256 .743 1464 430	90.5 -6 257 .744 1399 431	89.3 0 257 .745 1344 431	88.2 257 .745 1296 431	87.3 257 .745 1253 431	86.4 257 .745 1253 431	85.6 257 .745 1213 431	84.8 257 .745 1178 431	84.1 257 .745 1143 431	83.4 257 .745 1112 431	82.8 257 .744 1053 431	82.1 256 .742 1024 430	81.3 254 .737 989 427	80.4 251 .729 954 422	79.5 247 .719 917 416
33 (-26)	90.2 -3 263 .745 1437 433	89.1 2 263 .745 1386 433	88.2 263 .745 1340 433	87.2 263 .745 1298 433	86.4 263 .745 1260 433	85.7 263 .745 1224 433	85.0 263 .745 1191 433	84.3 263 .745 1159 433	83.6 263 .745 1129 433	83.0 263 .744 1101 433	82.4 262 .742 1071 432	81.6 260 .737 1036 428	80.8 257 .729 1000 424	79.9 253 .720 963 419	79.0 249 .709 925 412
32 (-23)	89.0 5 269 .745 1430 435	88.1 269 .745 1386 435	87.2 269 .745 1345 435	86.5 269 .745 1308 435	85.8 269 .745 1272 435	85.1 269 .745 1239 435	84.5 269 .745 1207 435	83.8 269 .745 1178 435	83.2 269 .744 1149 435	82.6 268 .742 1118 433	81.9 266 .737 1083 430	81.1 263 .729 1047 426	80.2 259 .720 1010 421	79.4 255 .710 972 415	78.4 251 .698 931 407

MAX TAT NOT SHOWN WHERE MAX CRUISE % N₁ CAN BE SET IN ISA + 30°C CONDITIONS
INCREASE/DECREASE %N₁ REQUIRED BY 1% PER 5°C ABOVE/BELOW STANDARD TAT
INCREASE/DECREASE FUEL FLOW 2% PER 10°C ABOVE/BELOW STANDARD TAT
INCREASE/DECREASE TAS BY 1KT PER 1°C ABOVE/BELOW STANDARD TAT

Figure 16.5

PERFORMANCE
CLASS A: EN ROUTE

16.5 WIND-ALTITUDE TRADE

Flying at the optimum altitude may not give the most favourable wind conditions, but moving from the optimum altitude will increase the fuel consumption. However if the gain from the more favourable wind exceeds the loss from the change of altitude, it would be beneficial to make the change. Figure 16.6 shows a method of determining the change of windspeed needed to maintain the range.

WIND - ALTITUDE TRADE

LONG RANGE CRUISE

PRESSURE ALTITUDE 1000FT	CRUISE WEIGHT 1000 KG																
	68	66	64	62	60	58	56	54	52	50	48	46	44	42	40	38	36
37								14	8	1	0	0	0	0	0	0	0
35					9	5	2	0	0	2	6	8	10	12	13	14	14
33		9	5	2	1	0	1	3	7	12	17	20	23	26	27	29	30
31	2	1	0	1	3	4	9	14	19	25	31	35	38	41	43	44	45
29	1	2	5	8	12	17	22	28	33	39	45	50	54	57	59	60	61
27	8	12	15	17	25	30	35	41	48	55	61	66	69	71	74	75	77
25	19	24	28	34	39	44	51	57	63	70	78	82	86	89	92	94	96
23	32	37	42	48	53	59	65	72	79	86	94	99	103	106	109	112	113

Figure 16.6

Example : What wind would be necessary at 27000 ft. to maintain the range for an aircraft at 31000 ft. with a headwind of 20 kt. (-20) If the Mass is 50000 kg.

From the table : Wind at 27000 ft.= 55 kt
Wind at 31000 ft.= 25 kt
Difference   = +30 kt

Wind required = -20 +30 = +10 kt (tailwind)

PERFORMANCE   CLASS A: EN ROUTE

16.6 LIMITATIONS ON CRUISE SPEED AND ALTITUDE

The choice of cruise speed and altitude may be limited by:

a)   thrust available

b)   buffet onset (manoeuvre capability)

16.7 THRUST LIMIT

The speed achieved with the maximum cruise thrust will vary with aeroplane mass, altitude and temperature. In high performance conditions, speeds could be achieved in excess of V_{MO}/M_{MO}, but this would not be permissible. Figure 16.7 shows a typical speed capability graph.

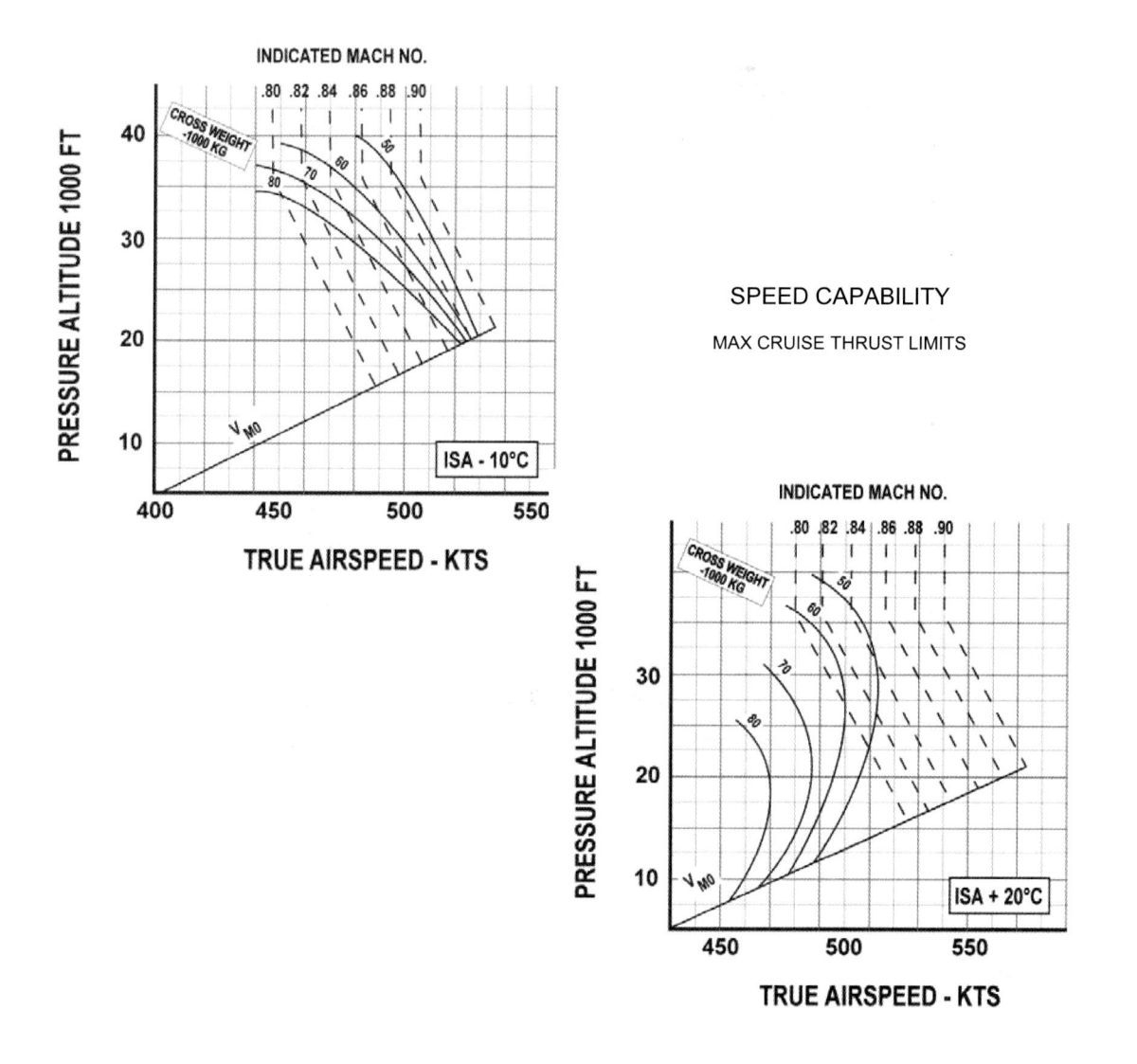

Figure 16.7 - Maximum Achievable Speed with Max. Cruise Thrust

PERFORMANCE
CLASS A: EN ROUTE

16.8 BUFFET LIMIT

Buffet occurs when the boundary layer air flow begins to separate, and this can be caused by:

a) the angle of attack approaching the stalling angle

b) the formation of shock waves

In flight these situations occur at low speed (high angle of attack) and at high speed (onset of shockwave development). Buffet onset therefore determines both low and high speed limits:

Figure 16.8 shows the buffet boundary in terms of C_L and Mach No.

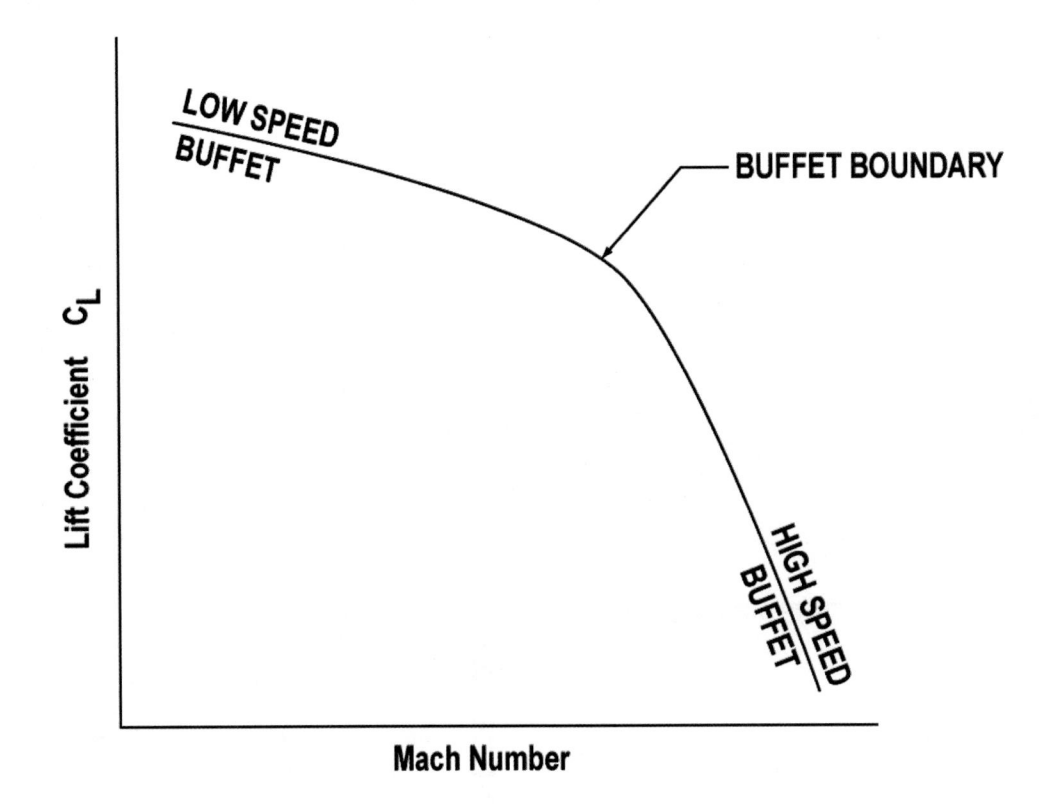

Figure 16.8 Buffet Boundary in Terms of C_L and M.No.

The lift coefficient is determined by aeroplane mass, pressure altitude, Mach number, and "g". For practical application the limits must be shown in terms of these variables. Figure 16.9 shows a presentation of the speed range available and altitude for onset of buffet, for aeroplane mass and "g".

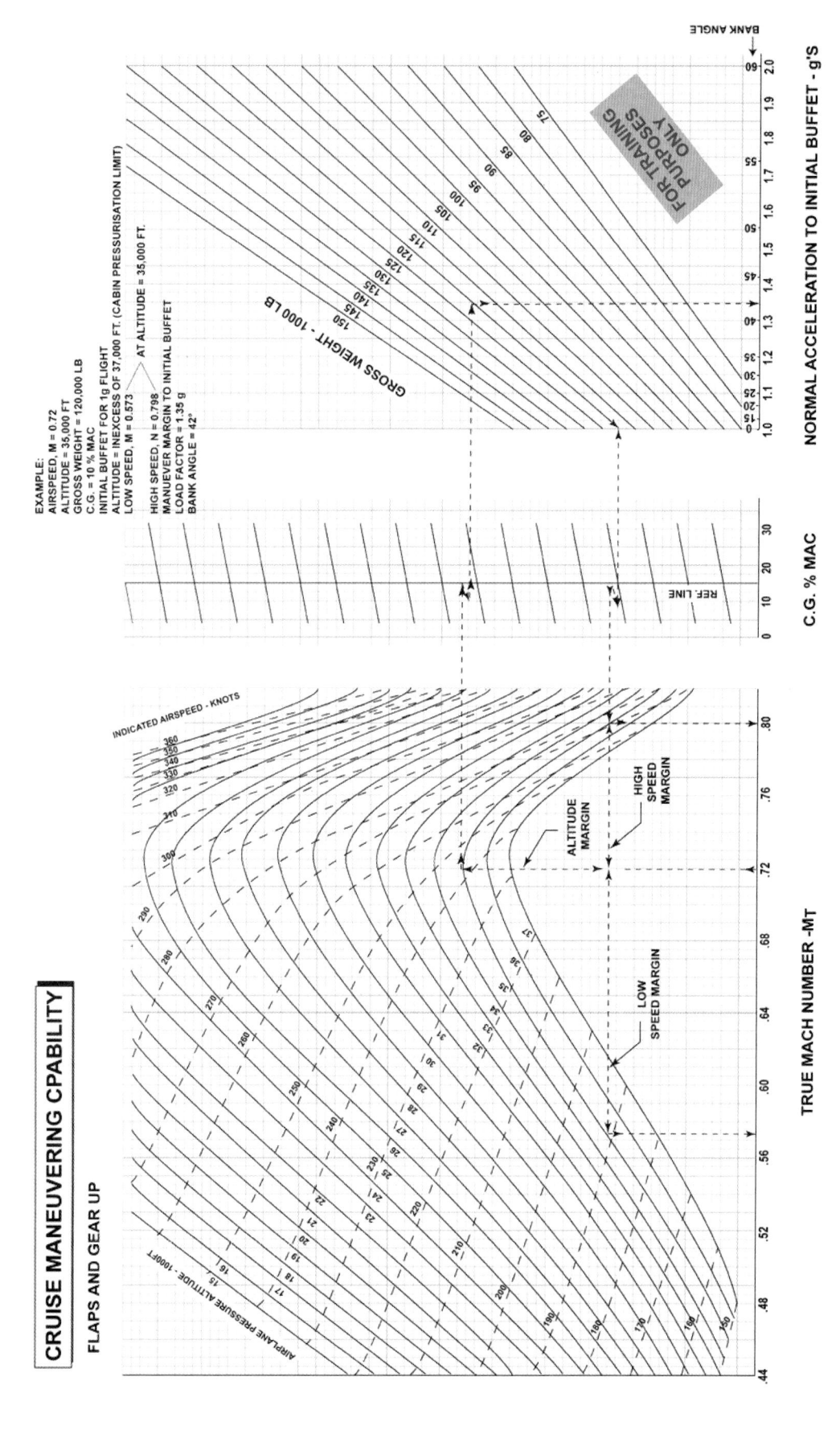

Figure 16.9 Buffet Onset Boundary in Terms of Weight, Altitude, M.No., IAS, "g" and C.G

PERFORMANCE CLASS A: EN ROUTE

16.9 CRUISE FOR MAXIMUM ECONOMY

Cruising to give the maximum miles per pound of fuel does not necessarily give the most economic operation. Data may be given to enable the lowest cost operation to be planned before despatch, and in flight, to assess whether a change from the planned flight procedure will give a more economical operation.

A 'cost index' can be defined :

$$\text{Cost Index} = \frac{\text{Operating cost / hour}}{\text{Fuel cost / pound}} \quad \text{(Excluding fuel cost)}$$

16.10 CRUISE WITH ONE ENGINE INOPERATIVE

The performance capability of an aircraft after the failure of one engine during the cruise will depend on the number of engines installed, a four engined aircraft will have 75% of its power remaining, a twin engined aircraft will have only 50% remaining. For a twin engined aircraft at its cruising altitude, in the event of an engine failure, it will normally be necessary to reduce speed and descend to a lower altitude. Cruising at a lower altitude will increase the fuel consumption and reduce the range.

After engine failure the operating engine should be set to the N_1 for Maximum Continuous Thrust and the speed brought to the optimum drift down speed.
Figure 16.10 shows typical data for the optimum speed for drift down and the expected level-off altitude.

WEIGHT 1000 KG		OPTIMUM DRIFT DOWN SPEED KIAS	LEVEL OFF ALTITUDE FT		
START DRIFT DOWN	LEVEL OFF		ISA +10°C & BELOW	ISA + 15°C	ISA + 20°C
68	65	242	17600	16300	14900
64	61	236	19500	18400	17100
60	57	229	21500	20500	19300
56	53	221	23500	22600	21500
52	50	213	25600	24800	23900
48	46	205	27700	27000	26200
44	42	197	29800	29400	28600
40	38	188	31900	31600	31100

Figure 16.10

PERFORMANCE   **CLASS A: EN ROUTE**

Figure 16.11 shows the range capability, including the drift down distance, and can be used to determine the fuel and time required to reach the destination/alternate, or, for the fuel remaining, the distance that could be flown.

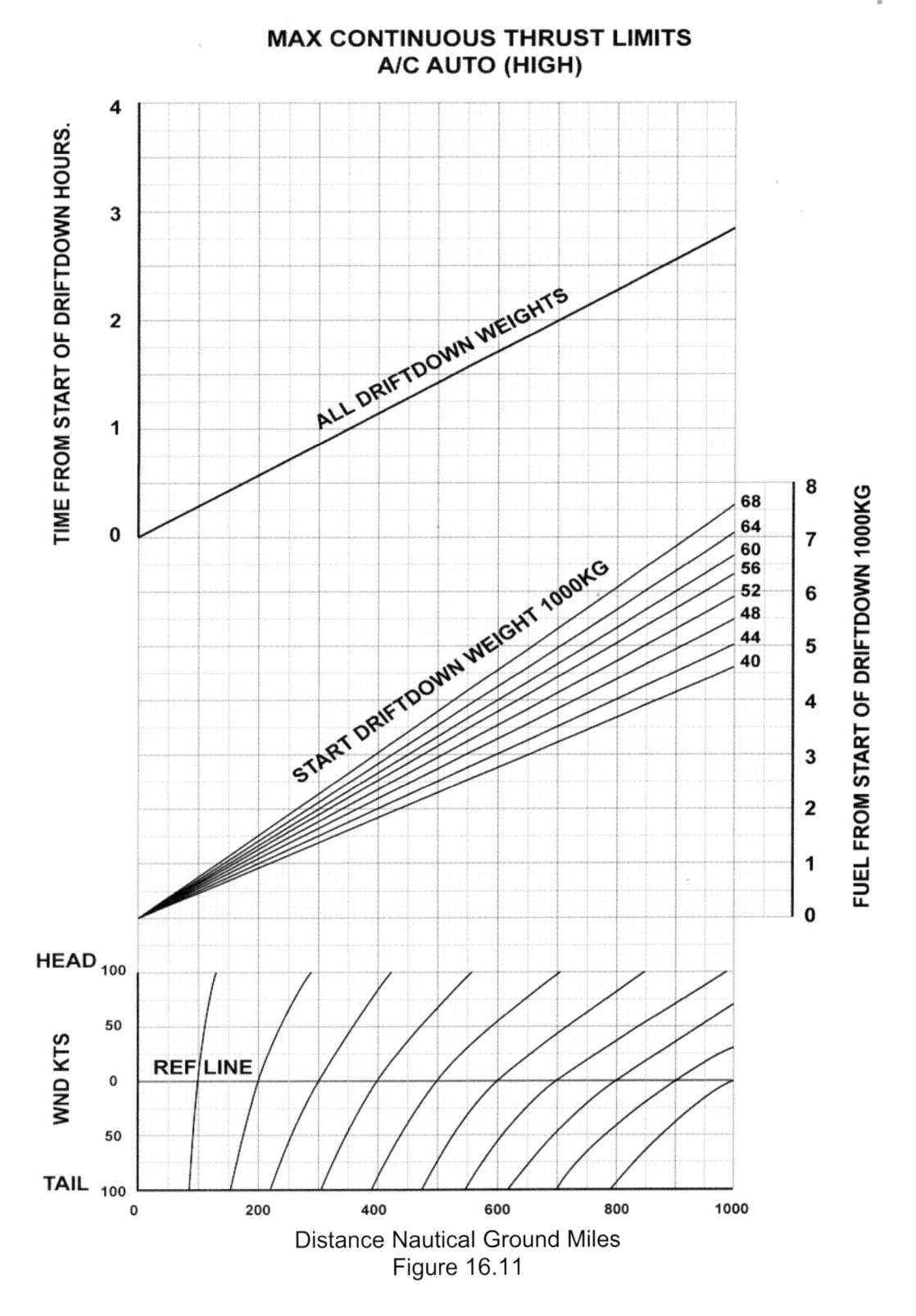

Figure 16.11

Data will be given for the one engine inoperative cruise, presented in the same way as the all engine cruise data.

16.11 RANGE LIMIT

Because of the increased fuel consumption at the lower altitudes at which it is necessary to fly following an engine failure, twin engined aircraft are restricted in the distance at which they are permitted to fly from an aerodrome which is adequate for landing. For twin engined aircraft in Performance Class A with either approved passenger seating of 20 or more or a maximum take-off mass of 45360 kg. or more, the distance is limited to that flown in 60 minutes at the one engine inoperative cruise speed in still air, ISA, unless they have ETOPS approval.

16.12 EXTENDED RANGE TWIN OPERATIONS (ETOPS)

The distance limit determined by the 60 minute requirement can be extended if sufficient reliability of engines and systems can be demonstrated. The aircraft may then be given ETOPS approval for an increased time limit. Typically this has been for 120 minutes but some aircraft have limits of 180 minutes.

16.13 ETOPS DATA

If the aircraft has ETOPS approval, additional data will be given in the Operation Manual. The data will include :

a) Area of operation (Diversion distance)

b) Critical fuel reserves.

c) Net level-off altitude

16.14 AREA OF OPERATION

The area of operation is the region within which the operator is authorised to conduct extended range operations. The distance to the diversion aerodrome from any point along the intended route must be covered within the approved time, using the single engine cruise speed, in still air, for ISA conditions.

PERFORMANCE

1 ENGINE INOP
MAX CONTINUOUS THRUST LIMITS
DIVERSION DISTANCE NM

CLASS A: EN ROUTE

AREA OF OPERATION (DIVERSION DISTANCE)

SPEED M/KIAS	WEIGHT AT DIVERSION 1000 KG	TIME MINUTES														
		60	70	80	90	100	110	120	130	140	150	160	170	180	190	200
0.70/280	35	411	481	550	619	688	757	826	895	964	1033	1102	1171	1240	1309	1378
	40	411	478	546	614	682	750	818	886	854	1022	1089	1157	1225	1293	1361
	45	406	473	539	606	673	739	806	873	939	1006	1072	1139	1206	1272	1339
	50	401	466	531	596	662	727	792	857	923	988	1053	1118	1184	1249	1314
	55	394	458	522	586	649	713	777	841	904	968	1032	1095	1159	1223	1287
	60	387	449	511	574	636	698	760	822	885	947	1009	1071	1133	1196	1258
	65	379	440	500	561	621	682	742	803	863	924	984	1045	1105	1166	1226
	70	372	431	489	548	607	666	724	783	842	900	959	1018	1077	1135	1194
0.74/290	35	423	492	561	630	699	768	838	907	976	1045	1114	1183	1252	1322	1391
	40	419	487	555	624	692	760	828	897	965	1033	1101	1170	1238	1306	1374
	45	415	482	549	616	683	751	818	885	952	1019	1087	1154	1221	1288	1355
	50	409	475	541	607	673	739	805	871	937	1003	1069	1135	1201	1267	1333
	55	403	468	532	597	662	726	791	856	920	985	1050	1114	1179	1244	1308
	60	396	459	522	585	649	712	775	838	902	965	1028	1091	1154	1218	1281
	65	388	450	511	573	635	696	758	820	881	943	1005	1066	1128	1190	1252
	70	381	441	501	561	621	681	741	801	861	921	981	1041	1101	1161	1222
0.74/310	35	425	495	564	633	702	771	840	909	978	1047	1116	1186	1255	1324	1393
	40	424	492	561	630	698	767	835	904	972	1041	1109	1178	1247	1315	1384
	45	421	489	556	624	692	760	828	896	963	1031	1099	1167	1235	1302	1370
	50	417	484	551	618	685	752	818	885	952	1019	1086	1153	1220	1287	1354
	55	412	478	544	609	675	741	807	873	939	1004	1070	1136	1202	1268	1334
	60	406	471	535	600	665	729	794	859	923	988	1052	1117	1182	1246	1311
	65	400	463	526	590	653	717	780	843	907	970	1033	1097	1160	1223	1287
	70	395	457	519	581	643	705	767	829	891	953	1015	1077	1139	1201	1263
0.74/330	35	427	495	564	632	700	769	837	906	974	1043	1111	1179	1248	1316	1385
	40	426	494	562	630	698	766	835	903	971	1039	1107	1175	1244	1312	1380
	45	424	492	559	627	695	763	830	898	966	1034	1101	1169	1237	1304	1372
	50	421	488	555	622	690	757	824	891	958	1025	1092	1159	1226	1293	1360
	55	417	484	550	616	683	749	815	881	948	1014	1080	1147	1213	1279	1345
	60	413	479	544	609	675	740	805	871	936	1001	1067	1132	1197	1263	1328
	65	408	472	537	601	665	730	794	858	923	987	1051	1115	1180	1244	1308
	70	404	467	530	594	657	720	783	847	910	973	1036	1100	1163	1226	1289
LRC	35	371	431	492	553	613	674	734	794	854	914	974	1034	1094	1153	1212
	40	376	438	499	561	622	683	744	805	866	827	987	1048	1108	1168	1228
	45	381	443	505	567	629	691	753	814	876	937	998	1059	1120	1181	1241
	50	385	448	510	573	635	697	759	821	883	944	1006	1067	1128	1189	1250
	55	387	450	512	575	637	699	761	823	885	946	1008	1069	1130	1191	1252
	60	388	451	513	576	638	700	762	824	885	947	1008	1069	1130	1191	1252
	65	389	451	513	576	638	699	761	822	884	945	1006	1067	1127	1188	1248
	70	391	453	515	576	638	700	761	822	883	944	1004	1065	1125	1185	1245

ISA BASED ON DRIFTDOWN STARTING AT OR NEAR OPTIMUM ALTITUDE.
3-K88A 737-400/CFM56-3C-1
APR 01/91 23.60.03

Figure 16.12

PERFORMANCE   **CLASS A: EN ROUTE**

16.15 CRITICAL FUEL RESERVES.

This is the fuel required to reach the diversion aerodrome, based on the worst case of three scenarios:

a)   Engine failure with descent at the selected speed and cruise at the selected level.

b)   Depressurisation, with emergency descent to FL 100 and cruise at the long range cruise speed with all engines operating.

c)   Depressurisation and engine failure, with emergency descent to FL 100 and cruise at the long range cruise speed.

Figure 16.13 shows the data for the critical fuel with all engines operating, and gives the various fuel allowances which must be made. The performance deterioration factor is 5% unless a value has been established by the operator.

Similar data is given for the critical fuel with one engine inoperative.

The highest fuel given by consideration of the three cases must be taken.

If the fuel on board the aircraft at the point on the route considered is less than the critical fuel reserve, then the fuel loaded must be increased to make up the difference.

PERFORMANCE CLASS A: EN ROUTE

CRITICAL FUEL RESERVES
LONG RANGE CRUISE

BASED ON: EMERGENCY DESCENT TO 10000 FT.
LEVEL CRUISE AT 10000 FT.   250 KIAS DESCENT TO 1500 FT.
15 MINUTE HOL AT 1500 FT.
ONE MISSED APPROACH; APPROACH AND LAND.
5% ALLOWANCE FOR WIND ERRORS. INCLUDES APU FUEL BURN

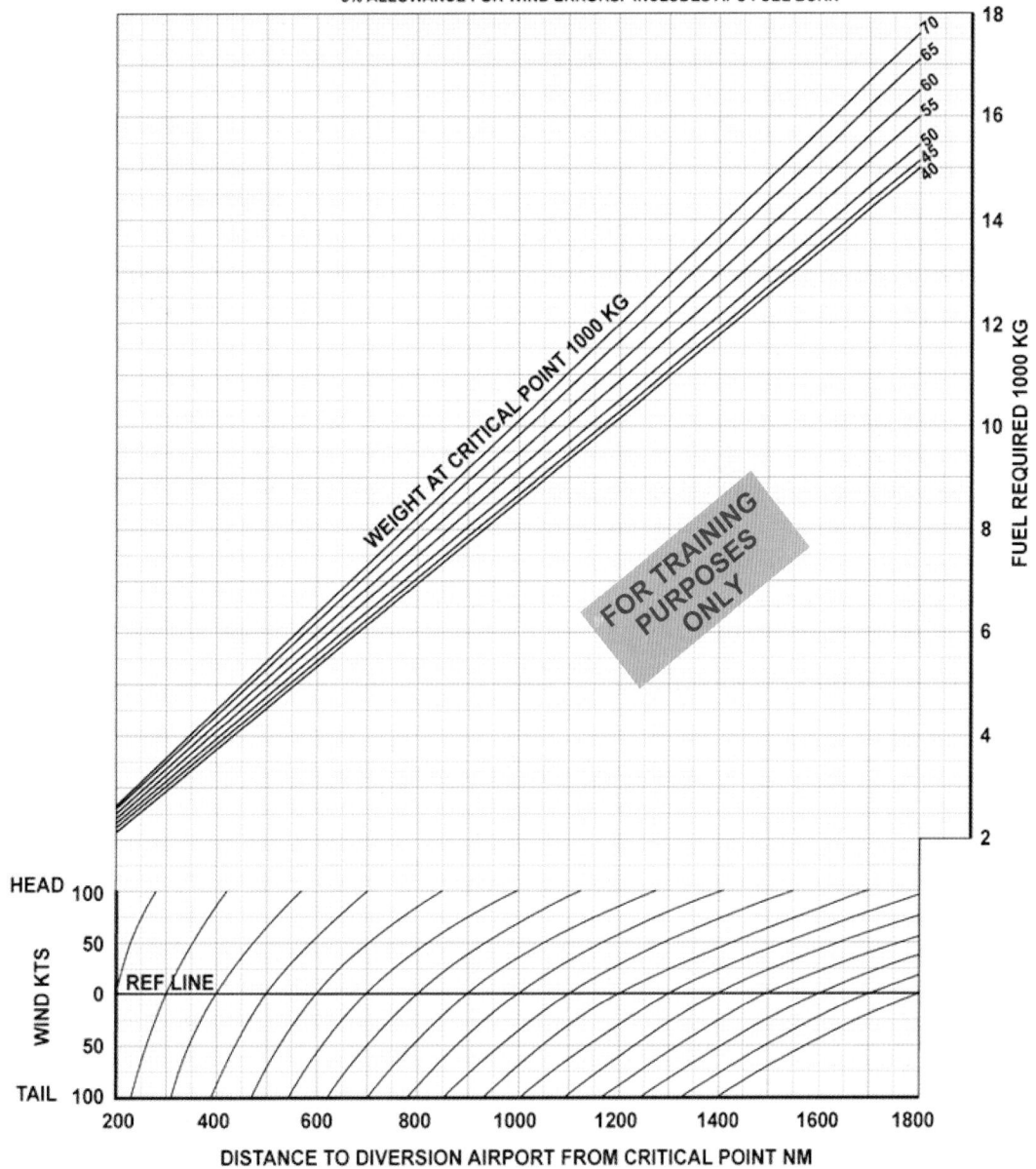

FOR TRAINING PURPOSES ONLY

INCREASE FUEL REQUIRED BY 0.5% FOR EACH 10°C HOTTER THAN ISA CONDITIONS.
IF ICING CONDITIONS EXIST, INCREASE FUEL REQUIRED BY 18% TO ACCOUNT FOR ENGINE AND WING A/I ON AND ICE ACCUMULATION ON UNHEATED SURFACES.
ALLOWANCE FOR PERFORMANCE DETERIORATION NOT INCLUDED.
COMPARE THE FUEL REQUIRED FROM THIS CHART WITH CRITICAL FUEL RESERVES FOR ONE ENGINE INOPERATIVE. USE THE HIGHER OF THE TWO.

Figure 16.13

PERFORMANCE   **CLASS A: EN ROUTE**

16.16 NET LEVEL OFF ALTITUDE

This gives the maximum mass for terrain clearance. (See En-Route Requirements para. 16.19)

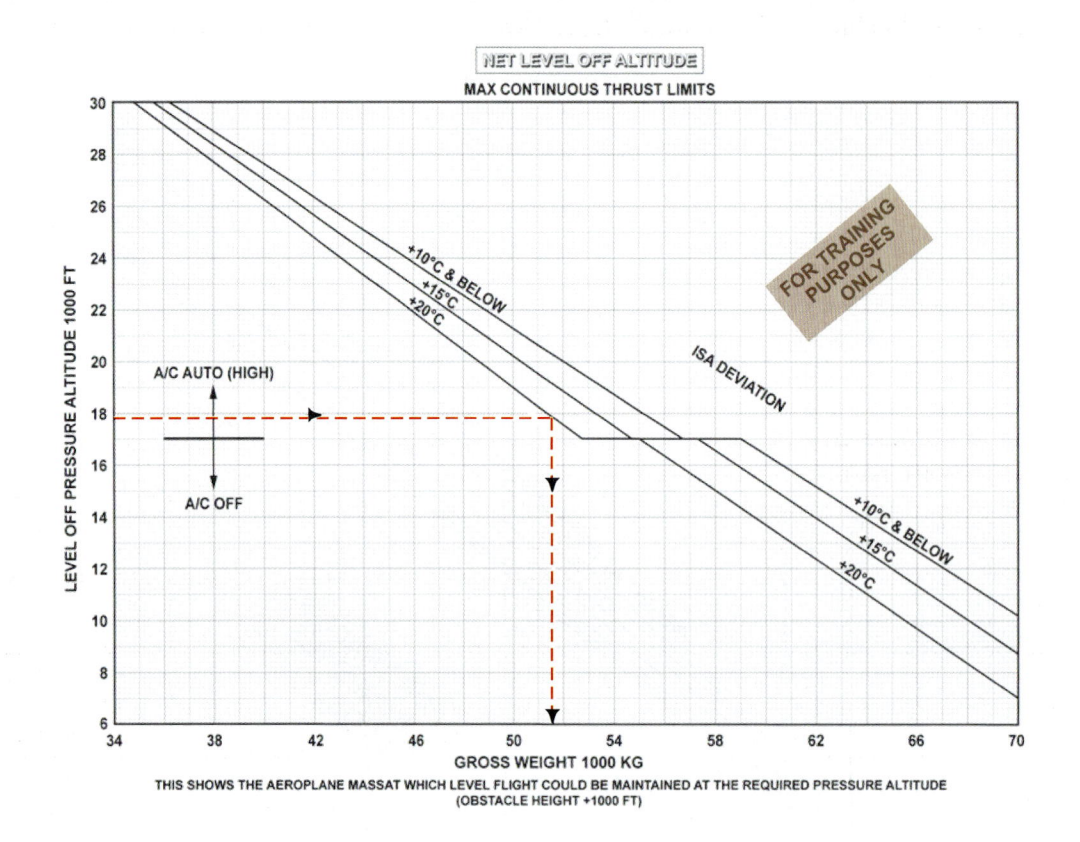

Figure 16.14 Net Level Off Altitude

PERFORMANCE   CLASS A: EN ROUTE

16.17 ENDURANCE.

If the aeroplane is required to hold, then maximum endurance is required. It was seen that this occurs at Vmd for a jet aircraft, although the recommended holding speed may be faster to give better speed stability. The following table shows a typical presentation of holding data.

HOLDING

% N_1
IAS KNOTS
FF PER ENG

FLAPS UP

PRESSURE ALTITUDE FT	WEIGHT 1000 KG														
	66	64	62	60	58	56	54	52	50	48	46	44	42	40	38
37000					92.3 239 1370	89.9 237 1270	88.2 234 1200	86.7 231 1130	85.5 228 1080	84.4 225 1040	83.3 221 990	82.3 217 950	81.2 212 900	80.2 210 870	79.3 210 840
35000		92.8 251 1510	90.6 249 1410	88.9 246 1330	87.5 243 1260	86.3 241 1210	85.3 238 1160	84.3 234 1110	83.3 230 1070	82.4 226 1030	81.4 222 980	80.4 218 940	79.3 213 900	78.3 210 860	77.4 210 830
30000	86.1 263 1420	85.3 259 1370	84.6 256 1330	83.8 252 1280	83.0 248 1240	82.1 245 1200	81.3 241 1150	80.4 236 1110	79.5 232 1070	78.5 227 1030	77.5 222 980	76.5 217 940	75.4 212 900	74.4 210 870	73.5 210 840
25000	82.3 265 1420	81.5 261 1380	80.8 257 1330	80.0 253 1290	79.2 248 1250	78.3 244 1210	77.5 239 1160	76.5 235 1120	75.6 230 1080	74.6 225 1040	73.6 220 1000	72.6 215 960	71.5 210 920	70.6 210 890	69.7 210 860
20000	78.5 263 1420	77.7 259 1380	77.0 255 1340	76.2 251 1290	75.4 246 1250	74.5 242 1210	73.6 237 1170	72.7 232 1130	71.7 227 1090	70.7 223 1050	69.6 218 1010	68.5 212 970	67.5 210 930	66.6 210 900	65.7 210 880
15000	74.7 261 1440	74.0 257 1400	73.2 253 1350	72.3 248 1310	71.4 244 1270	70.5 239 1230	69.6 235 1190	68.6 230 1150	67.6 225 1110	66.6 220 1070	65.4 215 1030	64.2 210 990	63.3 210 960	62.4 210 930	61.5 210 900
10000	70.8 259 1460	70.0 254 1410	69.2 250 1370	68.3 246 1330	67.4 241 1290	66.4 237 1250	65.4 232 1210	64.4 228 1170	63.3 223 1130	62.2 219 1090	61.1 214 1050	60.1 210 1010	59.3 210 990	58.5 210 960	57.9 210 940
5000	66.7 256 1480	65.8 252 1430	64.9 248 1390	64.0 244 1350	63.1 240 1310	62.2 235 1270	61.3 231 1230	60.4 227 1190	59.5 223 1150	58.6 219 1110	57.7 214 1070	56.7 210 1040	56.0 210 1010	55.3 210 980	54.7 210 960
1500	63.7 255 1500	62.9 251 1450	62.1 247 1410	61.3 243 1370	60.5 239 1330	59.7 235 1290	58.9 231 1260	58.1 228 1220	57.2 224 1180	56.4 220 1140	55.5 216 1110	54.5 211 1070	53.7 210 1040	53.1 210 1010	52.4 210 990

FUEL FLOW IS BASED ON A RACETRACK PATTERN.
FOR HOLDING IN STRAIGHT AND LEVEL FLIGHT REDUCE FUEL FLOW VALUES BY 5 PERCENT.

PERFORMANCE   **CLASS A: EN ROUTE**

16.18 EN-ROUTE REQUIREMENTS

The Regulations give requirements for obstacle clearance in the event of engine failure en-route, and for two engines inoperative in the case of three and four engined aeroplanes.

16.19 EN ROUTE GRADIENT MARGINS

The performance must be based on net performance, that is, the gross performance must be reduced by a specified margin. For one engine inoperative these margins are:

a)   1.1% for 2 engined aeroplanes

b)   1.4% for 3 engined aeroplanes

c)   1.6% for 4 engined aeroplanes

for two engines inoperative the margins are:

a)   0.3% for 3 engined aeroplanes

b)   0.5% for 4 engined aeroplanes

16.20 OBSTACLE CLEARANCE REQUIREMENTS EN-ROUTE JAR-OPS 1.500

In the event of an engine failure en-route the aeroplane must:

1. Either   a)   have a positive net gradient at 1000ft above all terrain and obstacles along the route within 5 nm on either side if the intended track.

or   b)   continue the flight to an aerodrome where a landing can be made in accordance with the landing regulations, the net flight path clearing vertically by at least 2000 ft all terrain and obstacles along the route within 5 nm on either side of the intended track.

2.   The net flight path must have a positive gradient at 1500 ft above the aerodrome where the landing is to be made after engine failure.

In assessing compliance with 1 (b) above :-

    (i)   the engine must be assumed to fail at the most critical point along the route.
    (ii)   account must be taken of the effect of wind.
    (iii)   fuel jettisoning is permitted.
    (iv)   landing requirements must be met.

PERFORMANCE — CLASS A: EN ROUTE

The width margins in paragraphs 1 (a) and 1 (b) must be increased to 10 nm if the navigational accuracy does not meet the 95% containment level.

16.21 PRESENTATION OF DATA

En-route data may be given in the form of the en-route gradient, from which the ceiling and drift down profile can be produced.

Alternatively the ceiling and drift down profiles may be presented.

Figure 16.14 shows the aeroplane mass at which level flight could be maintained at the required pressure altitude (obstacle height + 1000 ft).

Figure 16.15 shows the drift down profiles. These are profiles of height against distance for a range of "equivalent weights". The equivalent weight allows for the effect of temperature, a higher temperature being equivalent to a higher weight as far as gradient is concerned.

For a given required pressure altitude (obstacle height + 2000 ft) and equivalent weight, the maximum distance from the obstacle at engine failure can be determined.

16.22 PROCEDURE

For the highest obstacles along the route, determine whether the aircraft can maintain level flight with one engine inoperative, at 1000 ft above the obstacles. (Fig. 16.14)

If this is not possible, a drift down procedure must be evaluated, based on engine failure at the most critical point, and must show clearance of obstacles during the drift down by at least 2000 ft.

PERFORMANCE   **CLASS A: EN ROUTE**

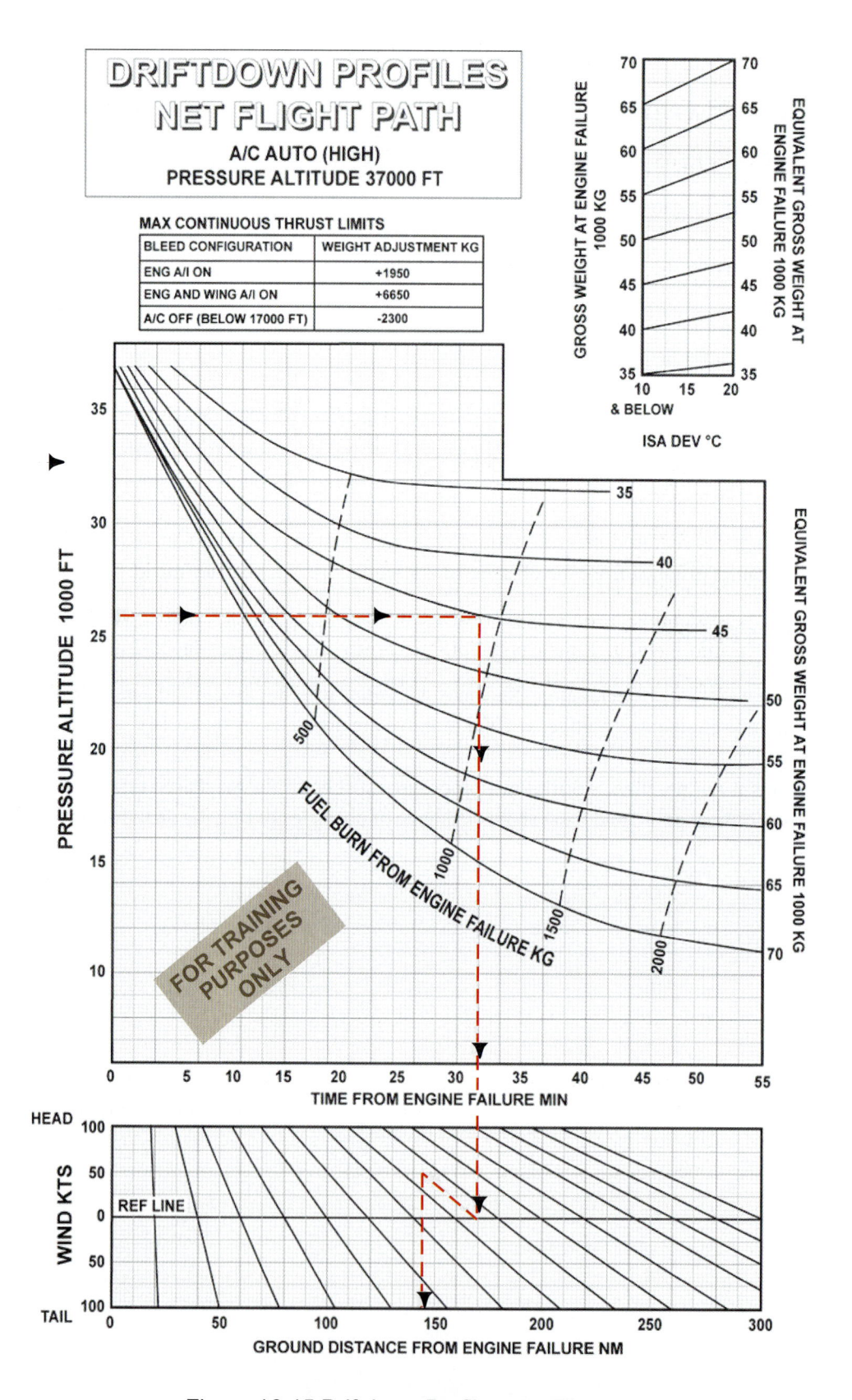

Figure 16.15 Driftdown Profiles Net Flight Path

16.23 TWO ENGINES INOPERATIVE

For aircraft with three or more engines, there are regulations for two engines inoperative.

JAR-OPS 1.505

a) An operator shall ensure that at no point along the intended track will an aeroplane having three or more engines be more than 90 minutes, at the all-engines long range cruising speed at standard temperature in still air, away from an aerodrome at which the performance requirements applicable at the expected landing mass are met unless it complies with sub-paragraphs (b) to (f) below.

b) The two engines inoperative en-route flight net path data must permit the aeroplane to continue the flight, in the expected meteorological conditions, from the point where two engines are assumed to fail simultaneously, to an aerodrome at which it is possible to land and come to a complete stop when using the prescribed procedure for a landing with two engine inoperative. The net flight path must clear vertically, by at least 2000 ft all terrain and obstructions along the route within 9.3 km (5 nm) on either side of the intended track. At altitudes and in meteorological conditions requiring ice protection systems to be operable, the effect of their use on the net flight path data must be taken into account. If the navigational accuracy does not meet the 95% containment level, an operator must increase the width margin given above to 18.5 km (10 nm).

c) The two engines are assumed to fail at the most critical point of that portion of the route where the aeroplane is more than 90 minutes, at the all engines long range cruising speed at standard temperature in still air, away from an aerodrome at which the performance requirements applicable at the expected landing mass are met.

d) The net flight path must have a positive gradient at 1500 ft above the aerodrome where the landing is assumed to be made after the failure of two engines.

e) Fuel jettisoning is permitted to an extent consistent with reaching the aerodrome with the required fuel reserves, if a safe procedure is used.

f) The expected mass of the aeroplane at the point where the two engines are assumed to fail must not be less than that which would include sufficient fuel to proceed to an aerodrome where the landing is assumed to be made, and to arrive there at least 1500 ft directly over the landing area and thereafter to fly level for 15 minutes.

PERFORMANCE

CLASS A: EN ROUTE

PAPER 1

1. For an aircraft climbing at a constant Indicated Air Speed. the True airspeed (i) and the Mach number (ii) will :

	(i)	(ii)
a)	increase	increase
b)	increase	decrease
c)	decrease	increase
d)	decrease	decrease

2. An aircraft is descending at a constant Mach number. The IAS (i) and the TAS (ii) will :

	(i)	(ii)
a)	increase	decrease
b)	decrease	increase
c)	increase	increase
d)	decrease	decrease

3. The Long Range Cruise procedure is to fly :

    a) at the speed which gives the maximum possible range
    b) at a speed above the maximum range speed, which gives 99% of the maximum possible range
    c) at a speed below the maximum range speed, which gives 95% of the maximum range
    d) at a speed which gives the minimum fuel flow.

4. It may be preferable to cruise at an altitude other than the optimum altitude for maximum range:

    a) because a lower altitude may give a higher V_{MO}
    b) because a higher altitude may give a larger buffet margin
    c) if a lower altitude has a more favourable wind
    d) because a higher altitude may give a higher Specific Range

5. Which of the following statements is correct with regard to cruising speed.

    a) Maximum cruise speed is always limited by maximum cruise thrust available.
    b) Maximum achievable cruise speed is not affected by weight.
    c) Maximum cruise speed may be limited by maximum cruise thrust available, M_{MO}, or V_{MO}
    d) Maximum cruise speed is always limited by a Mach number limitation

PERFORMANCE   **CLASS A: EN ROUTE**

6. Buffet may occur :

   a) only at very high speed
   b) only at very low speed
   c) only at very high or very low speed
   d) at any speed depending on the combination of angle of attack and Mach number.

Refer to Appendix A for questions 7 and 8

7. For an aircraft flying at 34000 ft, Mach 0.73 , mass 60000 kg, C of G 15 % MAC, what 'g' can be applied before buffet occurs ?

   a) 1.15
   b) 1.3
   c) 1.25
   d) 1.5

8. What speed range would be available between low speed and high speed buffet, when flying at 36000 ft. in 1g flight at a mass of 62000 kg, C of G 5 % MAC.

   a) 0.656 to 0.76
   b) 0.68 to 0.76
   c) 0.645 to 0.77
   d) 0.68 to 0.77

9. **Refer to Appendix B**

   For an aircraft holding at 5000 ft. at a mass of 56000 kg., if 6000 kg. of fuel is available for holding, the maximum time available to hold would be :

   a) 2 hr. 36 min.
   b) 2.36 hr.
   c) 4.7 hr.
   d) 62.2 min

10. **Refer to CAP 698 Fig. 4.24**

    Find the ground distance to descend from 37000 ft. to 20000 ft. ISA + 20°C at a mass of 60000 kg.if the wind component is 25 knots headwind

    a) 110 n.m.
    b) 95 n.m.
    c) 105 n.m.
    d) 125 n.m.

PERFORMANCE   **CLASS A: EN ROUTE**

11. Which of the following statements is correct with regard to the en route regulations for twin engined aircraft after engine failure :

    a) If the flight is continued to a landing aerodrome, obstacles must be cleared by 1000 ft.
    b) the gross gradient must be reduced by 0.9%
    c) the effect of fuel jettisoning on the aircraft weight is not permitted.
    d) the aircraft must have a positive climb gradient at 1500 ft. above the landing aerodrome

12. For a non- ETOP's approved twin engined aircraft,

    a) it must not be more than 60 minutes in still air, ISA, from a suitable aerodrome
    b) it must not be more than 60 n.m. from a suitable aerodrome.
    c) it must not be more than 60 minutes in the forecast wind, ISA. from a suitable aerodrome
    d) it must not be more than 90 minutes in still air, ISA, from a suitable aerodrome

Appendix A

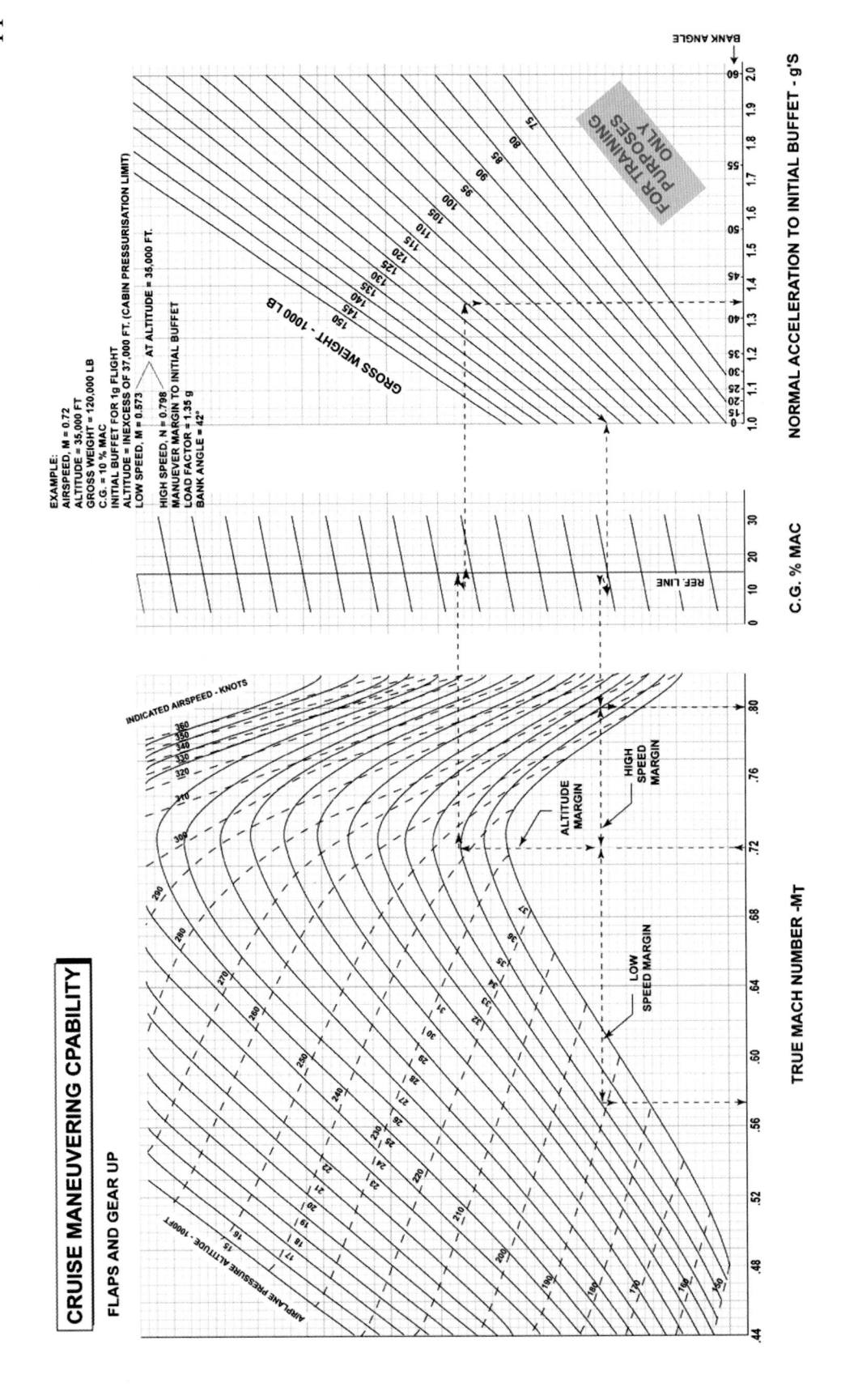

CRUISE MANEUVERING CAPABILITY

FLAPS AND GEAR UP

EXAMPLE:
AIRSPEED, M = 0.72
ALTITUDE = 35,000 FT
GROSS WEIGHT = 120,000 LB
C.G. = 10 % MAC
INITIAL BUFFET FOR 1g FLIGHT
ALTITUDE = INEXCESS OF 37,000 FT. (CABIN PRESSURISATION LIMIT)
AT ALTITUDE = 35,000 FT.
LOW SPEED, M = 0.573
HIGH SPEED, N = 0.798
MANUEVER MARGIN TO INITIAL BUFFET
LOAD FACTOR = 1.35 g
BANK ANGLE = 42°

FOR TRAINING PURPOSES ONLY

TRUE MACH NUMBER -MT

C.G. % MAC

GROSS WEIGHT - 1000 LB

NORMAL ACCELERATION TO INITIAL BUFFET - g'S

BANK ANGLE

PERFORMANCE CLASS A: EN ROUTE

Appendix B

BOEING 737
OPERATIONS MANUAL

HOLDING

%N₁
IAS KNOTS
FF PER ENG

FLAPS UP

PRESSURE ALTITUDE FT	WEIGHT 1000 KG														
	66	64	62	60	58	56	54	52	50	48	46	44	42	40	38
37000					92.3 239 1370	89.9 237 1270	88.2 234 1200	86.7 231 1130	85.5 228 1080	84.4 225 1040	83.3 221 990	82.3 217 950	81.2 212 900	80.2 210 870	79.3 210 840
35000		92.8 251 1510	90.6 249 1410	88.9 246 1330	87.5 243 1260	86.3 241 1210	85.3 238 1160	84.3 234 1110	83.3 230 1070	82.4 226 1030	81.4 222 980	80.4 218 940	79.3 213 900	78.3 210 860	77.4 210 830
30000	86.1 263 1420	85.3 259 1370	84.6 256 1330	83.8 252 1280	83.0 248 1240	82.1 245 1200	81.3 241 1150	80.4 236 1110	79.5 232 1070	78.5 227 1030	77.5 222 980	76.5 217 940	75.4 212 900	74.4 210 870	73.5 210 840
25000	82.3 265 1420	81.5 261 1380	80.8 257 1330	80.0 253 1290	79.2 248 1250	78.3 244 1210	77.5 239 1160	76.5 235 1120	75.6 230 1080	74.6 225 1040	73.6 220 1000	72.6 215 960	71.5 210 920	70.6 210 890	69.7 210 860
20000	78.5 263 1420	77.7 259 1380	77.0 255 1340	76.2 251 1290	75.4 246 1250	74.5 242 1210	73.6 237 1170	72.7 232 1130	71.7 227 1090	70.7 223 1050	69.6 218 1010	68.5 212 970	67.5 210 930	66.6 210 900	65.7 210 880
15000	74.7 261 1440	74.0 257 1400	73.2 253 1350	72.3 248 1310	71.4 244 1270	70.5 239 1230	69.6 235 1190	68.6 230 1150	67.6 225 1110	66.6 220 1070	65.4 215 1030	64.2 210 990	63.3 210 960	62.4 210 930	61.5 210 900
10000	70.8 259 1460	70.0 254 1410	69.2 250 1370	68.3 246 1330	67.4 241 1290	66.4 237 1250	65.4 232 1210	64.4 228 1170	63.3 223 1130	62.2 219 1090	61.1 214 1050	60.1 210 1010	59.3 210 990	58.5 210 960	57.9 210 940
5000	66.7 256 1480	65.8 252 1430	64.9 248 1390	64.0 244 1350	63.1 240 1310	62.2 235 1270	61.3 231 1230	60.4 227 1190	59.5 223 1150	58.6 219 1110	57.7 214 1070	56.7 210 1040	56.0 210 1010	55.3 210 980	54.7 210 960
1500	63.7 255 1500	62.9 251 1450	62.1 247 1410	61.3 243 1370	60.5 239 1330	59.7 235 1290	58.9 231 1260	58.1 228 1220	57.2 224 1180	56.4 220 1140	55.5 216 1110	54.5 211 1070	53.7 210 1040	53.1 210 1010	52.4 210 990

FUEL FLOW IS BASED ON A RACETRACK PATTERN.
FOR HOLDING IN STRAIGHT AND LEVEL FLIGHT REDUCE FUEL FLOW VALUES BY 5 PERCENT.

PERFORMANCE **CLASS A: EN ROUTE**

ANSWERS

Questions	Answers
1	A
2	C
3	B
4	C
5	C
6	D
7	B
8	C
9	B
10	B
11	D
12	A

CHAPTER SEVENTEEN - CLASS A: LANDING

Contents

		Page
17.1	LANDING CONSIDERATIONS	17 - 1
17.2	LANDING CLIMB REQUIREMENTS	17 - 1
17.3	LANDING DISTANCE REQUIREMENTS	17 - 4
17.4	RUNWAY SELECTION	17 - 4
17.5	NON COMPLIANCE	17 - 5
17.6	WET RUNWAYS	17 - 5
17.7	PRESENTATION OF DATA	17 - 5
	CLASS A AIRCRAFT: LANDING	17 - 7

© Oxford Aviation Services Limited

PERFORMANCE — CLASS A: LANDING

17.1 LANDING CONSIDERATIONS

The maximum mass for landing is the lesser of:

a) the climb limit mass

b) the field length limit mass

c) the structural limit mass

17.2 LANDING CLIMB REQUIREMENTS

a) **Landing climb. All engines operating JAR 25.119.**

A gradient of not less than 3.2% with:

i) All engines operating at the power available 8 seconds after initiation of movement of the thrust control from the minimum flight idle to the take off position

ii) Landing configuration.

iii) A climb speed which is:

1. not less than $1.13\ V_{SR}$ (may be $1.15\ V_{SR}$ for 4 engine aircraft if application of power results in significant reduction in stalling speed)

2. not less than V_{MCL}

3. not more than $1.23\ V_{SR}$

iv) Aerodrome altitude.

v) Ambient temperature expected at the time of landing.

PERFORMANCE   CLASS A: LANDING

b) **Discontinued Approach Climb. One engine inoperative. JAR 25.121**

A climb gradient not less than:

2.1% for 2 engined aircraft

2.4% for 3 engined aircraft

2.7% for 4 engined aircraft

With:
- i) The critical engine inoperative and the remaining engines at the available take-off thrust.
- ii) Speed: Normal approach speed but not greater than 1.41 V_{SR}.
- iii) Landing gear retracted
- iv) Flaps approach configuration, provided that the approach flap Vs does not exceed 110% of landing flap Vs
- v) Aerodrome altitude
- vi) Ambient temperature

The more limiting of the landing climb and the approach gradient requirements will determine the maximum mass for altitude and temperature at the landing aerodrome. Figure 17.1 shows a typical presentation of this data.

c) **Discontinued Instrument Approach Climb. JAR-OPS 1.510**

For instrument approaches with decision heights below 200ft, an operator must verify that the approach mass of the aeroplane, taking into account the take-off mass and the fuel expected to be consumed in flight, allows a missed approach gradient of climb, with the critical engine failed and with the speed and configuration used for go-around of at least 2.5%, or the published gradient, whichever is the greater.

PERFORMANCE   CLASS A: LANDING

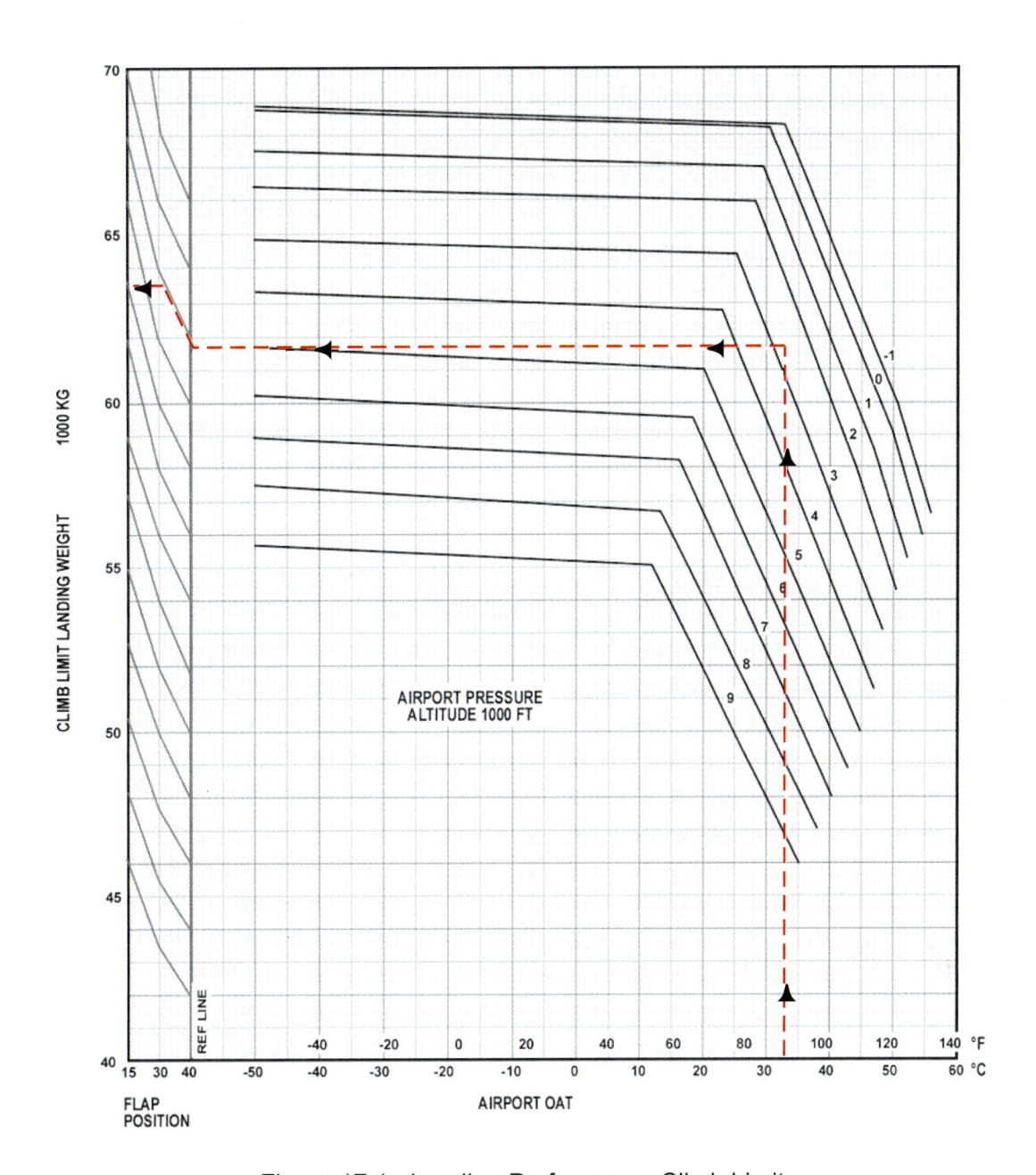

Figure 17.1. Landing Performance Climb Limit

PERFORMANCE   CLASS A: LANDING

17.3 LANDING DISTANCE REQUIREMENTS JAR-OPS 1.515

The landing distance required on a dry runway for destination and alternate aerodromes, from 50 ft to a full stop must not exceed:

a)   60% of the Landing Distance Available for turbo-jet aeroplanes

b)   70% of the Landing Distance Available for turbo-prop aeroplanes

(Short landing and steep approach procedures may be approved based on lower screen heights, but not less than 35 ft)

The landing distance required is based on:

a)   the aeroplane in the landing configuration

b)   the speed at 50ft not less than 1.23 V_{SR}

c)   aerodrome pressure altitude

d)   standard day temperature

e)   factored winds (50% headwind, 150% tailwind)

f)   the runway slope if greater than ± 2%

17.4 RUNWAY SELECTION.

Landing must be considered both in still air and in the forecast wind.

a)   Still air: The most favourable runway in still air may be selected.

b)   Forecast wind: The runway most likely to be used in the forecast wind.

The lower of the two masses obtained from a) and b) above must be selected as the limiting mass for the field lengths available.

PERFORMANCE   CLASS A: LANDING

17.5 NON COMPLIANCE

a) If the still air requirement cannot be met at an aerodrome with a single runway, that is, landing can only be made if there is an adequate wind component, the aircraft may be dispatched if 2 alternate aerodromes are designated at which full compliance is possible.

b) If the forecast wind requirement cannot be met, the aeroplane may be dispatched if an alternate is designated at which all the landing requirements are met.

17.6 WET RUNWAYS

a) If the runway is forecast to be wet at the estimated time of arrival the Landing Distance Available must be at least 115% of the dry runway landing distance required.

17.7 PRESENTATION OF DATA

Figure 17.2 shows a typical presentation of landing distance data.

Example: Find the maximum planned landing weight at a destination aerodrome given the following details;

Aerodrome pressure altitude : 2000 ft
Expected air temperature: 20°C
Flap position: 30°
Runway A: Length 4750ft, level, dry, forecast wind component 5 kt head
Runway B : Length 4500 ft, level, dry, forecast wind component 20 kt head
Solution: Using the data in CAP 698
Field length consideration :   Still Air   Forecast Wind
              R/W A   48500kg   50500 kg
              R/W B   45800 kg   51500 kg
Most favourable runway in still air : 48500 kg
Most favourable runway in forecast wind : 51500 kg
Selected weight 48500 kg
Landing climb considerations: 64000 kg
Maximum structural landing weight : 54900 kg
∴ Maximum planned landing weight = 48500 kg

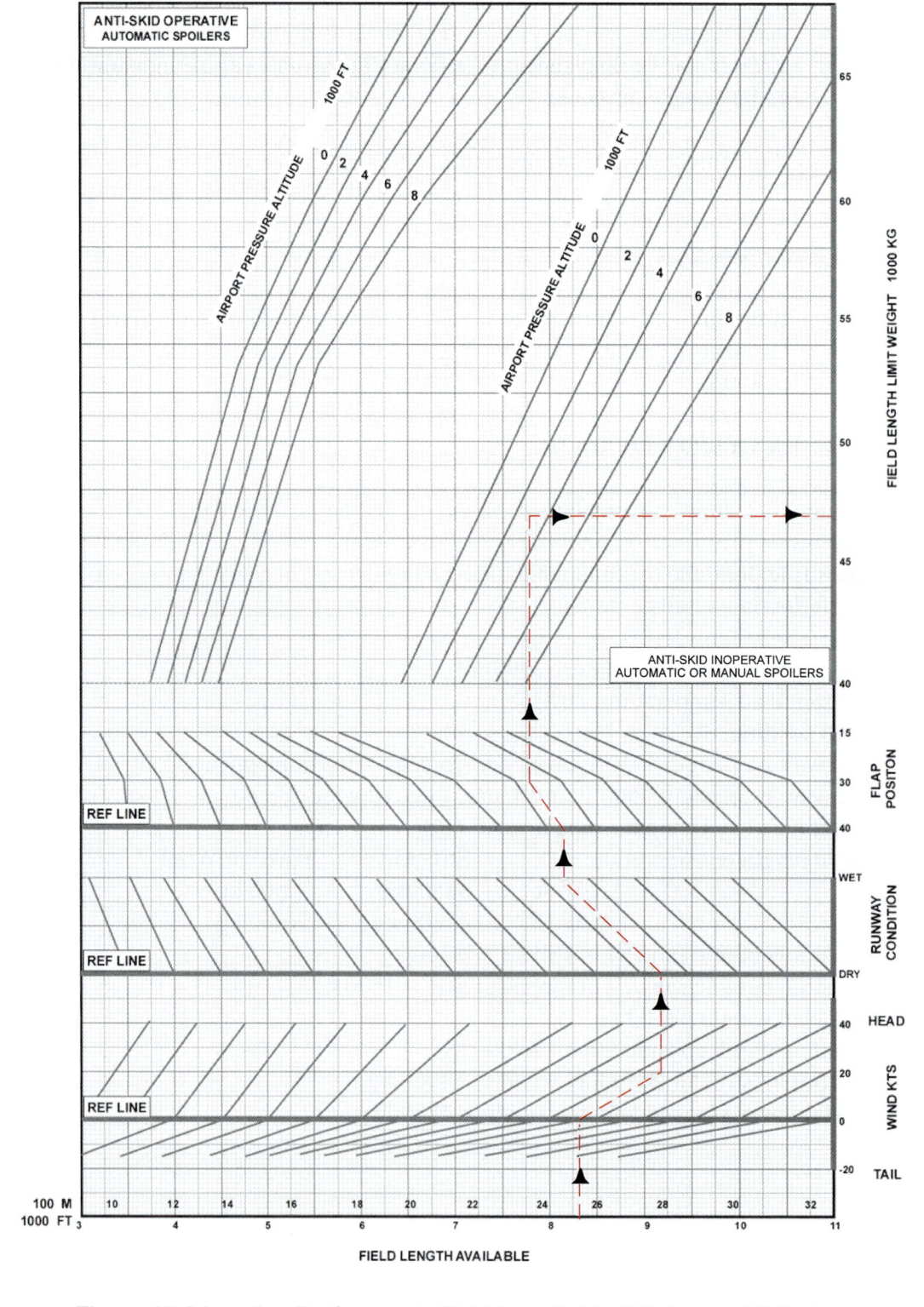

Figure 17.2 Landing Performance Field Length Limit Category 'A' Brakes

PERFORMANCE CLASS A: LANDING

PAPER 1

1. To meet the balked landing requirements an aircraft must achieve a climb gradient of :

    a) 3.2% in the landing configuration with the critical engine inoperative.
    b) 2.1% in the approach configuration with all engines operating.
    c) 3.2% in the landing configuration with all engines operating.
    d) 2.1% in the landing configuration with all engines operating.

    Refer to CAP 698 Fig. 4.29 for questions 2 and 3

2. The climb limited landing mass at an aerodrome with a pressure altitude of 4000 ft. and a temperature of 10°C with a flap setting of 30° and A/C packs off would be :

    a) 61800 kg.
    b) 63110 kg.
    c) 60800 kg.
    d) 62110 kg.

3. At a pressure altitude of 5000 ft. for temperatures below 20°C, the climb limited landing mass is approximately constant. This is due to :

    a) the climb speed becoming limited by V_{MCL}
    b) the engines becoming flat rated.
    c) the engine acceleration time exceeding 8 seconds
    d) allowance being made for anti-ice operation.

4. For a class A jet aircraft, the landing distance required must not exceed :

    a) 50% of the landing distance available
    b) 60% of the landing distance available
    c) 70% of the landing distance available
    d) 90% of the landing distance available.

Refer to CAP 698 Fig. 4.28 for questions 5, 6, 7 and 8

5. Find the maximum planned landing mass determined by the field length available, at an aerodrome with a pressure altitude of 3000 ft. The single runway has a landing distance available of 5000 ft .and is dry. The forecast wind component is 20 knots. The flap setting is 30° and the anti-skid is operative.

   a) 60000 kg.
   b) 54800 kg.
   c) 50000 kg.
   d) 57000 kg.

6. What landing distance must be available to satisfy the landing requirements, for an aircraft with a mass of 60000 kg. at a pressure altitude of 2000 ft. ISA in still air. The flap setting is 30° and the runway is wet.

   a) 4800 ft.
   b) 6250 ft.
   c) 6670 ft.
   d) 7000 ft.

7. Find the maximum planned landing weight for the aerodrome data given below, given that the climb and structural requirements are not limiting.

	Pressure altitude	2000 ft.		
			Length	Forecast Wind Component
	Runway A	Dry, level	5600 ft.	0
	Runway B	Dry, level	5000 ft.	30 kt. Headwind

   Flap setting : 40°
   Anti-skid : operative

   a) 54000 kg.
   b) 60200 kg.
   c) 58800 kg.
   d) 56400 kg.

PERFORMANCE   CLASS A: LANDING

8. For an aircraft mass of 54000 kg. what is the maximum tailwind at which the landing requirement could be met at a pressure altitude of 2000 ft. if the landing distance available is 6000 ft. The runway is dry. The flap setting is 30°

   a)   2 knots
   b)   8 knots
   c)   15 knots
   d)   a headwind would be required.

9. Which of the following statements is true with regard to runway slope in the calculation of landing weight.

   a)   Slope is only taken into account if it is downhill.
   b)   Slope should be taken into account if it exceeds 1.0%
   c)   Slope should be taken into account if it exceeds 2.0%
   d)   Slope is never taken into account.

10. Landing is planned at an aerodrome with a single runway. At the expected arrival weight the landing distance available is adequate with the forecast wind, but inadequate in still air.

   a)   Weight must be reduced until the still air requirement is met.
   b)   The screen height may be reduced to 30 ft. if approved by the aerodrome authority
   c)   the aircraft may be despatched at the original weight if an alternate at which all the landing requirements are met, is available.
   d)   the aircraft may be despatched at the original weight if two alternates at which all the landing requirements are met, are available.

PERFORMANCE   **CLASS A: LANDING**

ANSWERS

Questions	Answers
1	C
2	B
3	B
4	B
5	C
6	D
7	C
8	B
9	C
10	D

CHAPTER EIGHTEEN – SPECIMEN QUESTIONS

Contents

Page

SPECIMEN QUESTIONS…………………...……………………………………..18 – 1

ANSWERS TO SPECIMEN QUESTIONS…..……………………..……………18 – 30

SPECIMEN EXAM PAPER…..…………………………………..……………...18 – 31

ANSWERS TO SPECIMEN EXAM PAPER ………………………..………….18 – 38

EXPLANATIONS TO SPECIMEN EXAM PAPER ……….………..…….……18 – 39

AIRCRAFT PERFORMANCE — SPECIMEN QUESTIONS

SPECIMEN QUESTIONS

1. What happens to the speed for Vx and Vy with increasing altitude?

    a) Both remain constant.
    b) Vx remains constant and Vy increases.
    c) Vx increases and Vy remains constant.
    d) Vx remains constant and Vy decreases.

2. The affect of a contaminated runway on the field limit mass :

    a) Decreased weight, increased V1, increased VR.
    b) Decreased weight, same V1, increased VR.
    c) Decreased weight, same V1, same VR.
    d) Decreased weight, decreased V1, decreased VR.

3. When operating with anti-skid inoperative:

    a) Both landing and take off performance will be affected.
    b) Only landing performance will be affected.
    c) Only take off performance will be affected.
    d) Neither take off or landing performance will be affected.

4. When comparing Vx to Vy:

    a) Vx will always be greater than Vy.
    b) Vy will always be greater than or equal to Vx.
    c) Vy will always be greater than Vx.
    d) Vx will sometimes be greater than Vy, but sometimes be less than Vy.

5. Referring to Fig. 4.24. Why does the curve for an equivalent weight of 35 000 kg, only start 4 mins after engine failure?

    a) All the curves start at the same point higher up.
    b) At that altitude the engine takes longer to spool down after failure.
    c) At that weight the aircraft has a higher TAS and therefore more momentum.
    d) At that weight the aircraft takes longer to slowdown to the optimum drift down speed.

6. With which conditions would one expect Vmc to be the lowest?

    a) Cold temp, low altitude, low humidity.
    b) Hot temp, low pressure altitude, high humidity.
    c) Hot temp, high pressure altitude, high humidity.
    d) Cold temp, high altitude, low humidity.

7. Give the correct order for the following:

    a) Vmcg, VR, V1, V2
    b) Vmcg, V1, VR, V2
    c) V1, Vmcg, VR, V2
    d) Vmcg, V1, Vmca, VR, V2

AIRCRAFT PERFORMANCE   SPECIMEN QUESTIONS

8. If the C of G moves aft from the most forward position:

   a) The range and the fuel consumption will increase.
   b) The range and the fuel consumption will decrease.
   c) The range will increase and the fuel consumption will decrease.
   d) The range will decrease and the fuel consumption will increase.

9. Descending below the tropopause from FL370 to FL250 at a steady M#, then FL250 to FL100 at a constant CAS. What happens to descent angle?

   a) Increase   Increase
   b) Increase   Constant
   c) Decrease   Decrease
   d) Constant   Constant

10. With a constant weight and M#, a higher altitude will require:

    a) Lower CL.
    b) Lower CD.
    c) Higher AoA.
    d) No change.

11. When approaching a wet runway, with the risk of hydroplaning, what technique should the pilot adapt?

    a) Positive touch down, full reverse and brakes as soon as possible.
    b) Smoothest possible touch down, full reverse and only brakes below VP.
    c) Positive touch down, full reverse and only brakes below VP.
    d) Normal landing, full reverse and brakes at VP.

12. An aircraft with a grad of 3.3%, flying at an IAS of 85 KTS. At a P.ALT of 8500' with a temp of +15°c will have a ROC of:

    a) 284'/Min
    b) 623'/Min
    c) 1117'/Min
    d) 334'/Min

12. An aircraft with a mass of 110000kg is capable of maintaining a grad of 2.6%. With all the atmospheric variables remaining the same, with what mass would it be able to achieve a grad of 2.4?

    a) 119167 kg
    b) 101530 kg
    c) 110000 kg
    d) 121167 kg

14. Give the correct sequence:

    a) Vs, Vx, Vy
    b) Vx, Vs, Vy
    c) Vs, max range speed, max endurance speed.
    d) Max endurance speed, Vs, Max range speed.

15. Flying at an altitude close to coffin corner gives:

    a) Max speed.
    b) Less manoeuvrability.
    c) Greater 1 engine inoperative Range.
    d) Greater 1 engine inoperative Endurance.

16. The main reason for using the stepped climb technique is to:

    a) Decrease sector times.
    b) Increase endurance.
    c) Adhere to ATC procedures.
    d) Increase range.

17. Ignoring the effect of compressibility, what would CL do with an increase in altitude?

    a) Increase.
    b) Decrease.
    c) Remain the same.
    d) Increase, then decrease.

18. In climb limited mass calculations, the climb gradient is a ratio of:

    a) Height gained over distance travelled through the air.
    b) Height gained over distance travelled across the ground.
    c) TAS over rate of climb.
    d) TGS over rate of climb.

19. In a twin engined jet aircraft with six passenger seats, and a maximum certified take off mass of 5650kg. What is the required en-route obstacle clearance, with one engine inoperative during drift down towards the alternate airport?

    a) 2000ft
    b) 1500ft
    c) 1000ft
    d) 50ft or half the wingspan

20. When does THRUST = DRAG?

    a) Climbing at a constant IAS.
    b) Descending at a constant IAS.
    c) Flying level at a constant IAS.
    d) All of the above.

21. When take off mass is limited by VMBE, an increase in the uphill slope will:

    a) Have no affect.
    b) Require a decrease in the mass.
    c) Allow an increase in the mass.
    d) Decrease the TODR.

22. SFC will:

    a) Increase if C of G is moved further forward of the C of P.
    b) Decrease if C of G is moved further forward of the C of P.
    c) Not be affected by C of G position.
    d) Only be affected by C of G position, if it is behind the C of P.

23. Reference point zero refers to:

    a) Point where the aircraft lifts off the ground.
    b) Point where the aircraft reaches V2.
    c) Point where the aircraft reaches 35ft.
    d) Point where gear is selected up.

24. To maintain the same angle of attack and altitude at a higher gross weight an aeroplane needs:
    a) Less airspeed and same power.
    b) Same airspeed.
    c) More airspeed and less power.
    d) More airspeed and more power.

25. The coefficient of lift may be increased by lowering the flaps or:
    a) Increase CAS.
    b) Reduce nose up elevator trim.
    c) Increase angle of attack.
    d) Increase TAS.

26. An aircraft is certified to land with flaps at either 25 or 35 degrees of flap. If the pilot selects the higher setting there will be:
    a) Increased landing distance and reduced go-around performance.
    b) Increased landing distance and improved go-around performance.
    c) Reduced landing distance and improved go-around performance.
    d) Reduced landing distance and reduced go-around performance.

27. Which conditions are most suited to a selection of lower flap for take off?
    a) Low airfield elevation, close obstacles, long runway, high temperature.
    b) Low airfield elevation, no obstacles, short runway, low temperature.
    c) High elevation, no obstacles short runway, low temperature.
    d) High airfield elevations, distant obstacles, long runway, high ambient temperature.

28. During the certification of an aeroplane, the take off distance with all engines operating and the take off distance with one engine inoperative are
    1547m
    1720m
    What is the distance used in the aircraft certification?

    a) 1547m
    b) 1779m
    c) 1720m
    d) 1798m

AIRCRAFT PERFORMANCE   SPECIMEN QUESTIONS

29. V2 MIN is determined by: (excluding Vmca)

   a) 1.08VSR for 4 engine turboprops with 1.13VSR for 2 and 3 engined turboprops.
   b) 1.2VS for all turbojets.
   c) 1.2VSR for all turboprops and 1.15VSR for all turbojets.
   d) 1.15Vs for all aeroplanes.

30. If the flap setting is changed from 10 degrees to 20 degrees V2 will:

   a) Not change.
   b) Decrease if not limited to Vmca.
   c) Increase.
   d) Increase or decrease depending on weight.

31. For a turbojet aeroplane the second segment of the climb begins when:

   a) Accelerating from V2 to flap retraction speed begins.
   b) Landing gear is fully retracted.
   c) Flap retraction begins.
   d) Flaps are fully retracted.

32. For a turbojet aeroplane the third segment of climb begins when:

   a) Acceleration to flap retraction speed begins (min 400ft).
   b) Landing gear is fully retracted.
   c) Acceleration from VLOF to V2 begins.
   d) Flaps are fully retracted.

33. The buffet onset boundary chart tells the pilot:

   a) Critical mach number for various masses and altitudes.
   b) Values for low speed stall and mach buffet onset for various masses and altitudes.
   c) Mach number for low speed buffet and shock buffet for various masses and altitudes.
   d) Maximum operating MMO for various masses and altitudes.

34. Two identical turbojets are at the same altitude and same speed and have the same specific fuel consumption. Plane 1 weighs 130,000kg and fuel flow is 4,300kg/hr. If plane 2 weighs 115,000kg what is the fuel flow:

   a) 3804kg/hr
   b) 4044kg/hr
   c) 3364kg/hr
   d) 3530kg/hr

35. The speed for minimum power required in a turbojet will be:

   a) Slower than the speed for minimum drag.
   b) Faster than the speed for minimum drag.
   c) Slower in a climb and faster in the decent.
   d) Same as speed for minimum drag.

36. In wet conditions, what extra percentage over the dry gross landing distance must be available for a turbojet?

    a) 43%
    b) 92%
    c) 67%
    d) 15%

37. In dry conditions, when landing at an alternate airport in a turbojet by what factor should the landing distance available be divided by to give landing distance?

    a) 0.6
    b) 1.0
    c) 1.67
    d) 1.43

38. What landing distance requirements need to be met at an alternate airfield compared to a destination airfield for a turboprop?

    a) Less than destination.
    b) More than destination.
    c) Same as destination.
    d) None applicable.

39. For a twin engined aircraft, which can use either 5 or 15 degrees flap setting, using MRJT fig 4.4 what is the maximum field limited take off mass:
    Pressure Altitude   7000'
    OAT   -10°C
    Length available   2400m
    Slope   Level
    Wind   Calm

    a) 55000kg
    b) 56000kg
    c) 44000kg
    d) 52000kg

40. Absolute Ceiling is defined by:

    a) Altitude where theoretical rate of climb is zero.
    b) Altitude at which rate of climb is 100fpm.
    c) Altitude obtained when using lowest steady flight speed.
    d) Altitude where low speed Buffet and high speed Buffet speeds are coincident.

41. VREF for a Class B aircraft is defined by:

    a) 1.3Vs
    b) 1.2Vs
    c) 1.3 Vmcl
    d) 1.2 Vmcl

42. VR for a jet aircraft must be faster than, the greater of:

    a) 1.05 Vmca and V1
    b) Vmca and 1.1V1
    c) VMBE and V1
    d) V1 and 1.1 Vmca

43. Landing on a runway with 5mm wet snow will:

    a) Increase landing distance.
    b) Decrease landing distance.
    c) Not affect the landing distance.
    d) Give a slightly reduced landing distance, due to increased impingement drag.

44. Take off on a runway with standing water, with a depth of 0.5cm. Compared to a dry runway, field length limited mass will:

    a) Increase, with a reduced V1.
    b) Remain the same, with a reduced V1.
    c) Decrease, with an increased V1.
    d) Decrease, with a decreased V1.

45. A balanced field length is when:

    a) Distance taken to accelerate to V1 and distance to stop are identical
    b) TORA X 1.5
    c) V1 = VR
    d) ASDA equals TODA

46. Increase ambient temperature will result in:

    a) Increased field length limited mass.
    b) Decrease maximum brake energy limited mass.
    c) Increase climb limited mass.
    d) Increased obstacle limited mass.

47. Pitch angle during decent at a constant mach number will:

    a) Increase.
    b) Decrease.
    c) Increase at first then decrease.
    d) Stay constant.

48. At maximum range speed in a turbojet the angle of attack is:

    a) Same as L/D max.
    b) Less than L/D max.
    c) Maximum.
    d) More than L/D max.

49. If there is an increase in atmospheric pressure and all other factors remain constant, it should result in:

    a) Decreased take off distance and increased climb performance.
    b) Increased take off distance and increased climb performance.
    c) Decreased take off distance and decreased climb performance.
    d) Increased take off distance and decreased climb performance.

50. Climbing to cruise altitude with a headwind will:

    a) Increase time to climb.
    b) Decrease ground distance covered to climb.
    c) Decreased time to climb.
    d) Increased ground distance covered to climb.

51. Requirements for the third segment of climb are:

    a) Minimum acceleration altitude for one engine inoperative should be used.
    b) There is no climb gradient requirement during acceleration phase.
    c) Level acceleration with an equivalent gradient of 1.2%.
    d) Legal minimum altitude for acceleration is 1500'.

52. If the calculations for an aeroplane of 3250lb indicate a service ceiling of 4000m, what will the service ceiling be when the actual take off mass is 3000lb?

    a) Higher.
    b) Lower.
    c) Higher or lower, more calculations will have to be done.
    d) The same.

53. Why is there a requirement for an approach climb gradient?

    a) So that an aircraft falling below the glide path will be able to re-intercept it.
    b) Adequate performance for a go-around in the event of an engine failure.
    c) So that the aircraft will not stall when full flap is selected.
    d) To maintain minimum altitude on the approach.

54. The drift down is a procedure applied:

    a) After aircraft depressurisation.
    b) For a visual approach to a VASI.
    c) For an instrument approach at an airfield without an ILS.
    d) When the engine fails above the operating altitude for one engine inoperative.

55. A light twin engined aircraft is climbing from the screen height of 50ft, and has an obstacle 10,000m along the net flight path. If the net climb gradient is 10%, there is no wind and obstacle is 900m above the aerodrome elevation then what will the clearance be?

    a) The aircraft will not clear the object.
    b) 85m
    c) 100m
    d) 115m

AIRCRAFT PERFORMANCE   SPECIMEN QUESTIONS

56. Take off run required for a jet aircraft, with one engine inoperative is:

   a) Brake release point to midpoint between VLOF and 35ft.
   b) Brake release point to 35ft.
   c) Brake release point to 15ft.
   d) The same as for all engines.

57. A jet aircraft's maximum altitude is usually limited by:

   a) It's certification maximum altitude.
   b) It's pressurisation maximum altitude.
   c) The altitude at which low and high-speed buffet will occur.
   d) Thrust limits.

58. With respect to en-route diversions (using drift down graph), if you believe that you will not clear an obstacle do you:

   a) Drift down to clearance height and then start to jettison fuel.
   b) Jettison fuel from the beginning of the drift down.
   c) Asses remaining fuel requirements, then jettison fuel as soon as possible.
   d) Fly slight faster.

59. With respect to field length limit, fill in the blanks in the follow statement.

   "The distance to accelerate to _____, at which point an engine fails, followed by the reaction time of _____ and the ensuing deceleration to a full stop must be completed within the _____."

   a) VR, 2sec, TORA
   b) V1, 2sec, ASDA
   c) VEF, 2sec, TORA
   d) VGO, 2sec, ASDA

60. How does the power required graph move with an increase in altitude?

   a) Straight up.
   b) Straight down.
   c) Up and to the right.
   d) Straight across to the right.

61. What factors would cause V2 to be limited by Vmca?

   a) Flaps at high settings.
   b) With high pressure.
   c) With low temperature.
   d) Combination of the above.

62. In a climb, at a constant IAS / Mach No. 300 kts / 0.78 M. What happens at the change over point (29 500 ft, ISA) ?

   a) Accelerate from the IAS to the Mach number, and therefore rate of climb will decrease.
   b) No change in rate of climb since TAS remains constant.
   c) Find that rate of climb would start to increase, because TAS starts to increase.
   d) Find that rate of climb would start to decrease, because TAS would start to decrease.

AIRCRAFT PERFORMANCE SPECIMEN QUESTIONS

63. If not VMBE or Vmcg limited, what would V1 be limited by?

    a) V2
    b) Vmga
    c) VR
    d) VMU

64. What procedure is likely to require V1 to be reduced?

    a) Improved climb procedure.
    b) Reduced thrust take off.
    c) When ASDA is greater than TODA.
    d) Take off with anti-skid inoperative.

65. Which of the following is not affected by a tailwind:

    a) Landing climb limit mass.
    b) Obstacle limit mass.
    c) VMBE.
    d) Tyre speed limit mass.

66. When flying an aircraft on the back of the drag curve, maintaining a slower speed (but still faster than VS) would require:

    a) More flap.
    b) Less thrust due to less parasite drag.
    c) More thrust.
    d) No change.

67. During certification test flights for a turbojet aeroplane, the measured take off runs from brake release to a point equidistant between the point at which VLOF is reached and the point at which the aeroplane is 35ft above the take off surface are:

    1747 m with all engines operating.
    1950 m with the critical engine failure recognised at V1, other factors remaining unchanged.

    What is the correct value of the Take off Run?

    a) 1747 m
    b) 2243 m
    c) 1950 m
    d) 2009 m

68. Taking into account the following, what would be the minimum required head wind component for landing? (Using fig 2.4 in the CAP 698.)

    Factored landing distance of 1300 ft.
    ISA temperature at MSL.
    Landing mass of 3200 lbs.

    a) 10 Kts
    b) 5 Kts
    c) 0 Kts
    d) 15 Kts

69. Two identical aircraft at different masses are descending at idle thrust. Which of the following statements correctly describes their descent characteristics?

    a) There is no difference between the descent characteristics of the two aeroplanes.
    b) At a given angle of attack, the heavier aeroplane will always glide further than the lighter aeroplane.
    c) At a given angle of attack, the lighter aeroplane will always glide further than the heavier aeroplane.
    d) At a given angle of attack, both the vertical and the forward speeds are greater for the heavier aeroplane.

70. When flying in a headwind, the speed for max range should be :

    a) slightly decreased.
    b) slightly increased.
    c) unchanged.
    d) should be increased, or decreased depending on the strength of the wind.

71. VLO is defined as :

    a) Actual speed that the aircraft lifts off the ground.
    b) Minimum possible speed that the aircraft could lift off the ground.
    c) The maximum speed for landing gear operation.
    d) The long range cruise speed.

72. When flying at the optimum range altitude, over time the :

    a) Fuel consumption gradually decreases.
    b) Fuel consumption gradually increases.
    c) Fuel consumption initially decreases then gradually increases.
    d) Fuel consumption remains constant.

73. What happens to the field limited take off mass with runway slope ?

    a) It increases with a downhill slope.
    b) It is unaffected by runway slope.
    c) It decreases with a downhill slope.
    d) It increases with an uphill slope.

74. For a given aircraft mass, flying with a cost index greater that zero set will result in :

    a) a cruise at a slower M# than the best range M# for a given altitude.
    b) a cruise at the maximum endurance speed.
    c) climb at the slowest safe speed, taking into account stall and speed stability.
    d) a cruise at a faster M# than the M# giving best ANM/kg ratio for a given altitude.

75. Cruising with 1 or 2 engines inoperative at high altitude, compared to all engines operative cruise, range will :

    a) increase.
    b) decrease.
    c) not change.
    d) decrease with 1 engine inoperative, and increase with 2 engines inoperative.

76. Taking into account the values given below, what would be the maximum authorised brake release mass ?

Flap :	5°	10°	15°
Field limited mass :	49 850 kg	52 500 kg	56 850 kg
Climb limited mass :	51 250 kg	49 300 kg	45 500 kg

   a) 56 850 kg
   b) 49 300 kg
   c) 49 850 kg
   d) 51 250 kg

77. A turbo-prop aircraft with a maximum all up mass in excess of 5700 kg, is limited to :

   a) 10° angle of bank up to 400 ft.
   b) 15° angle of bank up to 400 ft.
   c) 20° angle of bank up to 400 ft.
   d) 25° angle of bank up to 400 ft.

78. With regards to the optimum altitude during the cruise, the aircraft is :

   a) always flown at the optimum altitude.
   b) always flown 2000 ft below the optimum altitude.
   c) may be flown above or below the optimum altitude, but never at the optimum altitude.
   d) flown as close to the optimum altitude as ATC will allow.

79. A tailwind on take off will not affect :

   a) climb limit mass.
   b) obstacle clearance.
   c) field limit mass.
   d) VMBE

80. When climbing at a constant M# through the troposphere, TAS :

   a) increases.
   b) decreases.
   c) remains constant.
   d) increases then decreases.

81. Concerning landing gear, which factors limit take off performance ?

   a) Brake temperature.
   b) Wheel rate of rotation, brake energy.
   c) Tyre temperature.
   d) Brake wear.

82. In a glide (power off descent) if pitch angle is increased, glide distance will :

   a) increases.
   b) decrease.
   c) remain constant.
   d) depend on the aircraft.

AIRCRAFT PERFORMANCE   SPECIMEN QUESTIONS

83. With which conditions would the aircraft need to be flown, in order to achieve maximum speed ?

   a) Thrust set for minimum drag.
   b) Best lift - drag ratio.
   c) Maximum thrust and maximum drag.
   d) Maximum thrust and minimum drag.

84. If a jet engine fails during take off, before V1 :

   a) the take off can be continued or aborted.
   b) the take off should be aborted.
   c) the take off should be continued.
   d) the take off may be continued if aircraft speed is, above VMCG and lies between VGO and VSTOP.

85. Up to which height in NADP 1 noise abatement procedure must V2 + 10-20 Kts be maintained ?

   a) 1500 ft.
   b) 3000 ft.
   c) 1000 ft.
   d) 500 ft.

86. At MSL, in ISA conditions.
   Climb gradient = 6 %
   What would the climb gradient be if :
   P.Altitude   1000 ft
   Temperature   17°C
   Engine anti-ice on.
   Wing anti-ice on.
   (- 0.2 % engine anti-ice, - 0,1 % wing anti ice, ± 0.2 % per 1000 ft P.altitude,
   ± 0.1 % per 1°C ISA deviation)

   a) 5.1 %
   b) 6.3 %
   c) 3.8 %
   d) 5.5 %

87. An aircraft with 180 minutes approval for ETOPS, must be :

   a) no more than 180 minutes from a suitable alternate, in still air, at the one engine inoperative TAS.
   b) no more than 180 minutes from a suitable alternate, in still air, at the all engine TAS.
   c) no more than 90 minutes from a suitable alternate, and 90 minutes from departure, at the one engine inoperative TAS
   d) no more than 180 minutes from a suitable alternate, at the one engine inoperative TGS.

88. In a balanced turn load factor is dependant on :

   a) radius of turn and aircraft weight.
   b) TAS and bank angle.
   c) radius of turn and bank angle.
   d) Bank angle only.

89. Putting in 16500 litres of fuel with an SG of 780 kg/m³, and writing 16500 kg of fuel on the load sheet will result in :

    a) TOD increasing and ASD decreasing, and the calculated V2 being too fast.
    b) TOD and ASD decreasing, and the calculated V2 being too fast.
    c) TOD and ASD remaining constant, if the calculated speeds are used.
    d) TOD and ASD increasing, if the calculated speeds are used.

90. If V1 is found to be lower than VMCG, which of the following statements will be true?

    a) VMCG must be reduced to equal V1.
    b) TOD will be greater than ASD.
    c) ASD will be greater than TOD.
    d) Take-off will not be permitted.

91. When gliding into a headwind airspeed should be :

    a) reduce airspeed to gust penetration speed.
    b) the same as the max. range glide speed in still air.
    c) lower than the max. range glide speed in still air.
    d) higher than the max. range glide speed in still air.

92. How does the slush thickness affect the V1 reduction required ?

    a) Greater reduction if thicker.
    b) Smaller reduction if thicker.
    c) No affect if mass is reduced.
    d) No affect at all.

93. Which denotes the stall speed in the landing configuration ?

    a) VSO.
    b) VS1.
    c) VS.
    d) VS1G.

94. When in a gliding manoeuvre, in order to achieve maximum endurance the aircraft should be flown at :

    a) the speed for max. lift.
    b) the speed for min. drag.
    c) the speed for max. lift / drag.
    d) the speed for min. power.

95. When descending below the optimum altitude at the long range cruise speed :

    a) Mach no. decreases.
    b) TAS increases.
    c) Mach no. remains constant.
    d) Mach no. increases.

AIRCRAFT PERFORMANCE   SPECIMEN QUESTIONS

96. What does density altitude signify ?

   a) Pressure altitude.
   b) Flight levels.
   c) ISA altitude.
   d) An accurate indication of aircraft and engine performance.

97. For a turbo-prop aircraft, the LDA at an aerodrome is 2200 m. If the conditions are indicated as wet, what would the equivalent dry LDA be ?

   a) 1451 m.
   b) 1913 m.
   c) 1538 m.
   d) 1317 m.

98. During aircraft certification, the value of VMCG is found with nose wheel steering inoperative. This is because :

   a) Nose wheel steering does not affect VMCG.
   b) VMCG must be valid in both wet and dry conditions.
   c) Nose wheel steering does not work after an engine failure.
   d) The aircraft may be operated even if the nose wheel steering is inoperative.

99. Referring to Fig 4.28.
   What would the landing distance required be for a MRJT aircraft with anti-skid inoperative if:
   Pressure altitude 2000 feet.
   Mass 50 000 kg
   Flaps for short field.
   15 Kts Tailwind
   Dry runway.

   a) 1700 m
   b) 2500 m
   c) 1900 m
   d) 3100 m

100. Which is true regarding a balanced field ?

   a) Provides largest gap between net and gross margins.
   b) Provides minimum field length required in the case of an engine failure.
   c) Take off distance will always be more than stopping distance.
   d) Distances will remain equal, even if engine failure speed is changed.

101. Climbing in the troposphere at a constant TAS :

   a) Mach no. increases.
   b) Mach no. decreases.
   c) CAS increases.
   d) IAS increases.

AIRCRAFT PERFORMANCE   SPECIMEN QUESTIONS

102. When a MRJT aircraft descends at the maximum range speed :

   a) IAS increases.
   b) CAS increases.
   c) Mach no. decreases.
   d) Mach no. increases.

103. What condition is found at the intersection of the thrust available and the drag curve :

   a) Un-accelerated flight in a climb.
   b) Accelerated climb.
   c) Un-accelerated level flight.
   d) Accelerated level flight.

104. Out of the four forces acting on the aircraft in flight, what balances thrust in the climb ?

   a) Drag.
   b) Weight
   c) W Sin θ
   d) Drag + W Sin θ

105. According to the information in a light aircraft manual, which gives two power settings for cruise, 65 % and 75 %. If you fly at 75 % in stead of 65 % :

   a) Cruise speed will be higher, fuel consumption will be higher.
   b) Cruise speed will be the same, fuel consumption will be the same.
   c) Cruise speed will be higher, fuel consumption will be lower.
   d) Cruise speed will be higher, fuel consumption will be the same.

106. With a downward sloping runway :

   a) V1 will increase.
   b) V1 will decrease.
   c) VR will increase.
   d) VR will decrease.

107. How is fuel consumption affected by the C of G position, in terms of ANM / kg?

   a) Increases with a forward C of G.
   b) Decreases with a aft C of G.
   c) Decreases with a forward C of G.
   d) Fuel consumption is not affected by the C of G position.

108. Rate of Climb 1000 ft/min
      TAS 198 kts
      What is the aircraft's gradient ?

   a) 5.08 %
   b) 3 %
   c) 4 %
   d) 4.98 %

109. The reduced thrust take off procedure may not be used when :

    a) Runway wet.
    b) After dark.
    c) Temperature varies by more than 10°C from ISA.
    d) Anti-skid unserviceable.

110. If the maximum take off mass is limited by tyre speed, what affect would a down sloping runway have ?

    a) No affect.
    b) Always increase the mass.
    c) Only increase the mass if not limited by any other limitation.
    d) Decrease the mass.

111. With an obstacle which is 160 m above the airfield elevation and 5000 m away from the end of the take off distance. (Screen height 50 ft) What would the obstacle clearance be with a gradient of 5% ?

    a) 105 m
    b) 90 m
    c) 250 m
    d) 265 m

112. Prior to take off the brake temperature needs to be checked, because :

    a) they indicate the state of the fusible plugs.
    b) if the brakes are already hot, they may fade / overheat during a RTO.
    c) they would work better if they are warm.
    d) they may need to be warmed up to prevent them from cracking during a RTO.

113. A turbo jet is flying at a constant M # in the cruise, how does SFC vary with OAT in Kelvin?

    a) Unrelated to T
    b) Proportional to T
    c) Proportional to 1/T
    d) Proportional to $1/T^2$

114. If an aircraft has a stall speed of 100 Kts, what would the speed on short finals have to be ?

    a) 100 Kts
    b) 115 Kts
    c) 130 Kts
    d) 120 Kts

115. When descending at a constant M #, which speed is most likely to be exceeded first ?

    a) Max operating speed.
    b) MMO.
    c) High speed buffet limit.
    d) VMO.

AIRCRAFT PERFORMANCE   SPECIMEN QUESTIONS

116. What is meant by 'Equivalent weight' on the drift down profile graph ?

   a) Weight compensated for fuel reduction prior to engine failure.
   b) Weight compensated for temperature of ISA +10°C and above.
   c) Weight compensated for density at different heights.
   d) Weight compensated for temperature at different heights.

117. What happens to the speeds, VX and VY, when lowering the aircraft's undercarriage ?

   a) VX increases, VY decreases.
   b) VX decreases, VY decreases.
   c) VX increases, VY increases.
   d) VX decreases, VY increases.

118. Maximum Endurance :

   a) can be achieved in level unaccelerated flight with minimum fuel consumption.
   b) can be achieved by flying at the best rate of climb speed in straight and level flight.
   c) can be achieved in a steady climb.
   d) can be achieved by flying at the absolute ceiling.

119. What factors affect descent angle in a glide ?

   a) Configuration and altitude.
   b) Configuration and angle of attack.
   c) Mass and attitude.
   d) Mass and configuration.

120. What is meant by balanced field available ?

   a) TORA = TODA
   b) ASDA = ASDR and TODA = TODR
   c) TODA = ASDA
   d) TORA = ASDA

121. For a piston-engined aeroplane at a constant altitude, angle of attack and configuration, an increased weight will require :

   a) more power but less speed.
   b) more power and the same speed.
   c) more power and more speed.
   d) the same power but more speed.

122. In the climb an aircraft has a thrust to weight ratio of 1:4 and a lift to drag ratio of 12: 1. While ignoring the slight difference between lift and weight in the climb, the climb gradient will be:

   a) 3.0 %
   b) 8.3 %
   c) 16.7 %
   d) 3.3 %

123. Which of the following will not decrease the value of VS?

    a) The C of G in an aft position within the C of G envelope.
    b) Increased altitude.
    c) Decreased weight.
    d) Increased flap setting.

124. All other factors being equal, the speed for minimum drag is:

    a) constant for all weights.
    b) a function of density altitude.
    c) proportional to weight.
    d) a function of pressure altitude.

125. Taking into account the values given below, what would be the maximum authorised brake release mass with a 10 Kt tailwind?

Flap :	5°	10°	15°
Field limited mass :	49 850 kg	52 500 kg	56 850 kg
Climb limited mass :	51 250 kg	49 300 kg	45 500 kg

    Assume 370kg per Kt of tailwind.

    a) 56 850kg
    b) 49 850kg
    c) 52 500kg
    d) 48 800kg

126. If a turn is commenced during the take off climb path :
    (i)   The load factor
    (ii) The induced drag
    (iii) The climb gradient

	(i)	(ii)	(iii)
a)	Increases	Decreases	Decreases
b)	Decreases	Increases	Increases
c)	Increases	Increases	Decreases
d)	Decreases	Decreases	Increases

127. What effect does an increase in weight have on V1 ?

    a) It will cause it to increase.
    b) It will cause it to decrease.
    c) It will have no effect.
    d) It will cause it to decrease by the same percentage as the weight increase.

128. VR for a Class A aeroplane must not be less than:

    a) 10 % above VMU.
    b) 5 % above VMCA.
    c) 5 % above VMCG.
    d) 10 % above VMCA.

129. As speed is reduced from VMD to VMP :

   a) power required decreases and drag decreases.
   b) power required decreases and drag increases.
   c) power required increases and drag increases.
   d) power required increases and drag decreases.

130. The maximum induced drag occurs at a speed of :

   a) VMD.
   b) VMP.
   c) VSO.
   d) VATO.

131. Profile drag is :

   a) inversely proportional to the square root of the EAS.
   b) directly proportional to the square of the EAS.
   c) inversely proportional to the square of the EAS.
   d) directly proportional to the square root of the EAS.

132. VMD for a jet aeroplane is approximately equal to :

   a) 1.3 VS.
   b) 1.7 VS.
   c) 1.6 VS.
   d) 2.1 VS.

133. The best EAS / Drag ratio is approximately :

   a) 1.3 VMD.
   b) 1.32 VMD.
   c) 1.6 VMD.
   d) 1.8 VMD.

134. The effect an increase of weight has on the value of stalling speed (IAS) is that VS .

   a) increases.
   b) decreases.
   c) remains constant.
   d) increases or decreases, depending on the amount of weight increase.

135. Which one of the following statements is true concerning the effect of changes of ambient temperature on an aeroplane's performance, assuming all other performance parameters remain constant ?

   a) An increase will cause a decrease in the landing distance required.
   b) An increase will cause a decrease in take off distance required.
   c) A decrease will cause an increase in the climb gradient
   d) A decrease will cause an increase in the take off ground run.

AIRCRAFT PERFORMANCE   SPECIMEN QUESTIONS

136. What percentages of the head wind and tail wind components are taken into account, when calculating the take off field length required ?

   a) 100% head wind and 100% tail wind.
   b) 150% head wind and 50% tail wind.
   c) 50% head wind and 100% tail wind.
   d) 50% head wind and 150% tail wind.

137. For a turbo jet aircraft planning to land on a wet runway, the landing distance available :

   a) may be less than 15% greater than the dry landing distance if the flight manual gives specific data for a wet runway.
   b) may be less than 15% greater than the dry landing distance if all reverse thrust systems are operative.
   c) may be less than 15% greater than the dry landing distance if permission is obtained from the relevant aerodrome authority.
   d) must always be at least 15% greater than the dry landing distance.

138. The effect of installing more powerful engines in a turbojet aircraft is :

   a) to increase the aerodynamic ceiling and increase the performance ceiling.
   b) to decrease the aerodynamic ceiling and increase the performance ceiling.
   c) to increase the performance ceiling but not affect the aerodynamic ceiling.
   d) to decrease both the aerodynamic and the performance ceilings.

139. In relation to runway strength, the ACN :

   a) may not exceed 90% of the PCN.
   b) may exceed the PCN by up to 10%.
   c) may never exceed the PCN.
   d) may exceed the PCN by a factor of 2.

140. An aerodrome has a clearway of 500m and a stopway of 200m.
   If the stopway is extended to 500m the effect will be :

   a) the maximum take off mass will increase, and V1 will decrease.
   b) the maximum take off mass will increase and V1 will remain the same.
   c) the maximum take off mass will remain the same and V1 will increase.
   d) the maximum take off mass will increase and V1 will increase.

141. An aircraft is climbing at a constant power setting and a speed of VX. If the speed is reduced and the power setting maintained the :

   a) Climb gradient will decrease and the rate of climb will increase.
   b) Climb gradient will decrease and the rate of climb will decrease.
   c) Climb gradient will increase and the rate of climb will increase.
   d) Climb gradient will increase and the rate of climb will decrease.

142. An aircraft is climbing in a standard atmosphere above the tropopause at a constant Mach number :

   a) the IAS decreases and TAS remain constant.
   b) the IAS and TAS remain constant.
   c) the IAS decreases and TAS decreases.
   d) the IAS remains constant and TAS increases.

143. An aircraft is climbing at a constant IAS, below the Mach limit. As height increases :

   a) drag decreases, because density decreases.
   b) drag remains constant, but the climb gradient decreases.
   c) drag increases, because TAS increases.
   d) drag remains constant and the climb gradient remains constant.

144. Optimum altitude can be defined as :

   a) the highest permissible altitude for an aeroplane type.
   b) the altitude at which an aeroplane attains the maximum specific air range.
   c) the altitude at which the ground speed is greatest.
   d) the altitude at which specific fuel consumption is highest.

145. If an aircraft is descending at a constant Mach number :

   a) the IAS will increase and the margin to low speed buffet will decrease.
   b) the IAS will increase and the margin to low speed buffet will increase.
   c) the IAS will decrease and the margin to low speed buffet will decrease.
   d) the IAS will decrease and the margin to low speed buffet will increase.

146. For a given flight level, the speed range determined by the buffet onset boundary chart will decrease with :

   a) decreased weight.
   b) decreased bank angle.
   c) a more forward CG position.
   d) increased ambient temperature.

147. Which of the following variables will not affect the shape or position of the drag vs. IAS curve, for speeds below MCRIT ?

   a) Aspect ratio.
   b) Configuration.
   c) Altitude.
   d) Weight.

148. Which of the following would give the greatest gliding endurance ?

   a) Flight at VMD.
   b) Flight at 1.32VMD.
   c) Flight at the best CL / CD ratio.
   d) Flight close to CL MAX.

149. The tyre speed limit is :

   a) V1 in TAS.
   b) Max VLOF in TAS.
   c) Max VLOF in ground speed.
   d) V1 in ground speed.

150. What gives one the greatest gliding time?

    a) Being light.
    b) A head wind.
    c) A tail wind.
    d) Being heavy.

151. For take off performance calculations, what is taken into account ?

    a) OAT, pressure altitude, wind, weight.
    b) Standard temperature, altitude, wind, weight.
    c) Standard altitude, standard temperature, wind, weight.
    d) Standard temperature, pressure altitude, wind, weight.

152. Which 3 speeds are effectively the same for a jet aircraft ?

    a) ROC, Range, minimum Drag.
    b) Range, Best angle of climb, minimum Drag.
    c) Best angle of climb, minimum Drag, Endurance.
    d) Best angle of climb, Range, Endurance.

153. The long range cruise speed is a speed that gives :

    a) a 1% increase in range and a decrease in IAS.
    b) a 1% increase in TAS.
    c) a 1% increase in IAS.
    d) gives 99% of best cruise range, with an increase in IAS.

154. When an aircraft takes-off at the mass it was limited to by the TODA:

    a) the end of the runway will be cleared by 35ft following an engine failure just before V1.
    b) the actual take-off mass equals the field length limited take-off mass.
    c) the distance from BRP to V1 is equal to the distance from V1 to the 35ft screen.
    d) the balanced take-off distance equals 115% of the all engine take-off distance.

155. Which of the following speeds give the maximum obstacle clearance in the climb?

    a) VY
    b) 1.2VS1
    c) VX
    d) VFE

156. The tangent, from the origin to the Power required curve gives?

    a) Minimum Drag coefficient.
    b) L/D Minimum
    c) D/L Maximum
    d) L/D Maximum

157. For a jet flying at a constant altitude, at the maximum range speed, what is the effect on IAS and Drag over time?

    a) Increase, Increases.
    b) Decrease, Constant.
    c) Constant, Decrease.
    d) Decrease, Decrease.

158. If an aircraft descends at a constant Mach #, what will the first limiting speed be?

    a) Max operating speed.
    b) Never exceed speed.
    c) Max operating Mach #.
    d) Shock stall speed.

159. For an aircraft gliding at it's best glide range speed, if AoA is reduced:

    a) glide distance will increase.
    b) glide distance will remain unaffected.
    c) glide distance will decrease.
    d) glide distance will remain constant, if speed is increased.

160. If an aircraft's climb schedule was changed from 280 / 0.74 M to 290 / 0.74 M, what would happen to the change over altitude?

    a) It would remain unchanged.
    b) It could move up or down, depending on the aircraft.
    c) It will move down.
    d) It will move up.

161. What happens to the cost index when flying above the optimum Long Range cruise speed?

    a) Cost index is not affected by speed.
    b) Cost index will increase with increased speed.
    c) Cost index will decrease with increased speed.
    d) It all depends on how much the speed is changed by.

162. For an aircraft flying at the Long Range cruise speed, (i) Specific Range and (ii) Fuel to time ratio:

    a) (i) Decreases   (ii) Increases
    b) (i) Increases   (ii) Increases
    c) (i) Decreases   (ii) Decreases
    d) (i) Increases   (ii) Decreases

163. By what percentage should V2 be greater than VMCA?

    a) 30%
    b) 10%
    c) 20%
    d) 15%

AIRCRAFT PERFORMANCE   SPECIMEN QUESTIONS

164. If a turbo-prop aircraft has a wet LDA of 2200m, what would the equivalent dry landing distance allowed be?

   a) 1540m
   b) 1148m
   c) 1913m
   d) 1339m

165. If a TOD of 800m is calculated at sea level, on a level, dry runway, with standard conditions and with no wind, what would the TOD be for the conditions listed below?

   2000 ft Airfield elevation
   QNH 1013.25mb
   Temp. of 21°C
   5 kts of tailwind
   Dry runway with a 2% up slope.

   (Assuming: ±20m/1000ft elevation, +10m/1kt of reported tailwind, ±5m/1°C ISA deviation and the standard slope adjustments)

   a) 836m
   b) 940m
   c) 1034m
   d) 1095m

166. At a constant mass and altitude, a lower airspeed requires:

   a) more thrust and a lower coefficient of lift.
   b) less thrust and a lower coefficient of lift.
   c) more thrust and a lower coefficient of drag.
   d) a higher coefficient of lift.

167. On a piston engine aeroplane, with increasing altitude at a constant gross mass, angle of attack and configuration, the power required:

   a) decreases slightly because of the lower air density.
   b) remains unchanged but the TAS increases.
   c) increases but the TAS remains constant.
   d) increases and the TAS increases.

168. Reduced take off thrust:

   a) can be used if the headwind component during take off is at least 10 kts.
   b) can be used if the take off mass is higher than the performance limited take off mass.
   c) is not recommended at very low temperatures.
   d) has the benefit of improving engine life.

169. Reduced take off thrust:

   a) can only be used in daylight.
   b) can't be used on a wet runway.
   c) is not recommended when wind shear is expected on departure.
   d) is not recommended at sea level.

AIRCRAFT PERFORMANCE   SPECIMEN QUESTIONS

170. May the anti-skid be considered in determining the take off and landing mass limits?

   a) Only landing.
   b) Only take off.
   c) Yes
   d) No

171. Climb limited take-off mass can be increased by:

   a) Lower V2
   b) Lower flap setting and higher V2
   c) Lower VR
   d) Lower V1

172. An operator shall ensure that the aircraft clears all obstacles in the net take-off flight path. The half-width of the Obstacle Accountability Area (Domain) at distance D from the end of the TODA is:

   a) 90m + (D / 0.125)
   b) 90m + (1.125 x D)
   c) 90m + (0.125 x D)
   d) (0.125 x D)

173. The take-off performance for a turbo-jet aircraft using 10° Flap results in the following limitations:

   Obstacle clearance limited mass: 4 630 kg
   Field length limited mass: 5 270 kg

   Given that it is intended to take-off with a mass of 5 000 kg, which of the following statements is true?

   a) With 5° flap the clearance limit will increase and the field limit will decrease.
   b) With 15° flap both will increase.
   c) With 5° flap both will increase.
   d) With 15° flap the clearance limit will increase and the field limit will decrease.

174. Induced drag?

   a) Increases with increased airspeed.
   b) Decreases with increased airspeed.
   c) Independent of airspeed.
   d) Initially increases and the decreases with speed.

175. Which of these graphs shows the relationship that thrust required has, with decreased weight?

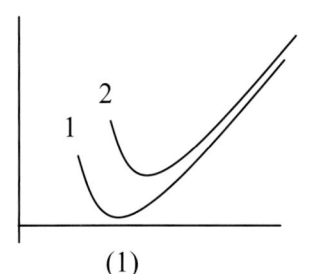

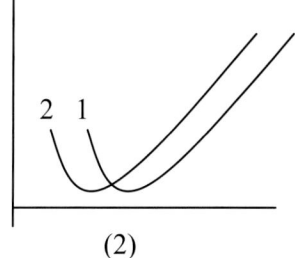

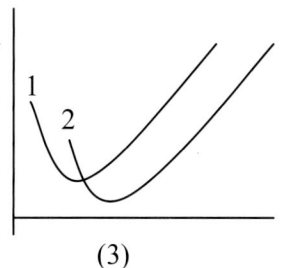

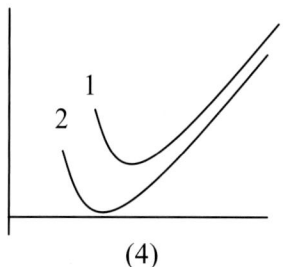

(1)   (2)   (3)   (4)

   a) 1
   b) 2
   c) 3
   d) 4

176. What is the formula for Specific Range?

   a) Ground speed divided by fuel flow.
   b) True airspeed divided by fuel flow.
   c) Fuel flow divided by SFC.
   d) Ground speed divided by SFC.

177. When does the first segment of the take-off climb begin?

   a) When V2 is reached.
   b) When 35 feet is reached.
   c) When flaps are up.
   d) When gear and flaps are up.

178. A headwind component:

   a) increases climb angle.
   b) decreases climb angle.
   c) increases best rate of climb.
   d) decreases rate of climb.

179. V1 is limited by:

   a) VMCG and VR.
   b) VMCA and VR.
   c) V2 and VR.
   d) 1.05 VMCA.

180. VR is:

   a) less than V1.
   b) more than V2.
   c) less than VMCG.
   d) equal to or more than V1.

181. What is the affect of an increase in pressure altitude?

    a) increased take-off distance with increased performance.
    b) decreased take-off distance and increased performance.
    c) increased take-off distance and decreased performance.
    d) decreased take-off distance with decreased performance.

182. What affects endurance?

    a) speed and weight.
    b) speed and fuel on board.
    c) speed, weight and fuel on board.
    d) none of the above.

183. What degrades aircraft performance?

    a) low altitude, low temperature, low humidity.
    b) high altitude, high temperature, high humidity.
    c) low altitude, high temperature, low humidity.
    d) high temperature, high altitude, low humidity.

184. If your take-off is limited by the climb limit mass, what is the effect of a headwind?

    a) No effect.
    b) Increased mass.
    c) Decreased mass.
    d) Dependant on the strength of the headwind.

185. What is the effect on Accelerate Stop Distance of the Anti-Skid system being inoperative?

    a) Increased.
    b) Decreased.
    c) Constant.
    d) Unable to be determined without further information.

186. What is V_{Ref}?

    a) $1.3 V_S$
    b) $1.13 V_{SR0}$
    c) $1.23 V_{SR0}$
    d) $1.05 V_{mcl}$

187. During planning V_{mcg} is found to be greater than V_1. If V_1 is adjusted to equal V_{mcg} and engine failure occurs at the new V_1, then:

    a) ASDR is smaller than TODR.
    b) ASDR is larger than TODR.
    c) ASDR is the same as TODR.
    d) The aircraft weight must be reduced in order to permit take off.

188. Refer to CAP698 Fig 2.1. What is the Gross TODR for an aircraft in the following conditions:

a/c TOM 1,591 kg
Field elevation 1,500ft (QNH 1013),
OAT is +18 Deg.C,
16 Kts Headwind Component,
1% downhill slope,
Paved, dry surface,
No stopway or clearway.

   a) 335m
   b) 744m
   c) 565m
   d) 595m

189. What is the Take Off Run defined as for a Class A aircraft:

   a) The distance from Brakes Release to V_{Lof}.
   b) The distance from Brakes Release to a point on the ground below which the aircraft has cleared a screen height of 35'.
   c) The distance from Brakes Release to a point on the ground below which the aircraft has cleared a screen height of 50'.
   d) The distance from Brakes Release to a point half way between where the aircraft leaves the ground and the point on the ground above which it clears a height of 35'.

190. Losing an engine during the take off above V_{mca} means the aircraft will be able to maintain:

   a) Altitude.
   b) Straight and level flight.
   c) Heading.
   d) Bank angle.

AIRCRAFT PERFORMANCE — SPECIMEN QUESTIONS

Answers To Specimen Questions

1. D	50. B	99. D	148. D
2. D	51. C	100. B	149. C
3. A	52. A	101. A	150. A
4. B	53. B	102. C	151. A
5. D	54. D	103. C	152. C
6. C	55. D	104. D	153. D
7. B	56. A	105. A	154. B
8. C	57. C	106. B	155. C
9. B	58. C	107. C	156. D
10. C	59. B	108. D	157. D
11. C	60. C	109. D	158. A
12. D	61. D	110. A	159. C
13. A	62. D	111. A	160. C
14. A	63. C	112. B	161. B
15. B	64. D	113. B	162. A
16. D	65. A	114. C	163. B
17. C	66. C	115. D	164. D
18. A	67. D	116. B	165. C
19. A	68. A	117. B	166. D
20. C	69. D	118. A	167. D
21. C	70. B	119. B	168. D
22. A	71. C	120. C	169. C
23. C	72. A	121. C	170. C
24. D	73. A	122. C	171. B
25. C	74. D	123. B	172. C
26. D	75. B	124. C	173. A
27. A	76. C	125. D	174. B
28. B	77. B	126. C	175. D
29. A	78. D	127. A	176. B
30. B	79. A	128. B	177. B
31. B	80. B	129. B	178. A
32. A	81. B	130. C	179. A
33. C	82. B	131. B	180. D
34. C	83. C	132. C	181. C
35. A	84. B	133. B	182. C
36. B	85. B	134. A	183. B
37. A	86. A	135. C	184. A
38. C	87. A	136. D	185. A
39. B	88. D	137. A	186. C
40. A	89. B	138. C	187. B
41. A	90. C	139. B	188. D
42. A	91. D	140. D	189. D
43. A	92. B	141. B	190. C
44. D	93. A	142. A	
45. D	94. D	143. B	
46. B	95. A	144. B	
47. B	96. D	145. B	
48. B	97. B	146. C	
49. A	98. B	147. C	

AIRCRAFT PERFORMANCE   SPECIMEN QUESTIONS

SPECIMEN EXAMINATION PAPER

40 Questions, 40 Marks
Time Allowed: 1 hour

1. A turbo-propeller aircraft is certified with a maximum take-off mass of 5600 kg and a maximum passenger seating of 10. This aircraft would be certified in:
    a. Class A
    b. Class B
    c. Class C
    d. Either class A or class B depending on the number of passengers carried.
    (1 mark)

2. How does the thrust from a fixed propeller change during the take-off run of an aircraft?
    a. It remains constant.
    b. It increases slightly as the aircraft speed builds up.
    c. It decreases slightly as the aircraft speed builds up.
    d. It only varies with changes in mass.
    (1 mark)

3. The take-off run is defined as:
    a. distance to V_1 and then to stop, assuing the engine failure is recognised at V_1
    b. distance from brake release to the point where the aircraft reaches V_2
    c. the horizontal distance from the start of the take-off roll to a point equidistant between V_{LOF} and 35 ft
    d. the distance to 35 ft with an engine failure at V_1 or 1.15 times the all engine distance to 35 ft.
    (1 mark)

4. What effect does a downhill slope have on the take-off speeds?
    a. It has no effect on V_1
    b. It decreases V_1
    c. It increases V_1
    d. It increases the IAS for take-off
    (1 mark)

5. Which of the following combinations most reduces the take-off and climb performance of an aircraft?
    a. high temperature and high pressure
    b. low temperature and high pressure
    c. low temperature and low pressure
    d. high temperature and low pressure
    (1 mark)

6. Density altitude is:
    a. the true altitude of the aircraft
    b. the altitude in the standard atmosphere corresponding to the actual conditions
    c. the indicated altitude on the altimeter
    d. used to calculate en-route safety altitudes
    (1 mark)

AIRCRAFT PERFORMANCE   SPECIMEN QUESTIONS

7. The take-off climb gradient:
    a. increases in a head wind and decreases in a tail wind
    b. decreases in a head wind and increases in a tail wind
    c. is independent of the wind component
    d. is determined with the aircraft in the take-off configuration

    (1 mark)

8. The effect of changing altitude on the maximum rate of climb (ROC) and speed for best rate of climb for a turbo-jet aircraft, assuming everything else remains constant, is:
    a. as altitude increases the ROC and speed both decrease
    b. as altitude increases the ROC and speed both increase
    c. as altitude increases the ROC decreases but the speed remains constant
    d. as altitude increases the ROC remains constant but the speed increases

    (1 mark)

9. A runway at an aerodrome has a declared take-off run of 3000 m with 2000 m of clearway. The maximum distance that may be allowed for the take-off distance is:
    a. 5000m
    b. 6000 m
    c. 3000 m
    d. 4500 m

    (1 mark)

10. An aircraft may use either 5° or 15° flap setting for take off. The effect of selecting the 5° setting as compared to the 15° setting is:
    a. take-off distance and take-off climb gradient will both increase
    b. take-off distance and take-off climb gradient will both decrease
    c. take-off distance will increase and take-off climb gradient will decrease
    d. take-off distance will decrease and take-off climb gradient will increase

    (1 mark)

11. The use of reduced thrust for take-off is permitted:
    a. if the field length limited take-off mass is greater than the climb limited take-off mass
    b. if the actual take-off mass is less than the structural limiting mass
    c. if the actual take-off mass is less than the field length and climb limited take-off masses
    d. if the take-off distance required at the actual take-off mass does not exceed the take-off distance available

    (1 mark)

12. Planning the performance for a runway with no obstacles, it is found that the climb limiting take-off mass is significantly greater than the field limiting take-off mass with 5° flap selected. How can the limiting take-off mass be increased?
    a. use an increased V_2 procedure
    b. increase the flap setting
    c. reduce the flap setting
    d. reduce the V_2

    (1 mark)

13. The maximum and minimum values of V_1 are limited by:
    a. V_R and V_{MCG}
    b. V_2 and V_{MCG}
    c. V_R and V_{MCA}
    d. V_2 and V_{MCA}

    (1 mark)

14. If the TAS is 175 kt and the rate of climb 1250 ft per minute, the climb gradient is approximately:
    a. 7%
    b. 14%
    c. 12%
    d. 10%

    (1 mark)

15. A pilot inadvertently selects a V_1 which is lower than the correct V_1 for the actual take-off weight. What problem will the pilot encounter if an engine fails above the selected V_1 but below the true V_1?
    a. the accelerate-stop distance required will exceed the distance available
    b. the climb gradient will be increased
    c. the take-off distance required will exceed that available
    d. there will be no significant effect on the performance

    (1 mark)

16. A turbo-jet is in a climb at a constant IAS what happens to the drag?
    a. It increases
    b. It decreases
    c. it remains constant
    d. it increases initially then decreases

    (1 mark)

17. Comparing the take-off performance of an aircraft from an aerodrome at 1000 ft to one taking off from an aerodrome at 6000 ft, the aircraft taking off from the aerodrome at 1000 ft:
    a. will require a greater take-off distance and have a greater climb gradient
    b. will require a greater take-off distance and have a lower climb gradient
    c. will require a shorter take-off distance and have a lower climb gradient
    d. will require a shorter take-off distance and have a greater climb gradient

    (1 mark)

18. Which is the correct sequence of speed?
    a. V_S, V_Y, V_X
    b. V_X, V_Y, V_X
    c. V_S, V_X, V_Y
    d. V_X, V_Y, V_S

    (1 mark)

19. Which of the following will increase the accelerate-stop distance on a dry runway?
    a. a headwind component
    b. an uphill slope
    c. temperatures below ISA
    d. low take-off mass, because of the increased acceleration

    (1 mark)

AIRCRAFT PERFORMANCE   SPECIMEN QUESTIONS

20. A turbo-jet aircraft is climbing at a constant Mach number in the troposphere. Which of the following statements is correct?
   a. TAS and IAS increase
   b. TAS and IAS decrease
   c. TAS decreases, IAS increases
   d. TAS increases, IAS decreases
   (1 mark)

21. The induced drag in an aeroplane:
   a. increases as speed increases
   b. is independent of speed
   c. decreases as speed increases
   d. decreases as weight decreases
   (1 mark)

22. The speed range between low speed and high speed buffet:
   a. decreases as altitude increases and weight decreases
   b. decreases as weight and altitude increase
   c. decreases as weight decreases and altitude increases
   d. increases as weight decreases and altitude increases
   (1 mark)

23. Thrust equals drag:
   a. in unaccelerated level flight
   b. in an unaccelerated descent
   c. in an unaccelerated climb
   d. in a climb, descent or level flight if unaccelerated
   (1 mark)

24. A higher mass at a given altitude will reduce the gradient of climb and the rate of climb. But the speeds:
   a. V_X and V_Y will decrease
   b. V_X and V_Y will increase
   c. V_X will increase and V_Y will decrease
   d. V_X and V_Y will remain constant
   (1 mark)

25. If the other factors are unchanged the fuel mileage (nm per kg) is:
   a. independent from the centre of gravity (CofG)
   b. lower with a forward CofG
   c. lower with an aft CofG
   d. higher with a forward CofG
   (1 mark)

26. Concerning maximum range in a turbo-jet aircraft, which of the following is true?
   a. the speed to achieve maximum range is not affected by the wind component
   b. to achieve maximum range speed should be increased in a headwind and reduced in a tailwind
   c. to achieve maximum range speed should be decreased in a headwind and increased in a tailwind
   d. The change is speed required to achieve maximum range is dependent on the strength of the wind component acting along the aircraft's flight path and may require either an increase or decrease for both headwind and tailwind
   (1 mark)

AIRCRAFT PERFORMANCE   SPECIMEN QUESTIONS

27. V_1 is the speed:
    a. above which take-off must be rejected if engine failure occurs
    b. below which take-off must be continued if engine failure occurs
    c. engine failure recognised below this speed, take-off must be rejected and above which take-off must be continued
    d. the assumed speed for engine failure

    (1 mark)

28. A constant headwind in the descent:
    a. Increases the angle of descent
    b. Increases the rate of descent
    c. Increases the angle of the descent flight path
    d. Increases the ground distance travelled in the descent

    (1 mark)

29. For a turbojet aircraft what is the reason for the use of maximum range speed?
    a. Greatest flight duration
    b. Minimum specific fuel consumption
    c. Minimum flight duration
    d. Minimum drag

    (1 mark)

30. Why are step climbs used on long range flights in jet transport aircraft?
    a. to comply with ATC flight level constraints
    b. step climbs have no significance for jet aircraft, they are used by piston aircraft
    c. to fly as close as possible to the optimum altitude as mass reduces
    d. they are only justified if the actual wind conditions differ significantly from the forecast conditions used for planning

    (1 mark)

31. The absolute ceiling of an aircraft is:
    a. where the rate of climb reaches a specified value
    b. always lower than the aerodynamic ceiling
    c. where the rate of climb is theoretically zero
    d. where the gradient of climb is 5%

    (1 mark)

32. In the take-off flight path, the net climb gradient when compared to the gross gradient is:
    a. greater
    b. the same
    c. smaller
    d. dependent on aircraft type

    (1 mark)

33. To answer this question use CAP698 SEP1 figure 2.1. Conditions: aerodrome pressure altitude 1000 ft, temperature +30°C, level, dry, concrete runway and 5 kt tailwind component. What is the regulated take-off distance to 50 ft for an aircraft weight 3500 lb if there is no stopway or clearway?
    a. 2800 ft
    b. 3220 ft
    c. 3640 ft
    d. 3500 ft

    (1 mark)

34. To answer this question use CAP698 MRJT figure 4.4. Conditions: Pressure altitude 5000 ft, temperature –5°C, balanced field length 2500 m, level runway, wind calm. What is the maximum field length limited take-off mass and optimum flap setting?
    a. 59400 kg 15°
    b. 60200 kg 5°
    c. 59400 kg, 5°
    d. 60200 kg 15°

    (1 mark)

35. The effect of a headwind component on glide range:
    a. the range will increase
    b. the range will not be affected
    c. the range will decrease
    d. the range will only be affected if incorrect speeds are flown

    (1 mark)

36. Refer to CAP398 MRJT figure 4.24. At a mass of 35000 kg, why does the drift down curve start at approximately 3 minutes at an altitude of 37000 ft?
    a. The origin of the curve lies outside the chart
    b. At this altitude it takes longer for the engines to slow down, giving extra thrust for about 4 minutes
    c. Because of inertia at the higher TAS it takes longer to establish the optimum rate of descent
    d. It takes about this time to decelerate the aircraft to the optimum speed for drift down.

    (1 mark)

37. A twin engine turbo-jet aircraft having lost one engine must clear obstacles in the drift down by a minimum of:
    a. 35 ft
    b. 1000 ft
    c. 1500 ft
    d. 2000 ft

    (1 mark)

38. The landing speed, V_{REF}, for a single engine aircraft must be not less than:
    a. $1.2V_{MCA}$
    b. $1.1V_{S0}$
    c. $1.05V_{S0}$
    d. $1.3V_{S0}$

    (1 mark)

39. What factor must be applied to the landing distance available at the destination aerodrome to determine the landing performance of a turbo-jet aircraft on a dry runway?
    a. 1.67
    b. 1.15
    c. 0.60
    d. 0.70

    (1 mark)

40. An aircraft is certified to use two landing flap positions, 25° and 35°. If the pilot selects 25° instead of 35° then the aircraft will have:
    a. an increased landing distance and reduced go-around performance
    b. an reduced landing distance and reduced go-around performance
    c. an increased landing distance and increased go-around performance
    d. an reduced landing distance and increased go-around performance

    (1 mark)

AIRCRAFT PERFORMANCE

ANSWERS TO SPECIMEN EXAMINATION PAPER

1	A	21	C
2	C	22	B
3	C	23	A
4	B	24	B
5	D	25	B
6	B	26	B
7	C	27	C
8	A	28	C
9	D	29	B
10	A	30	C
11	C	31	C
12	B	32	C
13	A	33	D
14	A	34	D
15	C	35	C
16	C	36	D
17	D	37	D
18	C	38	D
19	B	39	C
20	B	40	C

AIRCRAFT PERFORMANCE SPECIMEN QUESTIONS

EXPLANTIONS TO ANSWERS – SPECIMEN EXAM PAPER

a. The requirement for an aircraft to be certified in performance class B is that the maximum certified take-off mass must not exceed 5700kg AND the certified maximum number of passengers must not exceed 9. If both these conditions are not met then the aircraft will be certified in class A.

1. c. As speed increases then the angle of attack and hence the thrust of the propeller decrease.

2. c. see page 13-4

3. b. With a downhill slope the effort required to continue acceleration after V_{EF} is less than the effort required to stop the aircraft because of the effect of gravity. Hence take-off can be achieved from a lower speed, but stopping the aircraft within the distance available can only be achieved from a lower speed.

4. d. The lowest take-off performance will occur when air density is at its lowest, hence high temperature and low pressure.

5. b. see definitions chapter 1-2.

6. c. In determining take-off climb performance, the still air gradient is considered.

7. a. As altitude increases the speed for best gradient of climb remains constant, but the speed for best rate of climb decreases until the two speed coincide at the absolute ceiling.

8. d. TODA is limited by the lower of TORA plus clearway and 1.5 x TORA.

9. a. With a reduced flap setting the lift generated decreases and the stalling speed increases, so a greater take-off speed is required increasing the take-off distance. The reduced flap setting reduces the drag so the climb gradient increases.

10. c. The use of reduced thrust means that the take-off distance will be increased and the climb gradient reduced, so neither of these can be limiting.

11. b. Increasing the flap setting will reduce the TODR so the weight can be increased, but the climb gradient will reduce with the increased flap setting.

12. a. see CAP698 page 62, paragraph 2.5.1.

13. a. Gradient of climb can be approximated to rate of climb divided by TAS.

14. c. The decision to continue the take-off will be made, but because the speed is below the normal V_1, the distance required to accelerate will be greater, so the TODA may well be exceeded.

15. c. At a constant IAS drag will remain constant.

16. d. The air density is greater at the lower aerodrome so the performance will be better in terms of acceleration and gradient.

17. c. see chapter 3 page 5. All speeds must be greater than V_s.

18. b. Although a uphill slope will enhance the deceleration, the effect on the acceleration to V_1 will result in increased distances.

AIRCRAFT PERFORMANCE SPECIMEN QUESTIONS

19. b. Temperature is decreasing therefore the speed of sound and the TAS will decrease, and as altitude increases the IAS will also decrease.

20. c. As speed increases the induced drag reduces proportional to the square of the speed increase.

21. b. As altitude increases at a given weight the IAS for the onset of high speed buffet decreases and the IAS for the onset of low speed buffet remains constant, hence the speed range decreases. As weight increases at a given altitude the IAS for the onset of low speed buffet increases and there is a small decrease in the IAS for the onset of high speed buffet so once again the speed range decreases. (see the diagram in chapter 16, page 16-9)

22. a. In the climb and descent an element of gravity acts along the aircraft axis either opposing (climb) or adding to (descent) the thrust, so only in level unaccelerated flight will the two forces be equal.

23. b. As altitude increases the speed for best gradient of climb remains constant but the speed for the best rate of climb decreases. (see Q8)

24. b. As the CofG moves forward the download on the tail increases, so the lift required also increases. To create the extra lift either speed or angle of attack must increase. In either case more thrust is required and the fuel required per nm will increase.

25. b. The speed for maximum range in a jet aircraft is found where the line from the origin is tangential to the drag curve. With a head wind the origin moves to the right by the amount of the wind component so the line from the origin will be tangential at a higher speed, and to the left with a tail wind giving a lower speed.

26. c. V_1 is the decision speed. If engine failure is recognised below V_1 take-off must be abandoned and engine failure recognised above V_1 take-off must be continued.

27. c. the angle and rate of descent are independent of the wind, but the descent path is modified by the wind, the angle increasing with a head wind because of the reduced ground distance covered.

28. b. The maximum range speed is the speed at which the greatest distance can be flown, which means the lowest possible fuel usage. To achieve this implies we must have the lowest specific fuel consumption.

29. c. The most efficient way to operate a jet aircraft is to cruise climb, that is to set optimum cruise power and fly at the appropriate speed and allow the excess thrust as weight decreases to climb the aircraft.. For obvious safety reasons this is not possible, so the aircraft operates as close as possible to the optimum altitude by using the step climb technique.

30. c. The absolute ceiling is the highest altitude to which the aircraft could be climbed, where, at optimum climb speed, thrust = drag, so with no excess of thrust the rate (and gradient) of climb will be zero.

31. c. Net performance is gross (ie average) performance reduced by a regulatory amount to give a 'worst case' view. Therefore the worst case for the climb gradient will be a lower gradient than it is expected to achieve.

32. d. Gross take-off distance from the graph is 2800 ft. There is no stopway or clearway so the factor to apply is 1.25 to get the minimum TOR. (see CAP398, page 19)

33. d. The maximum take-off mass will be achieved with the higher flap setting because the take-off speeds will be reduced. As ever take care when using the graphs.

34. c. With a head wind component the glide performance will not be affected but the distance covered will reduce because of the reduced groundspeed.

35. d. The aircraft will be allowed to slow down to the optimum drift down speed before commencing descent which will take an amount of time dependent on weight and altitude.

36. d. This is the JAR regulatory requirement.

37. d. Again the JAR regulatory requirement. See chapter 8 paragraph 9.2.

38. c. The JAR regulatory requirement. See chapter 17 paragraph 17.3.

39. c. With a reduced flap setting the stalling speed and hence the V_{REF} will increase, so a greater landing distance will be required. With a reduced flap setting the drag will decrease, so the climb gradient will be better.